T0214744

Lecture Notes in Computer Science 12635

More information about this subseries at http://www.springer.com/series/7407

Ryuhei Uehara · Seok-Hee Hong ·
Subhas C. Nandy (Eds.)

WALCOM: Algorithms and Computation

15th International Conference and Workshops, WALCOM 2021
Yangon, Myanmar, February 28 – March 2, 2021
Proceedings

Springer

Editors
Ryuhei Uehara 🆔
Japan Advanced Institute of Science
and Technology
Ishikawa, Japan

Seok-Hee Hong
University of Sydney
Sydney, NSW, Australia

Subhas C. Nandy
Indian Statistical Institute
Kolkata, India

ISSN 0302-9743 ISSN 1611-3349 (electronic)
Lecture Notes in Computer Science
ISBN 978-3-030-68210-1 ISBN 978-3-030-68211-8 (eBook)
https://doi.org/10.1007/978-3-030-68211-8

LNCS Sublibrary: SL1 – Theoretical Computer Science and General Issues

This Springer imprint is published by the registered company Springer Nature Switzerland AG
The registered company address is: Gewerbestrasse 11, 6330 Cham, Switzerland

Preface

The 15th International Conference and Workshop on Algorithms and Computation (WALCOM 2021) was held at University of Information Technology, Yangon, Myanmar during February 28 – March 2, 2021. The conference covered diverse areas of algorithms and computation, that is, approximation algorithms, algorithmic graph theory and combinatorics, combinatorial algorithms, combinatorial optimization, computational biology, computational complexity, computational geometry, discrete geometry, data structures, experimental algorithm methodologies, graph algorithms, graph drawing, parallel and distributed algorithms, parameterized algorithms, parameterized complexity, network optimization, online algorithms, randomized algorithms, and string algorithms. The conference was organized in cooperation between University of Information Technology and Japan Advanced Institute of Science and Technology. Due to COVID-19, it was held online.

This volume of *Lecture Notes in Computer Science* contains 24 contributed papers that were presented at WALCOM 2021. There were 60 submissions from 21 countries. Each submission was reviewed by at least three Program Committee members with the assistance of external referees. Among them, the following paper was selected as the best paper: "On Compatible Matchings" by Oswin Aichholzer, Alan Arroyo, Zuzana Masárová, Irene Parada, Daniel Perz, Alexander Pilz, Josef Tkadlec, and Birgit Vogtenhuber. The volume also includes the abstract and the extended abstracts of three invited talks presented by Mohammad Kaykobad, Tetsuo Asano, and Erik D. Demaine. Two special issues, one of Theoretical Computer Science and one of the Journal of Graph Algorithms and Applications contained some selected papers among those presented at WALCOM 2021.

We wish to thank all who made this meeting possible: the authors for submitting papers, the Program Committee members and external referees (listed in the proceedings) for their excellent work, and our three invited speakers. We acknowledge the Steering Committee members for their continuous encouragement and suggestions. We also wish to express our sincere appreciation to the sponsors, local organizers, Proceedings Committee, and the editors of the *Lecture Notes in Computer Science* series and Springer for their help in publishing this volume. We especially thank Saw Sanda Aye, Wint Thida Zaw, Tin Htar Nwe, and their team at University of Information Technology for their tireless efforts in organizing this conference. Finally, we thank the EasyChair conference management system, which was very effective in handling the entire reviewing process.

March 2021

Seok-Hee Hong
Subhas C. Nandy
Ryuhei Uehara

Organization

Steering Committee

Tamal Dey	The Ohio State University, USA
Seok-Hee Hong	University of Sydney, Australia
Costas S. Iliopoulos	King's College London, UK
Giuseppe Liotta	University of Perugia, Italy
Petra Mutzel	University of Bonn, Germany
Shin-ichi Nakano	Gunma University, Japan
Subhas C. Nandy	Indian Statistical Institute, India
Md. Saidur Rahman	Bangladesh University of Engineering and Technology, Bangladesh
Ryuhei Uehara	Japan Advanced Institute of Science and Technology, Japan

Program Committee

Hee-Kap Ahn	Pohang University of Science and Technology, South Korea
Shamsuzzoha Bayzid	Bangladesh University of Engineering and Technology, Bangladesh
Guillaume Blin	University of Bordeaux, France
Hans Bodlaender	Utrecht University, The Netherlands
Gautam Das	Indian Institute of Technology Guwahati, India
Jesper Jansson	Hong Kong Polytechnic University, Hong Kong
Wing-Kai Hon	National Tsing Hua University, Taiwan R. O. C.
Seok-Hee Hong (Co-chair)	University of Sydney, Australia
Ralf Klasing	University of Bordeaux, France
Inbok Lee	Korea Aerospace University, South Korea
Giuseppe Liotta	University of Perugia, Italy
Takaaki Mizuki	Tohoku University, Japan
Debajyoti Mondal	University of Saskatchewan, Canada
Krishnendu Mukhopadhyaya	Indian Statistical Institute, India
Shin-ichi Nakano	Gunma University, Japan
Subhas C. Nandy (Co-chair)	Indian Statistical Institute, India
Solon Pissis	Centrum Wiskunde & Informatica, The Netherlands
Simon Puglisi	University of Helsinki, Finland
Tomasz Radzik	King's College London, UK
Atif Rahman	Bangladesh University of Engineering and Technology, Bangladesh

Mohammad Sohel Rahman	Bangladesh University of Engineering and Technology, Bangladesh
Kunihiko Sadakane	University of Tokyo, Japan
Slamin	University of Jember, Indonesia
William F. Smyth	McMaster University, Canada
Paul Spirakis	University of Liverpool, UK
Wing-Kin Sung	National University of Singapore, Singapore
Ryuhei Uehara (Co-chair)	Japan Advanced Institute of Science and Technology, Japan
Bruce Watson	Stellenbosch University, South Africa
Hsu-Chun Yen	National Taiwan University, Taiwan R. O. C.

Organizing Committee

Saw Sanda Aye	University of Information Technology, Myanmar
Ryuhei Uehara	Japan Advanced Institute of Science and Technology, Japan
Wint Thida Zaw	University of Information Technology, Myanmar
Myat Thida Mon	University of Information Technology, Myanmar
Swe Zin Hlaing	University of Information Technology, Myanmar
Ei Chaw Htoon	University of Information Technology, Myanmar
Swe Swe Kyaw	University of Information Technology, Myanmar
Aye Theingi	University of Information Technology, Myanmar
Win Win Myo	University of Information Technology, Myanmar

Additional Reviewers

Abu Reyan Ahmed
Hugo Akitaya
Eleni C. Akrida
Yuichi Asahiro
Evangelos Bampas
Giulia Bernardini
Stéphane Bessy
Anup Bhattacharya
Sujoy Bhore
Mohammad Tawhidul Hasan Bhuiyan
Arijit Bishnu
Hans-Joachim Böckenhauer
Sourav Chakraborty
Pratibha Choudhary
Minati De
Argyrios Deligkas
Arthur van Goethem
Barun Gorain

Arobinda Gupta
Mursalin Habib
Duc A. Hoang
Ramesh Jallu
Arindam Karmakar
Masashi Kiyomi
Grigorios Loukides
George Manoussakis
Themistoklis Melissourgos
George Mertzios
Othon Michail
Gopinath Mishra
Pawan Mishra
Daiki Miyahara
Dmitry Mokeev
Anisur Rahaman Molla
Hendrik Molter
Fabrizio Montecchiani

Jesper Nederlof
Yota Otachi
Dominik Pająk
Supantha Pandit
Vicky Papadopoulou
Vinod Reddy
Sasanka Roy
Anik Sarker
Buddhadeb Sau
Saket Saurabh

Md. Salman Shamil
Swakkhar Shatabda
Éric Sopena
Susmita Sur-Kolay
Michelle Sweering
Alessandra Tappini
Michail Theofilatos
Walter Unger
Mingyu Xiao
Tom van der Zanden

Sponsoring Institutions

University of Information Technology, Myanmar
Japan Advanced Institute of Science and Technology, Japan
Information Processing Society of Japan (IPSJ), Japan
The Institute of Electronics, Information and Communication Engineers (IEICE), Japan
Japan Chapter of European Association of Theoretical Computer Science (EATCS
    Japan), Japan

Understanding the Complexity of Motion Planning (Abstract of Invited Talk)

Erik D. Demaine

Computer Science and Artificial Intelligence Laboratory, Massachusetts Institute of Technology, Cambridge, MA 02139, USA
edemaine@mit.edu

Abstract. Motion planning of simple agents (robots, humans, cars, drones, etc.) is a major source of interesting algorithmic and geometric problems. We'll describe results for a variety of models ranging in number of agents and amount of available control, focusing on recent results. A single agent represents many single-player video games, as well as your daily life. With multiple agents, the amount of control can vary from moving one agent at a time (as in, e.g., Checkers or Chess, recently proved NP-hard even for a single move) to global control of all agents (as in the puzzle board game Tilt, recently proved PSPACE-complete) to simultaneous parallel control of all agents (recently proved to have a constant-factor approximation algorithm). Along the way, we'll describe a growing theory of "gadgets" aiming to characterize the complexity of motion planning problems.

Keywords: Gadgets · Motion planning · Computational complexity

Contents

Invited Talks

Majority Spanning Trees, Cotrees and Their Applications

Mohammad Kaykobad[1(✉)] and F. J. M. Salzborn[2]

[1] Department of Computer Science and Engineering,
BRAC University, Dhaka, Bangladesh
kaykobad@bracu.ac.bd
[2] Department of Applied Mathematics, University of Adelaide, Adelaide, Australia
fsalzbor@hotmail.com

Abstract. We show that in any digraph on an underlying connected graph with non-negative weights on its edges, there is a *Majority Spanning Tree* for which sum of weights of edges of a fundamental cutset, running along each edge of the spanning tree determining the cutset, is not less than sum of those running in opposite direction. Similarly, there is a *Majority Cotree*, each fundamental cycle of which has non-negative weight. We further prove simultaneous existence of majority spanning trees and majority cotrees in any non-negative weighted digraph. We have shown how these structures can be used to solved scheduling transports by minimizing sum of weighted connection times, ranking round-robin tournaments by minimizing number of upsets, in settling multiple debts and in construction of transport networks with unbalanced road capacity.

Keywords: Graph theory · Spanning tree · Cutset · Majority spanning tree · Cotree

1 Preliminaries

In this paper we consider weighted simple directed graphs $G = (V, E)$ whose underlying graph is connected. Let the weights $w_{ij} \geq 0$ for all $(i, j) \in E$. Furthermore, whenever we use the terms *spanning tree* and *cotree* we mean the corresponding structures in the underlying graph of G. We denote any arbitrary spanning tree of such underlying graphs by T and cotree by $\overline{T}$. Spanning trees have become a very important structure in solving problems of various fields like communication, optimization, networks and clustering. Minimum spanning trees are useful in designing telephone, electrical, hydraulic, TV cable, computer, road networks etc (see [1,3,9]). There are many algorithms like [6,8] and [10] for constructing minimum spanning trees of a graph.

Consider the example in Fig. 1. Let $T_1 = (V, E')$ be a spanning tree of the underlying graph of digraph G, with $E' = \{AB, BE, DA, FD, CF, CG\}$ (in

Supported by The Flinders University of South Australia.

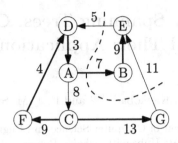

Fig. 1. A digraph with a majority spanning tree on bold edges.

bold) and $a_k = (i, j)$ be a directed edge in E. For ensuring disambiguity of related concepts, if $(i, j) \in E$ we refer the edge in T connecting vertices i and j as (i, j) and not (j, i). Notice that $T_1 \setminus \{a_k\}$ partitions V into two disjoint subsets V_i and V_j—each consisting of vertices of a connected component of $T_1 - \{a_k\}$ such that the vertices i and j belong to the subsets V_i and V_j respectively. Note that $V_i \cup V_j = V$ and $V_i \cap V_j = \emptyset$. Let

$$W(V_i, V_j) = \sum_{(l,m) \in K_k^+ | l \in V_i, m \in V_j} w_{lm}$$

and

$$W(V_j, V_i) = \sum_{(l,m) \in K_k^- | l \in V_j, m \in V_i} w_{lm}$$

where, $a_k = (i, j) \in T$, K_k^+ and K_k^- are respectively the sets of edges in the positive and negative orientations of the cutset K_k determined by the spanning tree edge a_k. Let

$$W[V_i, V_j] = W(V_i, V_j) - W(V_j, V_i)$$

Note that $[V_i, V_j]$ is said to be a *fundamental cutset* determined by the edge $a_k \in T_1$. In the same way an edge $a_k \in \overline{T_1}$ together with T_1 contains a unique cycle called *fundamental cycle*. In Fig. 1 if $(i, j) = (A, B)$ then $V_A = \{A, C, D, F, G\}, V_B = \{B, E\}, W(V_A, V_B) = w_{AB} + w_{GE} = 7 + 11 = 18$, whereas $W(V_B, V_A) = w_{ED} = 5$. So $W[V_A, V_B] = W(V_A, V_B) - W(V_B, V_A) = 18 - 5 = 13$.

Definition 1. *A spanning tree T of G is said to be a* majority spanning tree *(MST) if*

$$W[V_i, V_j] \geq 0, \forall (i, j) \in T$$

While every connected graph has a spanning tree, it is not clear whether every digraph has a majority spanning tree since it must satisfy some additional conditions. However, the spanning tree shown in Fig. 1 is indeed a majority spanning tree since the weights of each fundamental cutset is non-negative. For example, $W(CF) = w_{CF} + w_{GE} - w_{AC} = 9 + 11 - 8 = 12 \geq 0$. Note that a fundamental cutset need not have nonnegative weight. For example in Fig. 1, for the spanning tree $T' = T_1 \setminus \{CG\} \cup \{GE\}$, the weight of the fundamental cutset corresponding to edge GE is $W[V_G, V_E] = 11 - 13 < 0$. However, it can be easily seen that for the spanning tree T_1 in Fig. 1, weight of every cutset is non-negative. Hence T_1 is indeed a majority spanning tree of G.

We note here that for a spanning tree T of the underlying graph of a simple digraph G, $\overline{T} = G \setminus T$ is called a *cotree*. Each edge $a_k = (i, j) \in \overline{T}$ together with edges of T defines a unique cycle C_k or C_{ij}. While C_k is not necessarily directed, its positive orientation is assumed to be along the edge a_k in cycle C_k. Let us define the weight of a cycle as follows:

$$W(C_k) = \sum_{(l,m) \in C_k^+} w_{lm} - \sum_{(l,m) \in C_k^-} w_{lm}$$

where C_k^+ are the edges in C_k that are in the same orientation as $a_k = (i, j) \in \overline{T}$, and C_k^- are the set of edges in the reverse orientation.

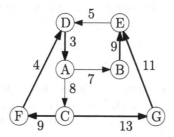

Fig. 2. A digraph with an MCT on thin edges.

Now we refer to Fig. 2, and find weight of a fundamental cycle. Here cotree $\overline{T_2} = \{ED, AB, AC\}$, thus for the fundamental cycle defined by ED, $C_{ED}^+ = \{ED, CG, GE\}$, $C_{ED}^- = \{FD, CF\}$, $W[C_{ED}] = (5 + 13 + 11) - (4 + 9) = 16$.

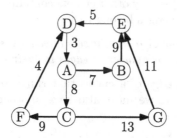

Fig. 3. A digraph with an MST whose complement is an MCT

Similar to majority spanning tree, we now introduce majority cotrees (MCT) as follows.

Definition 2. *A cotree* $\overline{T}$ *of* $G = (V, E)$ *is said to be a* majority cotree *if*

$$W(C_k) \geq 0, \forall a_k \in \overline{T}$$

It is not very clear that every directed graph with non-negative weights on its edges must have a majority spanning tree or a majority cotree. Note that the cotree drawn using thin arrows in Fig. 2 is indeed a majority cotree since $W[C_{ED}] = 16 \geq 0$ as shown earlier, and $W[C_{AC}] = 8 + 9 + 4 + 3 = 24 \geq 0$, $W[C_{AB}] = 7 + 9 + 3 + 4 + 9 - 13 - 11 = 8 \geq 0$.

Moreover, consider Fig. 3. In this figure $\overline{T_3} = \{ED, DA, AC\}$ is a majority cotree since $W[C_{ED}] = (5+13+11) - (4+9) = 16$, $W[C_{AC}] = 8+13+11-9-7 = 16 \geq 0$, $W[C_{DA}] = (3 + 7 + 9 + 9 + 4) - (13 + 11) = 8 \geq 0$. Not only that, $T_3 = \{AB, BE, GE, CG, CF, FD\}$ is indeed a majority spanning tree since all fundamental cutsets have non-negative values!

In the next section, we shall show that for all general digraphs with arbitrary non-negative weights on their edges there exist majority spanning trees, majority cotrees, and there is a majority spanning tree T whose complement $\overline{T} = G \setminus T$ is also a majority cotree.

2   Main Results

We start with the following important observations. Detailed proofs of the following results can be found in Kaykobad [5]. We note here that by cutsets and cycles we mean a set of edges not necessarily constituting a directed cutset or a directed cycle.

Observation 1. *Two edges that are in the same orientation in a cutset are in different orientations in any cycle containing them and vice versa.*

Again we refer to Fig. 1. Edges AB and GE are in the same orientation in the cutset $\{ED, AB, GE\}$, whereas they are in different orientations in cycle (A, B, E, G, C, F, D) or in any other cycle containing them. Edges ED and DA are in the same orientation in cycle (E, D, A, B) but in different orientation in cutset $\{ED, DA, AC, CG\}$ or any other cutsets containing them. This is true for any cycle and cutset these edges belong to.

Observation 2. *Let C and Q represent sets of edges respectively in a cycle and a cutset. Then there are even number of edges in their intersection.*

If a cycle contains an edge of a cutset that runs from one partition to the other, for cycle to complete we must also have another edge of the cutset to return to the same partition.

Observation 3. *If we subtract the weight of the minimum edge from all edges of a cutset in that orientation, and add the same value to weights of edges in the opposite orientation, then weights of the cycles do not change. Similarly, if a minimum positive weight is subtracted from weights of edges in a cycle in one orientation and the same weight is added to edges of the cycle in opposite orientation then the weights of cutsets will remain unchanged.*

For brevity for an edge AB of a digraph G, we write $AB = 7$ instead of $w_{AB} = 7$ so long as it does not create any confusion. In Fig 1, if we reduce weights along the cycle (A, B, D, E) by the minimum 3 then weights become $DA = 0, AB = 4, BE = 6, ED = 2$. Notice that in the process, weights of all cutsets remain unaffected. For example, consider the weight of the cutset defined by edge AB of spanning tree T_1. Its weight was $-5 + 7 + 11 = 13$. And now its weight is $-2 + 4 + 11 = 13$. On the other hand, if weights of edges in the cutset determined by the edge AB is reduced along the orientation of AB by the minimum edge weight 7 and increase weights of edges in the opposite orientation by 7 then the new changed weights are $ED = 12, AB = 0, GE = 4$. One of the affected cycles is (A, B, E, D). Its earlier weight was $7 + 9 + 5 + 3 = 24$, and the changed weight also remains $0 + 9 + 12 + 3 = 24$.

The above two results allow us to manipulate edge weights by changing weights of edges of a cycle while keeping weights of all cutsets fixed, or changing weights of edges of a cutset while keeping weights of all cycles fixed. In the following we shall use this technique to construct majority spanning trees and majority cotrees.

We first prove the following theorem by construction.

Theorem 1. *Every digraph $G = (V, E)$ with non-negative weights on its edges must have a majority spanning tree.*

Proof. So long as there is a cycle with positive weights on all its edges, subtract the weight of the minimum edge of the cycle from all edges in the same orientation in the cycle, and add the same weight to all edges in the opposite orientation of the cycle. Then at least one edge will be of weight 0. We carry out this process so long as there are cycles with all its weights positive. Ultimately positive weighted edges along with some 0-weight edges will constitute a spanning tree T that must be a majority spanning tree. Note that in this case, weight of the fundamental cutset defined by a spanning tree edge equals to the weight of that spanning tree edge — weights of edges of cotree being 0. In this process, no cutset weight has been changed. Now in each fundamental cutset, only weight of the edge of spanning tree determining the fundamental cutset is non-negative (other weights being 0). Hence weight of the corresponding fundamental cutset is at least non-negative.   □

The following theorem asserts the existence of a majority cotree in any digraph.

Theorem 2. *Every digraph $G = (V, E)$ with non-negative weights on its edges must have a majority cotree.*

Proof. The proof is similar to the proof of Theorem 1, but the edge weights now must be manipulated around cutsets, keeping weights of cycles unaffected. Then edges of positive weights possibly together with some edges of 0 weights constitute a cotree $\overline{T}$—which is a majority cotree. Furthermore, weight of each fundamental cycle equals the weight of the edge of the cotree determining the fundamental cotree.   □

Before we formulate the important result of simultaneous existence of majority spanning trees and majority cotrees, let us define the following structures.

Let r_T be a vector whose kth component equals weight of the fundamental cutset determined by the edge $a_k = (i, j) \in T$, that is,

$$r_k = \sum_{\{(l,m)|l \in V_i, m \in V_j\}} w_{lm} - \sum_{\{(l,m)|l \in V_j, m \in V_i\}} w_{lm}, \quad \forall a_k \in T \qquad (1)$$

Let Q be the signed fundamental cutset-edge incidence matrix. Row Q_k of Q is determined by edge $a_k \in T$. Then

$$q_{kj} = \begin{cases} 1, & \text{if } a_j \text{ has the same orientation as } a_k \text{ in the cutset} \\ -1, & \text{if } a_j \text{ has orientation opposite to } a_k \text{ in the cutset} \\ 0, & \text{otherwise} \end{cases} \qquad (2)$$

Let $s_{\overline{T}}$ be the vector whose pth component is the weight of the fundamental cycle determined by $a_p \in \overline{T}$. Then it can be calculated as

$$\sum_{(l,m) \in C_p^+} w_{lm} - \sum_{(l,m) \in C_p^-} w_{lm} = s_p, \quad \forall a_p \in \overline{T} \qquad (3)$$

Let L be the signed fundamental cycle-edge incidence matrix. Then pth component of L is determined by the edge $a_p \in \overline{T}$. Then

$$l_{pj} = \begin{cases} 1, & \text{if } a_j \text{ has the same orientation as } a_p \text{ in the cycle} \\ -1, & \text{if } a_j \text{ has orientation opposite to } a_p \text{ in the cycle} \\ 0, & \text{otherwise} \end{cases} \qquad (4)$$

From Observation 1, we know that if a cotree edge a_j is in the same orientation as that of the spanning tree edge a_k determining the cutset then $q_{kj} = 1$, whereas since in that case in the cycle determined by edge a_j, a_k is in the opposite orientation causing $l_{jk} = -1$. This relation holds true also in case cotree edge is in the opposite orientation. then made above we can see that $l_{pj} = -q_{jp}$. That is $L^t = -Q$. Moreover, removing identity matrices we get $L_{\overline{T}}^t = -Q_{\overline{T}}$.

As shown in Fig 3, both the spanning tree T_3 and the cotree $\overline{T}_3$ are both majority spanning tree and majority cotree respectively. The following result confirms simultaneous coexistence of majority spanning tree and majority cotree in any digraph with non-negative weights on edges.

Theorem 3. *Every digraph $G = (V, E)$ with non-negative weights on its edges has a majority spanning tree T such that $\overline{T}$ is a majority cotree, and vice versa.*

Proof. In order to prove the above theorem, we first formulate the following linear programming problem. Each equation corresponds to a signed fundamental cycle-edge incidence vector. Here right hand side equals to the weight of its corresponding cycle. Remember, variables corresponding to edges of the cotree form a basic solution since they do not appear in other equations/fundamental cycles.

In objective function, each variable corresponding to edge $a_k \in T$ appears with coefficients equal to weight of the fundamental cutset determined by the edge a_k.

The resulting linear programming problem is given below.

$$\max - s_{\overline{T}}^t x_{\overline{T}} \tag{5}$$
$$Q_{\overline{T}} x_{\overline{T}} \leq r_T$$
$$x_{\overline{T}} \geq 0$$

Its dual will be

$$\min r_T^t y_T \tag{6}$$
$$Q_{\overline{T}}^t y_T \geq -s_{\overline{T}}$$
$$y_T \geq 0$$

Equivalently,

$$\min r_T^t y_T \tag{7}$$
$$L_T y_T \leq s_{\overline{T}}$$
$$y_T \geq 0$$

While any feasible solution to (5) is a majority spanning tree, any feasible solution to (7) is a majority cotree. Right hand sides of both sets of inequalities are satisfied by equating the variables to weights of the corresponding edges, that is $x_{lm} = w_{lm}$ or $y_{lm} = w_{lm}$, where edge $(l, m) = a_k$. So both of them are feasible, and hence they have an optimal solution that corresponds to a majority spanning tree for (5) and majority cotree for (7).

This proves the theorem on simultaneous existence of majority spanning trees and majority cotrees in any digraph with non-negative weights on its edges. □

In the following section we present some of the applications of these concepts.

3 Applications

The structure of majority spanning tree has been used in Kaykobad [5] for scheduling trains by minimizing weighted sum of connection time over a railway network. Datta, Hossain and Kaykobad [4] and Kaykobad et al. [2] used the concepts for ranking players of a round-robin tournament. The concept of majority spanning trees has also been used in rank aggregation of meta-search engines [7].

3.1 Minimum Connection Time Problem

We address transportation systems where each trip is generated periodically like railway networks. Since most often each pair of destinations is not served by a single transport, there may be waiting time in connections between trips. We denote the trips of the network by vertices and the connections by edges. For the trip i, let x_i be its start time and t_i be its duration. The number of passengers taking the connection from trip i to trip j be p_{ij}. By w_{ij} we denote the waiting time for connection to trip j from trip i. Then the mathematical model for the problem is as follows:

$$\sum_{(i,j)\in E} p_{ij}w_{ij} \to \text{minimize} \tag{8}$$

$$x_j - x_i - t_i = w_{ij} \mod \tau, \ \forall (i,j) \in E$$

$$0 \leq w_{ij} < \tau, \ \forall (i,j) \in E$$

where τ is the period of occurrence of a trip. It can be shown that optimal scheduling will have 0 waiting times in connections that constitute a majority spanning tree of the digraph. However, it may be noted here that while majority spanning tree is a necessary condition for optimality, it is not sufficient. In fact, the problem has been shown to be NP-hard for general digraphs by Kaykobad [5].

3.2   Round-Robin Tournament Ranking

Consider the problem of ranking players of a round-robin tournament by minimizing the number of upsets, that is number of matches in which lowly ranked players have defeated highly ranked players. The results of such a tournament can be expressed in a digraph known as tournament digraph. In a *tournament digraph*, players correspond to vertices and each edge corresponds to the result of a match oriented from the winner to the defeated. Let R be a ranking of players, $V_R = V$ and $E_R = \{(i, i+1), 1 \leq i \leq n-1\}$ such that the rank of player corresponding to vertex $i+1$ is immediately below that of the player corresponding to vertex i. It is obvious that $G_R = (V_R, E_R)$ is a spanning tree of G. More accurately, $G_R = (V_R, E_R)$ is a Hamiltonian semipath.

Theorem 4. *Let R be any optimal ranking of a tournament represented by $G = (V, E)$. Then $G_R = (V_R, E_R)$ is a majority spanning tree of G.*

Let, $G_{ij}(R)$ be the subgraph of G induced by the set of vertices corresponding to players ranked from i to j as per rank R. Furthermore, let G_{ij}^R be the subgraph of $G_{ij}(R)$ having the same set of vertices, and only those edges that connect adjacently ranked players. Then,

Theorem 5. *For any optimal ranking R and $1 \leq i \leq j \leq n, G_{ij}^R$ must be a majority spanning tree of $G_{ij}(R)$.*

Let P_1 and P_2 be two disjoint set of players, ranked consecutively as shown in Fig. 4. if players in P_1 lose more games to players in P_2, then the cutset (denoted by dashed line) has negative value, and does not correspond to an MST. Thus players in P_2 deserve a better ranking, and hence swapping P_1 and P_2 results in a lesser number of upsets. MST algorithm continues to look for such violating cutsets and then swap the sets of players to reduce the number of upsets.

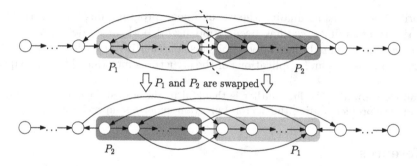

Fig. 4. MST algorithmic concept for round-robin tournament ranking.

3.3 Balancing One Way Roads

Imagine that due to the setting up of a number of important industries at various locations of a country, there has been a revised load on different sectors of the road network. As a result, the current road capacity can no longer satisfy transportation demand. Now we need to upgrade the capacity of certain segments of the road network to satisfy the new demand. Naturally we want to minimize the cost of reconstruction of the road network while satisfying the present demand of transportation. We can reconstruct a digraph with new demands of loads on different segments of the road network as weights of corresponding edges. Then for the optimal solution, the subgraph of the road network denoting the reconstructed road segments with updated capacity will be included in a majority spanning tree.

3.4 Settling Multiple Debts

Let us consider a set of borrowers and loan givers. Now we want to settle debts in a minimum amount of total transactions, or by minimizing total transaction cost. Transaction costs can be thought of being proportional to transaction amounts, with unit cost differing for different borrowers and loan giver pairs. This problem was first addressed in [11], and was modeled in a digraph with weights on directed edges from borrowers to loan givers equalling to the debt amounts. Edges corresponding to transactions in the optimal solution will be contained in a spanning tree of the underlying graph. Moreover, optimal set of transactions will correspond to a subset of edges of a majority spanning tree of the digraph where objective function will be to minimize total weighted transacted amount.

4 Conclusion

We have discovered a structure called majority spanning trees in general digraphs with non-negative weights on its edges. We have also proved the existence of majority cotree and the simultaneous existence of majority spanning trees and majority cotrees using duality theory of linear programming. The structure of majority spanning trees has been found useful both in modeling and solution of several seemingly unrelated problems. Further investigation may lead to enhanced properties and applications of the structures introduced in this paper.

Acknowledgment. The first author is thankful to the second author for proposing to study the problem during the doctoral program under his supervision.

References

1. Contreras, I., Fernández, E.: General network design: a unified view of combined location and network design problems. Eur. J. Oper. Res. **219**, 680–697 (2012)
2. Datta, A., Hossain, M., Kaykobad, M.: An improved MST algorithm for ranking players of a round robin tournament. Int. J. Comput. Math. **85**, 1–7 (2007)
3. Fortzab, B., Gouveiac, L., Joyce-Monizab, M.: Optimal design of switched ethernet networks implementing the multiple spanning tree protocol. Discrete Appl. Math. **234**, 114–130 (2018)
4. Kaykobad, M., Ahmed, Q.N.U., Khalid, A.S., Bakhtiar, R.: A new algorithm for ranking players of a round-robin tournament. Int. J. Comput. Oper. Res. **22**, 221–226 (1995)
5. Kaykobad, M.: Minimum connection time and some related complexity problems. Ph.D. thesis, The Flinders University of South Australia (1986)
6. Kruskal, J.B.: On the shortest spanning subtree of a graph and the traveling salesman problem **7**, 48–50 (1956)
7. Lam, K.W., Leung, C.H.: Rank aggregation for meta-search engines. In: 2004 Proceedings of the 13th International World Wide Web Conference on Alternate Track Papers & Posters, pp. 384–385. ACM, New York (2004)
8. Nešetřil, J., Milková, E.N.H.: Otakar Borůvka on minimum spanning tree problem: translation of both the 1926 papers, comments, history. Discrete Math. **233**, 3–36 (2001)
9. Nguyen Gia, N., Le, D.N.: A novel ant colony optimization-based algorithm for the optimal communication spanning tree problem. Int. J. Comput. Theory Eng. **5**, 509–513 (2012)
10. Prim, R.: Shortest connection networks and some generalizations. Bell Syst. Tech. J. **36**, 1389–1401 (1957)
11. Verhoeff, T.: Settling multiple debts efficiently - an invitation to computing science. Inform. Educ. **3**, 105–126 (2004)

A New Transportation Problem on a Graph with Sending and Bringing-Back Operations

Tetsuo Asano[1,2(✉)]

[1] Kanazawa University, Kanazawa, Japan
`asano@staff.kanazawa-u.ac.jp`
[2] Japan Advanced Institute of Science and Technology, Nomi, Japan

Abstract. This paper considers a transportation problem which is different from the conventional model. Suppose we are given many storages (nodes) to store multiple kinds of commodities together with roads (edges) interconnecting them, which are specified as a weighted graph. Some storages have surplus and others have shortages. Problem is to determine whether there are transportations to eliminate all of shortages. For transportation we can use a vehicle with the loading capacity at each node. Each vehicle visits one of its neighbors with some commodities which are unloaded at the neighbor. Then, we load some other commodities there, and then bring them back to the original node. How to design such send-and-bring-back transportations to eliminate all shortages is the problem. When we define a single round of transportations to be a set of those transportations at all nodes, whether there is a single round of valid transportations that eliminate all of shortages is our concern. After proving NP-completeness of the problem we present a linear time algorithm for a special case where an input graph is a forest.

Keywords: Forest · Linear program · Multi-commodity transportation problem · NP-completeness · Sending and bringing-back

1 Introduction

In this paper we consider a new type of a transportation problem which is different from the conventional ones but based on a very natural model. A typical transportation problem in operations research is to find the minimum cost of transporting a single commodity from m sources to n destinations along edges in a given network. This type of problems can be solved using linear programming. Many variations have been considered [2].

The transportation problem to be considered in this paper is defined very naturally. We are given a weighted graph $G = (V, E, w)$, where V is a set of places (nodes) to store many different commodities and E is a set of roads (edges) interconnecting nodes. At each node we can store multiple kinds of commodities. Some storages have surplus and others have shortages. Formally, we specify

© Springer Nature Switzerland AG 2021
R. Uehara et al. (Eds.): WALCOM 2021, LNCS 12635, pp. 13–24, 2021.
https://doi.org/10.1007/978-3-030-68211-8_2

quantities of commodities associated with a node u by $(w_1(u), w_2(u), \ldots, w_h(u))$, where $w_k(u)$ represents a quantity of the k-th commodity. If $w_k(u) > 0$ then $w_k(u)$ units of the k-th commodity are stored at node u. On the other hand, $w_k(u) < 0$ means that $|w_k(u)|$ units of the k-th commodity are needed at u.

A vehicle is available at each node for transportation. A transportation between two adjacent nodes u and v using a vehicle at u is done as follows. We load some amounts of commodities at node u, which are specified by $(s_1(u), s_2(u), \ldots, s_h(u))$. After visiting a target node v and unloading them at node v, we load commodities at v, which are specified by $(b_1(u), b_2(u), \ldots, b_h(u))$, and bring them back to the node u. All of vehicles start their nodes simultaneously. The first round of transportations finishes when the transportations are completed at all nodes.

One of our goals is to find a single round of transportations that eliminate all shortages. Of course, it is not realistic to transport a negative quantity of commodity and also to load more than there exists at a node. We also assume that each vehicle has some loading capacity, C. All the vehicles leave their nodes at the same time. Each transportation must be independent of their order. In other words, transportations between adjacent nodes are not synchronous and quantities of commodities are determined before the start time of all transportations. Without such constraint, it would be possible to send commodities no matter how far away it is (by sending some load from node u_1 to u_2 and then we send the load from u_2 to u_3 and so on). This assumption is referred to as the **order independence assumption**, which is a rigid principle in this paper. When we consider a single round of transportations, a transportation from non-negative to non-negative weight or from non-positive to non-positive weight is not effective and thus excluded.

The problem of measuring the Earth Mover's distance is related to this problem, which is a problem of finding the most economic way of transporting soil placed in many different places to holes to fill in. Earth movers are used for transporting the soil. It is possible to use whatever number of earth movers. Problem is to minimize the total cost of transportations. This problem looks hard, but a polynomial-time algorithm using network flow algorithm is known. It is similar to the problem in this paper, but it is different in the following senses: (1) we transport not only one kind of commodity but many kinds, (2) in our case transportations are restricted only between adjacent nodes, (3) we can send commodities from a node u to its adjacent node v and also bring commodities at v back to u, and (4) we discard transportation distances and weight of commodities to carry. Refer to the paper [4] for more detail about the earth mover's distance.

Our first question is, given a weighted graph $G = (V, E, w)$, whether there is a single round of transportations that eliminate all of shortages. Our second question is to find the minimum number of rounds that eliminate all of shortages. The second question looks much harder than the first one. In this paper we first consider the problem in one dimension and show that the first problem can be solved in linear time. The problem in two dimensions is much harder. After proving NP-completeness of the first problem above we present a linear-time algorithm for a special case where an input graph is a forest.

2   Problem Definition

Consider a weighted graph $G = (V, E, w)$, where V is a set of places (nodes) to store commodities and E a set of roads (edges) interconnecting those nodes. At each node we store different kinds of commodities. Some nodes have surplus and others have shortages. Formally, We specify quantities of commodities associated with a node u by $(w_1(u), w_2(u), \ldots, w_h(u))$, where $w_k(u)$ represents a quantity of the k-th commodity. If $w_k(u) > 0$ then $w_k(u)$ units of commodity are stored at node u. On the other hand, $w_k(u) < 0$ means that $|w_k(u)|$ units of the k-th commodity are needed at u. A vehicle for transportation is available at each node.

We specify a transportation using a vehicle at node u by a tuple $(u, \tau(u), (s_1(u), s_2(u), \ldots, s_h(u)), (b_1(u), b_2(u), \ldots, b_h(u)))$, where $\tau(u)$ is a target node and the amounts of commodities to be sent from u to $\tau(u)$ and those brought back from $\tau(u)$ to u are specified by $(s_1(u), s_2(u), \ldots, s_h(u))$ and $(b_1(u), b_2(u), \ldots, b_h(u))$, respectively. Since each vehicle has the loading capacity C, the sum of load must always be within C, that is, for each transportation we must have $\sum_{i=1}^{h} s_i(u) \leq C$ and $\sum_{i=1}^{h} b_i(u) \leq C$.

All of vehicles leave their nodes at the same time. Due to the order independence assumption, when we send some commodities from u to $\tau(u)$, the amounts of commodities sent from u must be within $(\max(0, w_1(u)), \max(0, w_2(u)), \ldots, \max(0, w_h(u)))$ and also those of commodities brought back from $\tau(u)$ must be within $(\max(0, w_1(\tau(u))), \max(0, w_2(\tau(u))), \ldots, \max(0, w_h(\tau(u))))$.

More formally, for each node u we must have

$$s_k(u) + \sum_{v s.t. \tau(v) = u} b_k(v) \leq \max(0, w_k(u)) \text{ for each } k = 1, 2, \ldots, h, \text{ and}$$

$$\max\left(\sum_{k=1}^{h} s_k(u), \sum_{k=1}^{h} b_k(u)\right) \leq C.$$

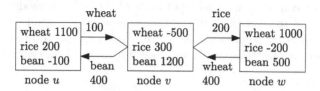

Fig. 1. An example of send-and-bring-back transportations. We send 100 units of wheat at node u, visit v, and bring-back 400 units of bean on the way back from v.

Figure 1 shows an example of transportations. Figure 2 gives another example. In the example, the transportation from node b to node a is characterized by $T_b = (b, a, (0, 30), (60, 0))$ which means we send 0 units of the first commodity and 30 units of the second commodity from the current node b to node a, and then bring-back 60 units of the first one and 0 units of the second one from a to b. Using these transportations we can eliminate all of shortages as shown in (b) in the figure.

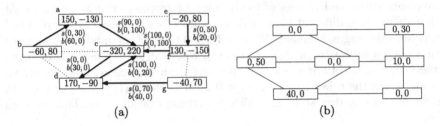

(a)   (b)

Fig. 2. Transportations necessary to eliminate all the shortages. (a) a single round of transportations (dotted lines are unused edges), and (b) the result after the transportations.

Under the definitions above, we consider the following two problems.

Problem 1: Given a weighted graph, determine whether there is a single round of transportations that eliminate all of shortages. Also, output such a set of transportations if any.

Problem 2: Given a weighted graph, find a single round of feasible transportations that minimize the largest shortage. Also, output such a set of transportations.

2.1 Formal Definitions and Basic Properties

An input instance to our problem is given by a graph $G = (V, E, w)$ in which each node u has h weights $w_1(u), w_2(u), \ldots, w_h(u)$. A transportation using a vehicle at node u is specified by a tuple $(u, \tau(u), (s_1(u), s_2(u), \ldots, s_h(u)), (b_1(u), b_2(u), \ldots, b_h(u)))$ which means that we load $s_i(u)$ units of the i-th commodity at node u for each $i \in [1, h]$ to send them to the target node $\tau(u)$, unload them at $\tau(u)$, and load $b_i(u)$ units of commodity at $\tau(u)$ for each $i \in [1, h]$ to bring them back to u. A single round of transportations is given by a set of transportations at all nodes.

Definition: (Basic Property) A single round of transportations for a graph $G = (V, E, w)$ given by $\mathcal{T} = \{(u, \tau(u), (s_1(u), s_2(u), \ldots, s_h(u)), (b_1(u), b_2(u), \ldots, b_h(u))) | u \in V\}$ is feasible if

(0) $s_i(u) \geq 0$ and $b_i(u) \geq 0$ for each $u \in V$ and $i \in [1, h]$,
(1) the node $\tau(u)$ is adjacent to node u,

(2) $\max(\sum_{i=1}^{h} s_i(u), \sum_{i=1}^{h} b_i(u)) \leq C$ for each $u \in V$, where C is the loading capacity,

(3) there is at least one $i \in [1, h]$ such that $w_i(u) \cdot w_i(\tau(u)) < 0$,

(4) for each $u \in V$, if $w_i(u) > 0$ and $w_i(\tau(u)) < 0$ then $s_i(u) \geq 0$ and $b_i(u) = 0$ for each $i \in [1, h]$,

(5) for each $u \in V$, if $w_i(u) < 0$ and $w_i(\tau(u)) > 0$ then $s_i(u) = 0$ and $b_i(u) \geq 0$ for each $i \in [1, h]$, and

(6) for each $u \in V$ such that $w_i(u) > 0$, $s_i(u) + \sum_{vs.t.\tau(v)=u} b_i(v) \leq w_i(u)$ for each $i \in [1, h]$.

The condition (1) above states that a target node must be selected among those adjacent to node u. The condition (2) states that the sum of load sent from u to $\tau(u)$ and the sum of load brought back from $\tau(u)$ to u must be bounded by the loading capacity. Whenever we transport some units of commodity, we must leave a node having surplus commodity and visit another node of shortage, which is stated in (3). The conditions (4) and (5) state that either $s_i(u)$ or $b_i(u)$ must be 0 for each $i \in [1, h]$. The last condition states that the sum of load of the i–th commodity carried out of a node u cannot exceed the original amount $w_i(u)$ of the commodity at u.

Lemma 1. *A feasible single round of transportations* $T = \{(u, \tau(u), (s_1(u), s_2(u), \ldots, s_h(u)), (b_1(u), b_2(u), \ldots, b_h(u)))|u \in V\}$ *eliminate all of shortages if* $w_i(u) + b_i(u) + \sum_{vs.t.\tau(v)=u} s_i(v) \geq 0$ *for each* $i \in [1, h]$ *and* $u \in V$ *such that* $w_i(u) < 0$.

Given a single round of transportations $T = \{(u, \tau(u), (s_1(u), s_2(u), \ldots, s_h(u)), (b_1(u), b_2(u), \ldots, b_h(u)))|u \in V\}$ for a graph $G = (V, E, w)$, we can define a directed graph $G^d = (V, E^d, w)$ where E^d contains a directed edge $(u, \tau(u))$ for each $u \in V$.

Lemma 2. *A single round of transportations* T *for a weighted graph* $G = (V, E, w)$ *is feasible only if the corresponding graph* G^d *has* **no two cycles** *(including oppositely directed edges between two nodes, called double edges) are contained in one connected component in its underlying undirected graph.*

3 One-Dimensional Transportation Problem

3.1 One-Commodity Problem Without Capacity Constraint

Generally each node stores many different kinds of commodities. First of all, consider only one kind of commodity in each node. Also we assume that a graph is a path, i.e., a linearly ordered array. Suppose n nodes on the horizontal line are numbered like $1, 2, \ldots, n$ from left to right. As input, we assume the quantity of the commodity at node i is given as $w(i)$, where $w(i) < 0$ means a shortage.

Consider a simple case where no loading capacity is assumed. Suppose we are given a sequence $(w(1), w(2), \ldots, w(n))$ of quantities of commodity. There are four cases to consider on the first two nodes. If $w(1) \geq 0$ and $w(2) \geq 0$ then we just discard node 1 since any transportation from node 1 to 2 has no effect. If $w(1) > 0$ and $w(2) < 0$ then we send $w(1)$ units of commodity using the vehicle at node 1 to reduce the shortage at node 2. If $w(1) + w(2) > 0$ then we have to reduce the amount of transportation to prevent the resulting surplus from sending to the right. If $w(1) < 0$ and $w(2) > 0$ then we use the vehicle at node 1 to bring back $\min(|w(1)|, w(2))$ units of commodity from node 2 to 1. If it results in non-negative quantity at node 1, then we proceed to the next step. Otherwise, the sequence is not feasible, i.e., there is no transportation schedule to eliminate the shortage at node 1. If $w(1) < 0$ and $w(2) < 0$ then there is no way to eliminate the shortage at node 1, and hence we just report that the sequence is not feasible and stop.

Lemma 3. *For any instance of a one-dimensional one-commodity transportation problem without constraint on loading capacity we can decide in linear time whether there is a single round of transportations that eliminate all of shortages.*

3.2 One-Commodity Transportation Problem with Loading Capacity

Next consider some loading capacity C on the total weight to carry by a vehicle. Input is specified by a sequence $(w(1), w(2), \ldots, w(n))$ of n values representing the quantities of commodities. If it contains 0 somewhere, say $w(i) = 0$, then to decide the feasibility we can separate the sequence into two subsequences $(1, \ldots, i-1)$ and $(i+1, \ldots, n)$ since there is no effective transportation from/to the node i of weight 0. If two consecutive values $w(i)$ and $w(i+1)$ are of the same sign, then we can separate the sequence into $(1, \ldots, i)$ and $(i+1, \ldots, n)$ since effective transportation occurs only between two nodes of different signs.

We assume that a sequence starts from a non-negative weight and ends also at a non-negative weight. If not, we add weight 0 at head and/or tail. We also assume that no two consecutive weights have the same sign.

Starting from the initial subsequence $(w(1), w(2), w(3))$, we find optimal solutions that can send the largest amount to the right in two settings. If it is not feasible then there is no feasible solution to the whole sequence. One solution is a feasible solution without constraint and the other with constraint in its last part. Recall that any solution can be represented by a graph on nodes $\{1, 2, 3\}$ with directed edges. We are interested in whether the last part, more exactly, the connected component of the last node 3 in this case in the underlying undirected graph contains double edges. If the last part contains double edges, then this causes some constraint to solutions of further right.

It is easy to compute the best solutions with/without using double edges in the last part. Let them be $\omega_u(3)$ and $\omega_c(3)$, where "u" and "c" represent "unconstrained" and "constrained", respectively. We extend the subsequence by two nodes. Suppose we have computed $\omega_u(k)$ and $\omega_c(k)$. To compute the best solution $\omega_u(k+2)$ there are two ways. One is to use $\omega_u(k)$ and to specify edges among $k, k+1$, and $k+2$ so that double edges are not included there. The other is to use $\omega_u(k)$ and to specify edges within $[k, k+2]$ so that double edges are not included there and also a cut is included between k and $k+1$ or between $k+1$ and $k+2$. The better value among them gives the value of $\omega_u(k+2)$.

Compute of $\omega_c(k+2)$ is symmetric. We compute one solution by using $\omega_u(k)$ and including double edges somewhere among $k, k+1$, and $k+2$. We compute the other one by using $\omega_c(k)$ and including double edges somewhere among $k, k+1$, and $k+2$. The better value among them gives the value of $\omega_c(k+2)$.

Finally, if none of $\omega_u(k+2)$ and $\omega_c(k+2)$ is defined then we have a conclusion that there is no feasible solution. Otherwise, we obtain an optimal feasible solution by taking better one among $\omega_u(n)$ and $\omega_c(n)$.

A formal description of the algorithm is as follows.

Algorithm for determining the feasibility of a given sequence.
instance: $(w(1), w(2), \ldots, w(n))$.

If the first and/or last elements are negative we insert 0 as the first and/or last elements, respectively.

output: True, if there is a single round of transportations to eliminate all of shortages, and **False** otherwise.

algorithm:

// Idea is to keep two solutions for each subsequence $(1, 2, \ldots, k)$, one without constraint and the other with constraint in its last part.

For the subsequence $(1, 2, 3)$ compute the best solutions, $\omega_u(3)$ in one case of no constraint and $\omega_c(3)$ in the other case of constraint, in their last part.

for $k = 3$ **to** n **step 2 do**{

    if $\omega_u(k)$ is defined **then**{

        $\omega_{uu}(k+2) = \text{extend}(\omega_u(k), \text{unconstrained})$.

        $\omega_{uc}(k+2) = \text{extend}(\omega_u(k), \text{constrained})$.

    } **else if** $\omega_c(k)$ is defined **then**{

        $\omega_{cu}(k+2) = \text{extend}(\omega_c(k), \text{unconstrained})$.

        $\omega_{cc}(k+2) = \text{extend}(\omega_c(k), \text{constrained})$.

    }

    Choose the better one among $\omega_{uu}(k+2)$ and $\omega_{cu}(k+2)$ as $\omega_u(k+2)$.

    Choose the better one among $\omega_{uc}(k+2)$ and $\omega_{cc}(k+2)$ as $\omega_c(k+2)$.

}

if at least one of $\omega_u(n)$ and $\omega_c(n)$ is defined and non-negative

then return True. else return False.

function extend$(\omega(k), \text{cons})${

    if cons = unconstrained **then**{

        extend the solution $\omega(k)$ by two nodes so that the extended solution is not constrained in its last part.

        if it is possible **then return** the extended solution.

        else return False.

    } **else** { // cons = constrained

        extend the solution $\omega(k)$ by two nodes so that the extended solution is constrained in its last part.

        if it is possible **then return** the extended solution.

else return False.

}

}

See Fig. 3 as an example. In this example, we have only the unconstrained solution $\omega_u(3)$. Extending it by two nodes, we have only constrained solution $\omega_c(5)$. Using it, we have constrained and unconstrained solutions, $\omega_u(7)$ and $\omega_c(7)$. Finally, we have only constrained solution $\omega_c(9)$.

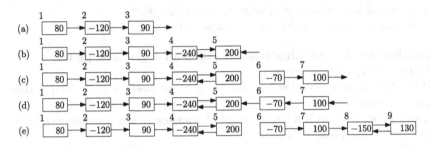

Fig. 3. A behavior of the algorithm. (a) only $\omega_u(3)$ exists, (b) only $\omega_c(5)$ exists, (c) unconstrained solution $\omega_u(7)$, (d) constrained solution $\omega_c(7)$, (e) only constrained solution $\omega_c(9)$ exists, which is a solution.

Lemma 4. *For any instance of a transportation problem with finite loading capacity C we can decide in linear time whether there is a single round of transportations that eliminate all of shortages.*

3.3   Optimization Problem

Now, consider the optimization problem to find the minimum shortage we can achieve for a given instance of one-commodity transportation problem. We do not know whether there is a polynomial-time algorithm, but we can design a pseudo-polynomial-time algorithm as follows. Given an instance of one-commodity transportation problem, let $-M$ be the largest shortage. Let $0 < t < M$ be arbitrary number between 0 and M. If we add t to every shortage at each node, we have a modified problem $\mathcal{P}(t)$. More exactly, for each negative weight $w(u) < 0$ we set $w(u) = \min(0, w(u) + t)$ so that no new positive weight is generated. We can decide the feasibility of the problem $\mathcal{P}(t)$ in linear time. If it is feasible then the minimum shortage is at most t. Otherwise, we can conclude that it is beyond t. Using the observation we can find the minimum shortage at any precision by using binary search. The number of iterations is $O(\log M)$.

Lemma 5. *Given a one-dimensional instance of a one-commodity transportation problem, we can determine the minimum shortage in linear time at any precision.*

3.4   Multi-commodity Transportation Problem

In a multi-commodity transportation problem each node u is characterized by an h-tuple of values $(w_1(i), w_2(i), \ldots, w_h(i))$. If no constraint on loading capacity is assumed, we can implement a transportation for each commodity independently. Therefore, we could apply the algorithm for one-commodity problems to solve a multi-commodity problem.

With constraint on loading capacity, however, we have a trouble. Recall that the algorithm for one-commodity case is based on the fact that a solution for the whole array contains at most two different solutions for the first three nodes. We have found two solutions, one unconstrained and the other constrained. But this is not true anymore for multi-commodity case. For we can design a simple example consisting of four nodes such that any feasible solution for the first three nodes fails to be feasible due to the fourth node.

4   Two-Dimensional Transportation Problem

4.1   NP-completeness

In this subsection we prove NP-completeness of the problem of deciding whether an instance of a general two-dimensional transportation problem is feasible or not, that is, whether there is a single round of transportations that eliminate all of shortages in the instance.

Lemma 6. *The problem of deciding whether, given a weighted graph, there is a single round of transportations with loading capacity C that eliminate all of shortages using sending and bringing-back operations is NP-complete.*

Proof Our proof is based on a reduction from integer partition problem, one of NP-complete problems [1].

Suppose the set $\{a_1, a_2, \ldots, a_{2n}\}, a_i > 0, i = 1, \ldots, 2n$ is an instance of integer partition, where $\sum_{i=1}^{2n} a_i = 2A$. Then, let U be a collection $\{u_1, u_2, \ldots, u_{2n}, u_{2n+1}, u_{2n+2}\}$ with $w(u_i) = A + a_i, i = 1, \ldots, 2n$ and $w(u_{2n+1}) = w(u_{2n+2}) = 2n^2A$, and let V be the pair $\{v_1, v_2\}$ with $w(v_1) = w(v_2) = -(2n^2 + n + 1)A$. Consider a bipartite graph $G = (U, V, E)$ where E consists of all edges between U and V. Assume that the loading capacity C is n^2A.

Consider a single round of transportations on the graph G, expressed as a graph G^d. First observation is that each of v_1 and v_2 is incident to exactly one of u_{2n+1} and u_{2n+2} in G^d. If v_1 is incident to none of them, then it is impossible to eliminate the shortage $-(2n^2+n+1)A$ at v_1 even if we send storages of all other nodes to v_1 since $w(v_1)+\sum_{i=1}^{2n} a_i = -(2n^2+n+1)A+2nA+2A < 0$. So, we can assume without loss of generality that v_1 is connected to u_{2n+1} and v_2 to u_{2n+2} in G^d. Moreover, the connection between them must be bi-directional, that is, we have to send n^2A units of commodity from u_{2n+1} to v_1 using the vehicle at u_{2n+1} and also to bring the same amount back to u_{2n+1} since otherwise there is no way to eliminate the large shortage at v_1. So, we can assume that we have bidirectional transportations (double edges in G^d) between u_{2n+1} and v_1 and also ones between u_{2n+2} and v_2. To have a feasible set of transportations we have to send commodities from nodes $u_1, u_2, \ldots, u_{2n}$ to either v_1 or v_2 using "send" operations using vehicles at those nodes.

Let U_1 and U_2 be the sets of nodes connected to v_1 and v_2, respectively. Suppose $\sum_{u \in U_1} w(u) < \sum_{u \in U_2} w(u)$. Since the total sum is given by $\sum_{i=1}^{2n+2} w(u_i) = 2n^2 A + 2nA + 2A = 2(n^2 + n + 1)A$, $\sum_{u \in U_1} w(u) \le (n^2 + n + 1)A$. To eliminate the shortage at v_1 we must have $w(v_1) + \sum_{u \in U_1} w(u) = -(n^2 + n + 1)A + (n^2 + n + 1)A = 0$. This implies that $\sum_{u \in U_1} w(u) = \sum_{u \in U_2} w(u) = (n^2 + n + 1)A$.

The above argument implies that if there is a single round of transportations that eliminate all of shortages then there is an integer partition for the set $\{a_1, a_2, \ldots, a_{2n}\}$. $\qquad\square$

4.2   One-Commodity Transportation Problem on a Forest

Consider a special case where we have a single kind of commodity and a graph G is a (undirected) forest composed of trees. We assume that end nodes of each edge have weights of different signs since an edge between two nodes of weights of the same sign has no meaning for transportation as far as we are interested in a single round of transportations.

Consider a simple case first where there is no constraint on the loading capacity. We can deal with each tree in a forest independently. Each tree has at least two **leaf nodes** (of degree 1). Starting from a leaf node u, we traverse a tree until we encounter a **branching node** v of degree ≥ 3. If there is no such branching node, it is just a path, for which we already had an algorithm in Sect. 3.

Such a path from a leaf node u to branching node v is called a **leaf path**. If a leaf node is adjacent to a leaf branching node, the leaf node is considered as a degenerated leaf path. A **leaf branching node** is a node v of degree ≥ 3 which is incident to at least $deg(v) - 1$ leaf paths, where $deg(v)$ is the degree of v. We create a one-dimensional instance by replacing v in P with an imaginary node $v(P)$ whose weight is 0 if $w(u_k) > 0$ and $+\infty$ otherwise. We can solve the one-dimensional problem using the algorithm in Sect. 3 which gives us an optimal value x in linear time. If it is infeasible then we have a conclusion that we have no feasible solution. Otherwise, we replace the path with a single node having weight determined by the value of x, that is, x if $w(u_k) > 0$ and $-x$ otherwise.

If we have a leaf branching node whose adjacent set consists of at most one internal node (of degree ≥ 2) and leaf nodes, then we just combine those leaf nodes with the branching node into a single node after a maximal transportation from every such leaf node to the branching node. We repeat the process while each leaf path is feasible. If a single node of weight ≥ 0 is left, then we have a feasible realization. Otherwise, the given tree is not feasible.

A formal description of the algorithm is given below.

Algorithm for determining the feasibility of a given sequence.
instance: $(w(1), w(2), \ldots, w(n))$.

If the first and/or last elements are negative we insert 0 as the first and/or last elements, respectively.

output: True, if there is a single round of transportations to eliminate all of shortages, and **False** otherwise.

algorithm:

// Idea is to keep two solutions for each subsequence $(1, 2, \ldots, k)$, one without constraint and the other with constraint in its last part.

For the subsequence $(1, 2, 3)$ compute the best solutions, $\omega_u(3)$ in one case of no constraint and $\omega_c(3)$ in the other case of constraint, in their last part.

for $k = 3$ **to** n **step 2 do**{

    if $\omega_u(k)$ is defined **then**{

        $\omega_{uu}(k + 2) = \text{extend}(\omega_u(k), \text{unconstrained})$.

        $\omega_{uc}(k + 2) = \text{extend}(\omega_u(k), \text{constrained})$.

    } **else if** $\omega_c(k)$ is defined **then**{

        $\omega_{cu}(k + 2) = \text{extend}(\omega_c(k), \text{unconstrained})$.

        $\omega_{cc}(k + 2) = \text{extend}(\omega_c(k), \text{constrained})$.

    }

    Choose the better one among $\omega_{uu}(k + 2)$ and $\omega_{cu}(k + 2)$ as $\omega_u(k + 2)$.

    Choose the better one among $\omega_{uc}(k + 2)$ and $\omega_{cc}(k + 2)$ as $\omega_c(k + 2)$.

}

if at least one of $\omega_u(n)$ and $\omega_c(n)$ is defined and non-negative **then return True. else return False.**

function extend$(\omega(k), \text{cons})${

    if cons = unconstrained **then**{

        extend the solution $\omega(k)$ by two nodes so that the extended solution is not constrained in its last part.

        if it is possible **then return** the extended solution.

        else return False.

    } **else** { // cons = constrained

        extend the solution $\omega(k)$ by two nodes so that the extended solution is constrained in its last part.

        if it is possible **then return** the extended solution.

        else return False.

    }

}

Lemma 7. *Assume no constraint on the loading capacity. Given a weighted forest, it is possible in linear time to determine whether there is a single round of transportations that eliminate all of shortages.*

Next, we consider the constraint on the loading capacity. In the case where no loading capacity is assumed, when we shorten a leaf path to a single node, it induces no constraints to the process of the remaining part. However, in this general case, we have to consider such a constraint. So, we modify the algorithm for a leaf path. We take a leaf path and shorten it to a single node, which becomes a child of a leaf branching node v. When we combine the node v with its children into a single node v', we traverse the tree again from the node v' until we encounter a branching node, which makes a leaf path again. The leaf path may have a constraint in its first part if one of its children is produced with constraint (double edges).

Taking the situations above into account, we consider for a leaf path four different cases depending on whether it has a constraint in its first part and also in its last part. Formally, for a leaf path P_i, we compute four different solutions $\omega_{uu}(P_i), \omega_{uc}(P_i), \omega_{cu}(P_i)$, and $\omega_{cc}(P_i)$, where $\omega_{uc}(P_i)$, for example, is the value of an optimal solution in the case where P_i is unconstrained in its first part but constrained in its last part. It is not so difficult to modify the algorithm for one-dimensional array so that it can compute the four values. The modified algorithm also runs in linear time. Although we omit the detail, we have the following lemma.

Lemma 8. *Given a forest weighted arbitrarily, it is possible in linear time to determine whether there is a single round of transportations that eliminate all of shortages.*

5   Concluding Remarks

We have many open questions. (1) Extension to a multi-commodity problem is not known even for one-dimensional problems. Is there any polynomial-time algorithm? (2) We had an efficient algorithm for minimizing the largest shortage in any precision, which runs in $O(n \log M)$ time where n and M are the size of the array and the largest shortage, respectively. It is efficient in practice, but it also depends on $\log M$. Is there any polynomial-time algorithm which does not depend on M? (3) A more general problem is to determine how many rounds of transportations are needed to eliminate all of shortages. It is more difficult since transportation from a node of positive weight to one of positive weight is effective in two rounds. (4) We assumed that only one vehicle is available, but what happens if two or more vehicles are available at some busy nodes? (5) If vehicles at some nodes are not available, how can we compensate them?

Many other open questions exist although we have no space to list them.

Acknowledgment. This work was supported by JSPS KAKENHI Grant Number JP20K11673. The author would like to thank David Kirkpatrick and Ryuhei Uehara for giving a version of the NP-completeness proof of the transportation problem in two dimensions.

References

1. Andrews, G.E., Eriksson, K.: Integer Partitions. Cambridge University Press, Cambridge (2004)
2. Appa, G.M.: The Transportation problem and its variants. Oper. Res. Q. **24**, 79–99 (1973)
3. Cormen, T.H., Leiserson, C.E., Rivest, R.L., Stein, C.: Introduction to Algorithms, 2nd edn. MIT Press and McGraw-Hill (2001)
4. Peleg, S., Werman, M., Rom, H.: A unified approach to the change of resolution: space and gray-level. IEEE Trans. Pattern Anal. Mach. Intell. **11**, 739–742 (1989)

Long Papers

Long Papers

Algorithms for Diameters of Unicycle Graphs and Diameter-Optimally Augmenting Trees

Haitao Wang and Yiming Zhao$^{(\boxtimes)}$

Department of Computer Science, Utah State University, Logan, UT 84322, USA
{haitao.wang,yiming.zhao}@usu.edu

Abstract. We consider the problem of computing the diameter of a unicycle graph (i.e., a graph with a unique cycle). We present an $O(n)$ time algorithm for the problem, where n is the number of vertices of the graph. This improves the previous best $O(n \log n)$ time solution [Oh and Ahn, ISAAC 2016]. Using this algorithm as a subroutine, we solve the problem of adding a shortcut to a tree so that the diameter of the new graph (which is a unicycle graph) is minimized; our algorithm takes $O(n^2 \log n)$ time and $O(n)$ space. The previous best algorithms solve the problem in $O(n^2 \log^3 n)$ time and $O(n)$ space [Oh and Ahn, ISAAC 2016], or in $O(n^2)$ time and $O(n^2)$ space [Bilò, ISAAC 2018].

Keywords: Diameter · Unicycle graphs · Augmenting trees · Shortcuts

1 Introduction

Let G be a graph of n vertices with a positive length on each edge. A *shortest path* connecting two vertices s and t in G is a path of minimum total edge length; the length of the shortest path is also called the *distance* between s and t. The *diameter* of G is the maximum distance between all pairs of vertices of G. G is a *unicycle graph* if it has only one cycle, i.e., G is a tree plus an additional edge.

We consider the problem of computing the diameter of a unicycle graph G. Previously, Oh and Ahn [10] solved the problem in $O(n \log n)$ time, where n is the number of vertices of G. We present an improved algorithm of $O(n)$ time. Using our new algorithm, we also solve the *diameter-optimally augmenting tree* (DOAT for short) problem, defined as follows.

Let T be a tree of n vertices such that each edge has a positive length. We want to add a new edge (called *shortcut*) to T such that the new graph (which is a unicycle graph) has the minimum diameter. We assume that there is an oracle that returns the length of any given shortcut in $O(1)$ time. Previously, Oh and Ahn [10] solved the problem in $O(n^2 \log^3 n)$ time and $O(n)$ space, and Bilò [1]

This research was supported in part by NSF under Grant CCF-2005323.

R. Uehara et al. (Eds.): WALCOM 2021, LNCS 12635, pp. 27–39, 2021.
https://doi.org/10.1007/978-3-030-68211-8_3

reduced the time to $O(n^2)$ but the space increases to $O(n^2)$. The problem has an $\Omega(n^2)$ lower bound on the running time as all $\Theta(n^2)$ possible shortcuts have to be checked in order to find an optimal shortcut [10]. Hence, Bilò's algorithm is time-optimal. In this paper, we propose an algorithm with a better time and space trade-off, and our algorithm uses $O(n^2 \log n)$ time and $O(n)$ space.

1.1   Related Work

The diameter is an important measure of graphs and computing it is one of the most fundamental algorithmic graph problems. For general graphs or even planar graphs, the only known way to compute the diameter is to first solve the all-pair-shortest-path problem (i.e., compute the distances of all pairs of vertices of the graph), which inherently takes $\Omega(n^2)$ time, e.g., [5,15]. Better algorithms exist for special graphs. For example, the diameter of a tree can be computed in linear time, e.g., by first computing its center [9]. If G is an outerplanar graph and all edges have the same length, its diameter can be computed in linear time [4]. The diameter of interval graphs (with equal edge lengths) can also be computed in linear time [11]. Our result adds the unicycle graph (with different edge lengths) to the linear-time solvable graph category.

The DOAT problem and many of its variations enjoy an increasing interest in the research community. If the tree T is embedded in a metric space (so that the triangle inequality holds for edge lengths), Große et al. [7] first solved the problem in $O(n^2 \log n)$ time. Bilò [1] later gave an $O(n \log n)$ time and $O(n)$ space algorithm, and another $(1 + \epsilon)$-approximation algorithm of $O(n + \frac{1}{\epsilon} \log \frac{1}{\epsilon})$ time and $O(n + \frac{1}{\epsilon})$ space for any $\epsilon > 0$. A special case where T is a path embedded in a metric space was first studied by Große et al. [6], who gave an $O(n \log^3 n)$ time algorithm, and the algorithm was later improved to $O(n \log n)$ time by Wang [12]. Hence, Bilò's work [1] generalizes Wang's result [12] to trees.

A variant of the DOAT problem which aims to minimize the *continuous diameter*, i.e., the diameter of T is measured with respect to all the points of the tree (including the points in the interior of the edges), has also been studied. If T is a path embedded in the Euclidean plane, De Carufel et al. [2] solved the problem in $O(n)$ time. If T is a tree embedded in a metric space, De Carufel et al. [3] gave an $O(n \log n)$ time algorithm. If T is a general tree, Oh and Ahn [10] solved the problem in $O(n^2 \log^3 n)$ time and $O(n)$ space.

The DOAT problem is to minimize the diameter. The problem of minimizing the radius was also considered. For the case where T is a path embedded in a metric space, Johnson and Wang [8] presented a linear time algorithm which adds a shortcut to T so that the radius of the resulting graph is minimized. The radius considered in [8] is defined with respect to all points of T, not just the vertices. Wang and Zhao [14] studied the same problem with radius defined with respect to only the vertices, and they gave a linear time algorithm.

1.2 Our Approach

To compute the diameter of a unicycle graph G, Oh and Ahn [10] reduces the problem to a geometric problem and then uses a one-dimensional range tree to solve the problem. We take a completely different approach. Let C be the unique cycle of G. We define certain "domination" relations on the vertices of C so that if a vertex v is dominated by another vertex then v is not important to the diameter. We then present a pruning algorithm to find all undominated vertices (and thus those dominated vertices are "pruned"); it turns out that finding the diameter among the undominated vertices is fairly easy. In this way, we compute the diameter of G in linear time.

For the DOAT problem on a tree T, Oh and Ahn [10] considered all possible shortcuts of T by following an Euler tour of T; they used the aforementioned 1D range tree to update the diameter for the next shortcut. Bilò's method [1] is to transform the problem to adding a shortcut to a path whose edge lengths satisfy a property analogous to the triangle inequality (called graph-triangle inequality) and then the problem on P can be solved by applying the $O(n \log n)$ time algorithm for trees in metric space [1]. Unfortunately, the problem transformation algorithm needs $\Theta(n^2)$ space to store the lengths of all possible $\Theta(n^2)$ shortcuts of T. The algorithm has to consider all these $\Theta(n^2)$ shortcut lengths in a global manner and thus it inherently uses $\Omega(n^2)$ space. Note that Bilò's method [1] does not need an algorithm for computing the diameter of a unicycle graph.

We propose a novel approach. We first compute a diametral path P of T. Then we reduce the DOAT problem on T to finding a shortcut for P. To this end, we consider vertices of P individually. For each vertex v_i of P, we want to find an optimal shortcut with the restriction that it must connect v_i, dubbed a v_i-shortcut. For this, we define a "domination" relation on all v_i-shortcuts and we show that those shortcuts dominated by others are not important. We then design a pruning algorithm to find all shortcuts that are not dominated by others; most importantly, these undominated shortcuts have certain monotonicity properties that allow us to perform binary search to find an optimal v_i-shortcut by using our diameter algorithm for unicycle graphs as a subroutine. With these effort, we find an optimal v_i-shortcut in $O(n \log n)$ time and $O(n)$ space. The space can be reused for computing optimal v_i-shortcuts of other vertices of P. In this way, the total time of the algorithm is $O(n^2 \log n)$ and the space is $O(n)$.

Outline. In the following, we present our algorithm for computing the diameter of a unicycle graph in Sect. 2. Section 3 solves the DOAT problem. Due to the space limit, many proofs are omitted but can be found in our full paper [13].

2 Computing the Diameter of Unicycle Graphs

In this section, we present our linear time algorithm for computing the diameter of unicycle graphs.

For a subgraph G' of a graph G and two vertices u and v from G', we use $\pi_{G'}(u, v)$ to denote a shortest path from u to v in G' and use $d_{G'}(u, v)$ to denote the length of the path. We use $\Delta(G)$ to denote the diameter of G. A pair of vertices (u, v) is called a *diametral pair* and $\pi_G(u, v)$ is called a *diametral path* if $d_G(u, v) = \Delta(G)$.

In the following, let G be a unicycle graph of n vertices. Our goal is to compute the diameter $\Delta(G)$ (along with a diametral pair). Let C denote the unique cycle of G.

2.1 Observations

Removing all edges of C (while keeping its vertices) from G results in several connected components of G. Each component is a tree that contains a vertex v of C; we use $T(v)$ to denote the tree. Let $v_1, v_2, \ldots, v_m$ be the vertices ordered clockwise on C. Let $\mathcal{T}(G) = \{T(v_i) \mid 1 \leq i \leq m\}$. Note that the sets of vertices of all trees of $\mathcal{T}(G)$ form a partition of the vertex set of G.

Consider a diametral pair (u^*, v^*) of G. There are two cases: (1) both u^* and v^* are in the same tree of $\mathcal{T}(G)$; (2) u^* and v^* are in two different trees of $\mathcal{T}(G)$. To handle the first case, we compute the diameter of each tree of $\mathcal{T}(G)$, which can be done in linear time. Computing the diameters for all trees takes $O(n)$ time. The longest diameter of these trees is the diameter of G. In the following, we focus on the second case.

Suppose $T(v_i)$ contains u^* and $T(v_j)$ contains v^* for $i \neq j$. Observe that the diametral path $\pi_G(u^*, v^*)$ is the concatenation of the following three paths: $\pi_{T(v_i)}(u^*, v_i)$, $\pi_C(v_i, v_j)$, and $\pi_{T(v_j)}(v_j, v^*)$. Further, u^* is the farthest vertex in $T(v_i)$ from v_i; the same holds for v^* and $T(v_j)$. On the basis of these observations, we introduce some concepts as follows.

For each vertex $v_i \in C$, we define a *weight* $w(v_i)$ as the length of the path from v_i to its farthest vertex in $T(v_i)$. The weights for all vertices on C can be computed in total $O(n)$ time. With this definition in hand, we have $\Delta(G) = \max_{1 \leq i < j \leq m}(w(v_i) + d_C(v_i, v_j) + w(v_j))$. We say that (v_i, v_j) is a *vertex-weighted diametral pair* of C if $T(v_i)$ contains u^* and $T(v_j)$ contains v^* for a diametral pair (u^*, v^*) of G. To compute $\Delta(G)$, it suffices to find a vertex-weighted diameter pair of C. We introduce a domination relation for vertices on C.

Definition 1. *For two vertices $v_i, v_j \in C$, we say that v_i dominates v_j if $w(v_i) > w(v_j) + d_C(v_i, v_j)$.*

The following lemma shows that if a vertex is dominated by another vertex, then it is not "important".

Lemma 1. *For two vertices v_i and v_j of C, if v_i dominates v_j, then v_j cannot be in any vertex-weighted diametral pair of C unless (v_i, v_j) is such a pair.*

2.2 A Pruning Algorithm

We describe a linear time *pruning algorithm* to find all vertices of C dominated by other vertices (and thus those dominated vertices are "pruned"). As will be seen later, the diameter can be easily found after these vertices are pruned.

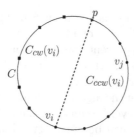

Fig. 1. Illustrating the definitions of $C_{ccw}(v_i)$ (the disks except v_i) and $C_{cw}(v_i)$ (the squares). We assume that p is a point on C that together with v_i partitions C into two half-cycles of equal length.

Let $|C|$ denote the sum of the lengths of all edges of C. For any vertex v_i of C, define $C_{ccw}(v_i)$ as the set of vertices v_j of C such that the path from v_i to v_j counterclockwise along C has length at most $|C|/2$ (Fig. 1); define $C_{cw}(v_i)$ as the set of vertices of C not in $C_{ccw}(v_i)$. We assume that v_i is in neither subset.

Lemma 2. *With $O(n)$ time preprocessing, given any two vertices v_i and v_j of C, we can do the following in $O(1)$ time: (1) compute $d_C(v_i, v_j)$; (2) determine whether v_j is in $C_{ccw}(v_i)$; (3) determine whether v_i and v_j dominate each other.*

Proof. We first compute the weight $w(v_i)$ for all vertices $v_i \in C$. This can be done in $O(n)$ time. Then, we compute the length $|C|$. Next, by scanning the vertices $v_1, v_2, \ldots, v_m$ on C, we compute an array $A[1, \ldots, m]$ with $A[i]$ equal to the length of the path from v_1 to v_i clockwise along C. Hence, for any $1 \leq i < j \leq m$, $A[j] - A[i]$ is the length of the path from v_i to v_j clockwise along C and $|C| - (A[j] - A[i]))$ is the length of the path from v_i to v_j counterclockwise along C. Note that $d_C(v_i, v_j) = \min\{A[j] - A[i], |C| - (A[j] - A[i])\}$.

Consider any two vertices v_i and v_j of C. Without loss of generality, we assume $i < j$. By comparing $A[j] - A[i]$ with $|C|/2$, we can determine whether v_j is in $C_{ccw}(v_i)$ in $O(1)$ time. As $w(v_i)$ and $w(v_j)$ are both available, whether v_i and v_j dominate each other can be determined in $O(1)$ time.   □

With Lemma 2 in hand, starting from v_1, our pruning algorithm processes the vertices of C from v_1 to v_m in order (see our full paper [13] for the pseudocode). The algorithm maintains a stack S, which is $\emptyset$ initially. Consider a vertex v_i. If $S = \emptyset$, then we push v_i into S. Otherwise, let v be the vertex at the top of S. If v is not in $C_{ccw}(v_i)$, then we also push v_i into S. Otherwise, we check whether v and v_i dominate each other. If they do not dominate each other, then we push v_i into S. Otherwise, if v_i dominates v, we pop v out of S, and then we continue to pop the new top element v of S out as long as the following three conditions are all satisfied: (1) $S \neq \emptyset$; (2) $v \in C_{ccw}(v_i)$; (3) v_i dominates v. Once one of the three conditions is not satisfied, we push v_i into S.

After v_m is processed, the first stage of the pruning algorithm is over. In the second stage, we process the vertices in the stack S in a bottom-up manner until a vertex not in $C_{cw}(v_1)$; the processing of a vertex is done in the same way as above (the vertex should be removed from S first). Specifically, let v_i be the vertex at the bottom of S. If v_i is not in $C_{cw}(v_1)$, then we stop the algorithm and return the vertices in the current stack S. Otherwise, we remove v_i from S and then apply the same processing algorithm as above in the first stage (i.e., begin with checking whether S is empty).

Intuitively, the first stage of the algorithm does a "full-cycle" scan on C while the second stage does a "half-cycle" scan (i.e., the half-cycle clockwise from v_1). With Lemma 2, the algorithm can be implemented in $O(n)$ time. The following lemma establishes the correctness of the algorithm.

Lemma 3. *Let S be the stack after the algorithm is over.*

1. *Each vertex of C that is not in S is dominated by a vertex in S.*
2. *No two vertices of S dominate each other.*

2.3   Computing the Diameter

In the following, we use S to refer to the stack after the pruning algorithm. Note that S cannot be empty. The following lemma shows how S can help to find a vertex-weighted diametral pair of C.

Lemma 4. *If $|S| = 1$, then any vertex-weighted diametral pair of C must contain the only vertex in S. Otherwise, for any vertex v of C that is not in S, v cannot be in any vertex-weighted diametral pair of C.*

If $|S| = 1$, we compute the diameter $\Delta(G)$ as follows. Let v be the only vertex in S. We find the vertex $u \in C \setminus \{v\}$ that maximizes the value $w(u) + d_C(u, v) + w(v)$, which can be done in $O(n)$ time with Lemma 2. By Lemma 4, (u, v) is a vertex-weighted diametral pair and $\Delta(G) = w(u) + d_C(u, v) + w(v)$.

If $|S| > 1$, by Lemma 4, $\Delta(G) = \max_{u,v \in S}(w(u) + d_C(u, v) + w(v))$. Lemma 5 finds a vertex-weighted diametral pair and thus computes $\Delta(G)$ in linear time.

Lemma 5. *A pair (u, v) of vertices in S that maximizes the value $w(u) + d_C(u, v) + w(v)$ can be found in $O(n)$ time.*

We thus obtain the following theorem, whose proof summarizes the entire algorithm and is in our full paper [13].

Theorem 1. *The diameter (along with a diametral pair) of a unicycle graph can be computed in linear time.*

3   The Diameter-Optimally Augmenting Trees (DOAT)

In this section, we solve the DOAT problem in $O(n^2 \log n)$ time and $O(n)$ space. The algorithm of Theorem 1 will be used as a subroutine.

3.1   Observations

We follow the same notation as in Sect. 2 such as $\pi_{G'}(s,t)$, $d_{G'}(s,t)$, $\Delta(G)$.

Let T be a tree of n vertices such that each edge of T has a positive length. For any two vertices u and v of T, we use $e(u,v)$ to refer to the shortcut connecting u and v; note that even if T already has an edge connecting them, we can always assume that there is an alternative shortcut (or we could also consider the shortcut as the edge itself with the same length). Let $|e(u,v)|$ denote the length of $e(u,v)$. Again, there is an oracle that can return the value $|e(u,v)|$ in $O(1)$ time for any shortcut $e(u,v)$. Denote by $T + e(u,v)$ the graph after adding $e(u,v)$ to T. The goal of the DOAT problem is to find a shortcut $e(u,v)$ so that the diameter of the new graph $\Delta(T + e(u,v))$ is minimized. Let $\Delta^*(T)$ be the diameter of an optimal solution. In the following we assume that $\Delta^*(T) < \Delta(T)$, since otherwise any shortcut would be sufficient.

For any shortcut $e(u,v)$, T has a unique path $\pi_T(u,v)$ between u and v. We make an assumption that $|e(u,v)| < d_T(u,v)$ since otherwise $e(u,v)$ can never be used (indeed, whenever $e(u,v)$ was used in a shortest path, we could always replace it with $\pi_T(u,v)$ to get a shorter path). This assumption is only for the argument of the correctness of our algorithm; the algorithm itself still uses the true value of $|e(u,v)|$ (this does not affect the correctness, because if $|e(u,v)| \geq d_T(u,v)$, then $e(u,v)$ cannot be an optimal shortcut). For the reference purpose, we refer to this assumption as the *shortcut length assumption*.

At the outset, we compute a diametral path P of T in $O(n)$ time. Let $v_1, v_2, \ldots, v_m$ be the vertices of P ordered along it. Removing the edges of P from T results in m connected components of T, each of which is a tree containing a vertex of P; we let $T(v_i)$ denote the tree containing v_i. For each v_i, we define a weight $w(v_i)$ as the distance from v_i to its farthest vertex in $T(v_i)$. Let $\mathcal{T} = \{T(v_i) \mid 1 \leq i \leq m\}$.

For any pair (i,j) of indices with $1 \leq i < j \leq m$, we define a *critical pair* of vertices (x,y) with $x \in T(v_i)$ and $y \in T(v_j)$ such that they minimize the value $d_{T(v_i)}(v_i, x') + |e(x', y')| + d_{T(v_j)}(y', v_j)$ among all vertex pairs (x', y') with $x' \in T(v_i)$ and $y' \in T(v_j)$.

The following lemma will be used later.

Lemma 6. *For any vertex v in any tree $T(v_k) \in \mathcal{T}$, it holds that $d_{T(v_k)}(v, v_k) \leq \min\{d_T(v_1, v_k), d_T(v_k, v_m)\}$. Also, $d_T(v_1, v_k) = d_P(v_1, v_k)$ and $d_T(v_k, v_m) = d_P(v_k, v_m)$.*

Lemma 7 shows why critical pairs are "critical".

Lemma 7. *Suppose $e(u^*, v^*)$ is an optimal shortcut with $u^* \in T(v_i)$ and $v^* \in T(v_j)$. Then, $i \neq j$ and any critical pair of (i,j) also defines an optimal shortcut.*

3.2   Reducing DOAT to Finding a Shortcut for P

In light of Lemma 7, we reduce our DOAT problem on T to finding a shortcut for the vertex-weighted path P as follows.

For an index pair (i, j) with $1 \leq i < j \leq m$, we define a shortcut $\bar{e}(v_i, v_j)$ connecting v_i and v_j with length $|\bar{e}(v_i, v_j)| = d_{T(v_i)}(v_i, x) + |e(x, y)| + d_{T(v_j)}(v_j, y)$, where (x, y) is a critical pair of (i, j). The diameter $\Delta(P + \bar{e}(v_i, v_j))$ is defined as $\max_{1 \leq k < h \leq m} \{ w(v_k) + d_{P + \bar{e}(v_i, v_j)}(v_k, v_h) + w(v_h) \}$. The diameter-optimally augmenting path (DOAP) problem on P is to find a shortcut $\bar{e}(v_i, v_j)$ so that the diameter $\Delta(P + \bar{e}(v_i, v_j))$ is minimized; we use $\Delta^*(P)$ to denote the minimized diameter.

With the help of Lemma 7, the following lemma shows that the DOAT problem on T can be reduced to the DOAP problem on P. Similar problem reductions were also used in [1,6].

Lemma 8. 1. *For any index pair (i, j) with $1 \leq i < j \leq m$, it holds that $\Delta(P + \bar{e}(v_i, v_j)) = \Delta(T + e(x, y))$, where (x, y) is a critical pair of (i, j).*
2. *If $\bar{e}(v_i, v_j)$ is an optimal shortcut for the DOAP problem on P, then $e(x, y)$ is an optimal shortcut for the DOAT problem on T, where (x, y) is a critical pair of (i, j).*
3. *$\Delta^*(T) = \Delta^*(P)$.*

By Lemma 8, we will focus on solving the DOAP problem on the vertex-weighted path P. Notice that the lengths of the shortcuts of P have not been computed.

3.3 Computing an Optimal Shortcut for P

To find an optimal shortcut for P, for each $i \in [1, m-1]$, we will compute an index $j(i)$ that minimizes the diameter $\Delta(P + \bar{e}(v_i, v_j))$ among all indices $j \in [i+1, m]$, i.e., $j(i) = \arg\min_{i+1 \leq j \leq m} \Delta(P + \bar{e}(v_i, v_j))$, as well as the diameter $\Delta(P + \bar{e}(v_i, v_{j(i)}))$. After that, the optimal shortcut of P is the one that minimizes $\Delta(P + \bar{e}(v_i, v_{j(i)}))$ among the shortcuts $\bar{e}(i, j(i))$ for all $i \in [1, m-1]$. We refer to the shortcuts for $\bar{e}(v_i, v_j)$ for all $j \in [i+1, m]$ as v_i-*shortcuts*. Therefore, our goal is to find an optimal v_i-shortcut $\bar{e}(i, j(i))$ for each $i \in [1, m-1]$.

Let n_i denote the number of vertices in $T(v_i)$, for each $1 \leq i \leq m$. Note that $n = \sum_{i=1}^{m} n_i$. Fix an index i with $1 \leq i \leq m-1$. In the following, we will present an algorithm that computes an optimal v_i-shortcut $\bar{e}(v_i, v_{j(i)})$ and the diameter $\Delta(P + \bar{e}(v_i, v_{j(i)}))$ in $O(n \cdot n_i + n \log n)$ time and $O(n)$ space. In this way, solving the DOAP problem on P takes $O(n^2 \log n)$ time and $O(n)$ space in total.

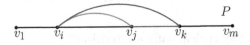

Fig. 2. $\bar{e}(v_i, v_j)$ dominates $\bar{e}(v_i, v_k)$ if $|\bar{e}(v_i, v_j)| + d_P(v_j, v_k) \leq |\bar{e}(v_i, v_k)|$, i.e., the length of the red path is less than or equal to the length of the blue path.

We introduce a domination relationship among v_i-shortcuts.

Definition 2. *For any two index j and k with $i < j < k \leq m$, we say that $\bar{e}(v_i, v_j)$ dominates $\bar{e}(v_i, v_k)$ if $|\bar{e}(v_i, v_j)| + d_P(v_j, v_k) \leq |\bar{e}(v_i, v_k)|$; e.g., see Fig. 2.*

The following lemma implies that if $\overline{e}(v_i, v_j)$ dominates $\overline{e}(v_i, v_k)$, then short-cut $\overline{e}(v_i, v_k)$ can be ignored or "pruned".

Lemma 9. *If $\overline{e}(v_i, v_j)$ dominates $\overline{e}(v_i, v_k)$, then $\Delta(P + \overline{e}(v_i, v_j)) \leq \Delta(P + \overline{e}(v_i, v_k))$.*

Let S_i be the set of all v_i-shortcuts, i.e., $S_i = \{\overline{e}(v_i, v_j) \| i + 1 \leq j \leq m\}$. In the following, we describe a pruning algorithm that computes a subset S of S_i such that no two shortcuts of S dominate each other and S contains at least one optimal v_i-shortcut. As will be seen later, these properties of S allow an efficient algorithm to find an optimal v_i-shortcut.

Before running the pruning algorithm, we compute the lengths of shortcuts of S_i by brute force, as follows. First, with $O(n)$ time preprocessing, given any two vertices u and v with $u \in T(v_i)$ and $v \in T(v_j)$ for $j \neq i$, we can compute $d_T(u, v)$ in constant time. Consider a tree $T(v_j)$ with $j \geq i + 1$. Computing the length of $\overline{e}(v_i, v_j)$ reduces to finding a critical pair of (i, j). To this end, we compute $d_{T(v_i)}(v_i, u) + |e(u, v)| + d_{T(v_j)}(v, v_j)$ for all vertices $u \in T(v_i)$ and all vertices $v \in T(v_j)$, which can be done in $O(n_i \cdot n_j)$ time (and $O(n)$ space). As such, computing the lengths of all shortcuts of S_i takes $O(n_i \cdot n)$ time.

Our pruning algorithm processes the shortcuts $\overline{e}(v_i, v_j)$ for all $j = i+1, \ldots, m$ one by one. A stack S is maintained and $S = \emptyset$ initially. Consider any $j \in [i + 1, m]$. If $S = \emptyset$, we push $\overline{e}(v_i, v_j)$ into S. Otherwise, let $\overline{e}$ be the shortcut at the top of S. If $\overline{e}$ and $\overline{e}(v_i, v_j)$ do not dominate each other, we push $\overline{e}(v_i, v_j)$ into S. Otherwise, if $\overline{e}$ dominates $\overline{e}(v_i, v_j)$, then we proceed on $j + 1$, i.e., $\overline{e}(v_i, v_j)$ is pruned. If $\overline{e}(v_i, v_j)$ dominates $\overline{e}$, then we pop $\overline{e}$ out of S (i.e., $\overline{e}$ is pruned). Next, we keep popping the top element out of S until either S becomes $\emptyset$ or $\overline{e}(v_i, v_j)$ does not dominate it; in either case we push $\overline{e}(v_i, v_j)$ into S.

As the lengths of the shortcuts of S_i are available, the algorithm runs in $O(n)$ time. The following lemma establishes the correctness of the algorithm.

Lemma 10. *After the algorithm, no two shortcuts of S dominate each other and S contains at least one optimal v_i-shortcut.*

Using the algorithm for Theorem 1 as a subroutine, Lemma 11 below provides a binary search algorithm that finds an optimal v_i-shortcut from S in $O(n \log n)$ time and $O(n)$ space.

Lemma 11. *An optimal v_i-shortcut in S can be found in $O(n \log n)$ time and $O(n)$ space.*

Proof. We first prove some properties that our algorithm relies on.

Consider the graph $P + \overline{e}(v_i, v_j)$ for an index j with $i < j \leq m$. Let $\Delta(i, j) = \Delta(P + \overline{e}(v_i, v_j))$. Suppose (v_a, v_b) is a diametral pair of $P + \overline{e}(v_i, v_j)$ with $a < b$. Then, $\Delta(i, j) = w(v_a) + d_{P+\overline{e}(v_i, v_j)}(v_a, v_b) + w(v_b)$.

We claim that if $a \in (1, i]$, then (v_1, v_b) is also a diametral pair. Indeed, since $a \neq 1$, by Lemma 6 and the definition of $w(v_a)$, we have $w(v_a) \leq d_P(v_1, v_a)$.

Hence, we can derive

$$w(v_a) + d_{P+\bar{e}(v_i,v_j)}(v_a, v_b) + w(v_b) \leq d_P(v_1, v_a) + d_{P+\bar{e}(v_i,v_j)}(v_a, v_b) + w(v_b)$$
$$= d_{P+\bar{e}(v_i,v_j)}(v_1, v_b) + w(v_b) \leq w(v_1) + d_{P+\bar{e}(v_i,v_j)}(v_1, v_b) + w(v_b).$$

Hence, (v_1, v_b) is also a diametral pair.

Similarly, we claim that if $b \in [j, m)$, then (v_a, v_m) is also a diametral pair. The claim can be proved by a similar argument as above.

Note that since $a < b$, $a \neq m$ and $b \neq 1$. Due to the above two claims, we assume that $a \in \{1\} \cup (i, m)$ and $b \in (1, j) \cup \{m\}$. Based on the values of a and b, we define the following five functions (illustrations can be found in our full paper [13]).

1. For the case $a = 1$ and $b = m$, we define

$$\alpha(i, j) = w(v_1) + d_P(v_1, v_i) + |\bar{e}(v_i, v_j)| + d_P(v_j, v_m) + w(v_m).$$

Hence, if $a = 1$ and $b = m$, we have $\Delta(i, j) = \alpha(i, j)$.
2. For the case $a = 1$ and $b \in (i, j)$, we define $\beta(i, j) = w(v_1) + \max_{i<b'<j}\left\{ \min\{d_P(v_1, v_{b'}), d_P(v_1, v_i) + |\bar{e}(v_i, v_j)| + d_P(v_j, v_{b'}) + w(v_{b'})\}\right\}$.
Hence, if $a = 1$ and $b \in (i, j)$, we have $\Delta(i, j) = \beta(i, j)$.
3. For the case $a \in (i, m)$ and $b = m$, we define

$$\gamma(i, j) = \max_{i<a'<m}\left\{ w(v_{a'}) + d_{P+\bar{e}(v_i,v_j)}(v_{a'}, v_m)\right\} + w(v_m).$$

Note that $d_{P+\bar{e}(v_i,v_j)}(v_{a'}, v_m)$ is equal to $\min\{d_P(v_{a'}, v_m), d_P(v_{a'}, v_i) + |\bar{e}(v_i, v_j)| + d_P(v_j, v_m)\}$ if $a' \in (i, j)$, and $d_P(v_{a'}, v_m)$ otherwise.
Hence, if $a \in (i, m)$ and $b = m$, we have $\Delta(i, j) = \gamma(i, j)$.
4. For the case $i < a < b < j$, we define $\delta(i, j) = \max_{i<a'<b'<j}\left\{ w(v_{a'}) + \min\{d_P(v_{a'}, v_{b'}), d_P(v_{a'}, v_i) + |\bar{e}(v_i, v_j)| + d_P(v_j, v_{b'})\} + w(v_{b'})\right\}$. Hence, if $a, b \in (i, j)$, we have $\Delta(i, j) = \delta(i, j)$.
5. For the case $a = 1$ and $b \in (1, i]$, we define

$$\lambda(i, j) = \max_{1<b'\leq i}\left\{ w(v_1) + d_P(v_1, v_{b'}) + w(v_{b'})\right\}.$$

Hence, if $a = 1$ and $b \in (1, i]$, we have $\Delta(i, j) = \lambda(i, j)$.

With these definitions, we have $\Delta(i, j) = \max\{\alpha(i, j), \beta(i, j), \gamma(i, j), \delta(i, j), \lambda(i, j)\}$. Hence, if j changes in $[i+1, m]$, the graph of $\Delta(i, j)$ is the upper envelope of the graphs of the five functions.

Recall that our goal is to find an optimal v_i-shortcut in S. Let I denote the set of the indices j of all shortcuts $\bar{e}(v_i, v_j)$ of S. We consider these indices of I

in order. We intend to show that $\Delta(i, j)$ is a unimodal function (first decreases and then increases) as j changes in I. To this end, we prove that each of the above five functions is a monotonically increasing or decreasing function as j changes in I. Note that each index of I is in $[i+1, m]$. To simplify the notation, we simply let $I = \{i + 1, i + 2, \ldots, m\}$, or equivalently, one may consider that our pruning algorithm does not prune any shortcut from S_i and thus $S = S_i$.

As no two shortcuts of S dominate each other, $\overline{e}(v_i, v_j)$ and $\overline{e}(v_i, v_{j+1})$ do not dominate each other for any $j \in (i, m)$, i.e., $|\overline{e}(v_i, v_j)| + d_P(v_j, v_{j+1}) > |\overline{e}(v_i, v_{j+1})|$ and $|\overline{e}(v_i, v_{j+1})| + d_P(v_j, v_{j+1}) > |\overline{e}(v_i, v_j)|$. Relying on this property, we can prove the following monotonicity properties of the five functions.

Monotonicity Properties: Let j be any index in (i, m). The following hold:

1. $\alpha(i, j) > \alpha(i, j + 1)$. Hence, $\alpha(i, j)$ is a decreasing function for $j \in (i, m)$.
2. $\beta(i, j) \leq \beta(i, j + 1)$. Hence, $\beta(i, j)$ is an increasing function for $j \in (i, m)$.
3. $\gamma(i, j) \geq \gamma(i, j + 1)$. Hence, $\gamma(i, j)$ is a decreasing function for $j \in (i, m)$.
4. $\delta(i, j) \leq \delta(i, j + 1)$. Hence, $\beta(i, j)$ is an increasing function for $j \in (i, m)$.
5. $\lambda(i, j) = \lambda(i, j + 1)$. Hence, $\lambda(i, j)$ is a constant function for $j \in (i, m)$.

On the basis of the above monotonicity properties, we present a binary search algorithm that finds an optimal v_i-shortcut in $O(n \log n)$ time and $O(n)$ space.

Our algorithm performs binary search on the indices $[l, r]$, with $l = i + 1$ and $r = m$ initially. In each step, we decide whether we will proceed on $[l, k]$ or on $[k, r]$, where $k = \lfloor \frac{l+r}{2} \rfloor$. To this end, we compute $\Delta(i, k)$ and $\Delta(i, k + 1)$. By Lemma 8, $\Delta(i, k) = \Delta(T + e(x, y))$ where (x, y) is a critical pair of (i, k). Since $T + e(x, y)$ is a unicycle graph, we compute $\Delta(T + e(x, y))$ in $O(n)$ time by Theorem 1. Therefore, $\Delta(i, k)$ can be computed in $O(n)$ time. Note that the algorithm of Theorem 1 also returns a diametral pair for $T + e(x, y)$, and we can decide which of the five cases for the functions α, β, γ, δ, and λ the diametral pair belong to. We do the same for $\Delta(i, k + 1)$. Assume that $\Delta(i, k) = f(i, k)$ and $\Delta(i, k + 1) = g(i, k + 1)$, for two functions f and g in $\{\alpha, \beta, \gamma, \delta, \lambda\}$. Then we have the following cases

1. $f = g$. We have the following subcases.
   - $f = g \in \{\beta, \delta\}$. In this case, our algorithm proceeds on the interval $[l, k]$. To see this, since both β and δ are monotonically increasing functions, we have $\Delta(i, j) \geq f(i, j) \geq f(i, k) = \Delta(i, k)$, for any $j \in (k, r]$. As such, the diameter $\Delta(i, j)$ would increase if we proceed on $j \in (k, r]$.
   - $f = g \in \{\alpha, \gamma\}$. In this case, we proceed on the interval $[k, r]$ because both functions are monotonically decreasing.
   - $f = g = \lambda$. In this case, we stop the algorithm and return $\overline{e}(i, k)$ as an optimal v_i-shortcut. To see this, $\Delta(i, j) \geq \lambda(i, j) = \lambda(i, k) = \Delta(i, k)$ for any $j \in [l, r]$. Hence, $\Delta(i, j)$ achieves the minimum at $j = k$ among all $j \in [l, r]$.
2. $f \neq g$. We have the following subcases.
   - One of f and g is λ. In this case, by a similar argument as before, we return $\overline{e}(v_i, v_k)$ as an optimal v_i-shortcut if $f = \lambda$, and return $\overline{e}(v_i, v_{k+1})$ as an optimal v_i-shortcut if $g = \lambda$.

- $\{f, g\} = \{\beta, \delta\}$. In this case, since both β and δ are increasing functions, by a similar argument as before, we proceed on the interval $[l, k]$.
- $\{f, g\} = \{\alpha, \gamma\}$. In this case, since both α and γ are decreasing functions, by a similar argument as before, we proceed on the interval $[k, r]$.
- One of f and g is in $\{\beta, \delta\}$ and the other is in $\{\alpha, \gamma\}$. In this case, one of $\overline{e}(v_i, v_k)$ and $\overline{e}(v_i, v_{k+1})$ is an optimal v_i-shortcut, which can be determined by comparing $\Delta(i, k)$ with $\Delta(i, k+1)$. To see this, without loss of generality, we assume that $f \in \{\beta, \delta\}$ and $g \in \{\alpha, \gamma\}$. Hence, $\Delta(i, j) \geq f(i, j) \geq f(i, k) = \Delta(i, k)$ for any $j \in [k+1, r]$, and $\Delta(i, j) \geq g(i, j) \geq g(i, k+1) = \Delta(i, k+1)$ for any $j \in [l, k]$. As such, $\min\{\Delta(i, k), \Delta(i, k+1)\} \leq \Delta(i, j)$ for all $j \in [l, r]$.

The algorithm will find an optimal v_i-shortcut in $O(\log n)$ iterations. As each iteration takes $O(n)$ time, the total time is $O(n \log n)$. The space is $O(n)$. □

Based on Lemma 11, we have the following result.

Theorem 2. *The DOAT problem on the tree T can be solved in $O(n^2 \log n)$ time and $O(n)$ space.*

References

1. Bilò, D.: Almost optimal algorithms for diameter-optimally augmenting trees. In: Proceedings of the 29th International Symposium on Algorithms and Computation (ISAAC), pp. 40:1–40:13 (2018)
2. Carufel, J.-L.D., Grimm, C., Maheshwari, A., Smid, M.: Minimizing the continuous diameter when augmenting paths and cycles with shortcuts. In: Proceedings of the 15th Scandinavian Workshop on Algorithm Theory, pp. 27:1–27:14 (2016)
3. Carufel, J.-L.D., Grimm, C., Schirra, S., Smid, M.: Minimizing the continuous diameter when augmenting a tree with a shortcut. In: Proceedings of the 15th Algorithms and Data Structures Symposium (WADS), pp. 301–312 (2017)
4. Farley, A., Proskurowski, A.: Computation of the center and diameter of outerplanar graphs. Discrete Appl. Math. **2**, 185–191 (1980)
5. Federickson, G.: Fast algorithms for shortest paths in planar graphs, with applications. SIAM J. Comput. **16**, 1004–1022 (1987)
6. Große, U., Gudmundsson, J., Knauer, C., Smid, M., Stehn, F.: Fast algorithms for diameter-optimally augmenting paths. In: Proceedings of the 42nd International Colloquium on Automata, Languages and Programming, pp. 678–688 (2015)
7. Große, U., Gudmundsson, J., Knauer, C., Smid, M., Stehn, F.: Fast algorithms for diameter-optimally augmenting paths and trees. arXiv:1607.05547 (2016)
8. Johnson, C., Wang, H.: A linear-time algorithm for radius-optimally augmenting paths in a metric space. In: Proceedings of the 16th Algorithms and Data Structures Symposium (WADS), pp. 466–480 (2019)
9. Megiddo, N.: Linear-time algorithms for linear programming in R^3 and related problems. SIAM J. Comput. **12**(4), 759–776 (1983)
10. Oh, E., Ahn, H.-K.: A near-optimal algorithm for finding an optimal shortcut of a tree. In: Proceedings of the 27th International Symposium on Algorithms and Computation (ISAAC), pp. 59:1–59:12 (2016)

11. Olariu, S.: A simple linear-time algorithm for computing the center of an interval graph. Int. J. Comput. Math. **34**, 121–128 (1990)
12. Wang, H.: An improved algorithm for diameter-optimally augmenting paths in a metric space. Comput. Geom.: Theory Appl. **75**, 11–21 (2018)
13. Wang, H., Zhao, Y.: Algorithms for diameters of unicycle graphs and diameter-optimally augmenting trees. arXiv:2011.09591 (2020)
14. Wang, H., Zhao, Y.: A linear-time algorithm for discrete radius optimally augmenting paths in a metric space. In: Proceedings of the 32nd Canadian Conference on Computational Geometry (CCCG), pp. 174–180 (2020)
15. Williams, R.: Faster all-pairs shortest paths via circuit complexity. SIAM J. Comput. **47**, 1965–1985 (2018)

On Short Fastest Paths in Temporal Graphs

Umesh Sandeep Danda[3], G. Ramakrishna[1(✉)], Jens M. Schmidt[2(✉)], and M. Srikanth[1]

[1] Indian Institute of Technology Tirupati, Tirupati, India
{rama,cs18d501}@iittp.ac.in
[2] Institute for Algorithms and Complexity, TU Hamburg, Hamburg, Germany
jens.m.schmidt@tuhh.de
[3] Hyderabad, India

Abstract. Temporal graphs equip their directed edges with a *departure time* and a *duration*, which allows to model a surprisingly high number of real-world problems. Recently, Wu et al. have shown that a *fastest* path in a temporal graph G from a given vertex s to a vertex z can be computed in near-linear time, where a *fastest* path is one that minimizes the arrival time at z minus the departure time at s.

Here, we consider the natural problem of computing a fastest path from s to z that is in addition *short*, i.e. minimizes the sum of durations of its edges; this maximizes the total amount of spare time at stops during the journey. Using a new dominance relation on paths in combination with lexicographic orders on the departure and arrival times of these paths, we derive a near-linear time algorithm for this problem with running time $O(n + m \log p(G))$, where $n := |V(G)|$, $m := |E(G)|$ and $p(G)$ is upper bounded by both the maximum in-degree and the maximum edge duration of G.

The dominance relation is interesting in its own right, and may be of use for several related problems like fastest paths with minimum fare, fastest paths with minimum number of stops, and other pareto-optimal path problems in temporal graphs.

1 Introduction

Temporal graphs capture various problems such as message dissemination in online social networks, epidemics spreading in complex networks and routing in scheduled public transportation networks [10]. This generality comes with a price: many standard graph parameters (such as the number of strongly connected components) are not known to admit polynomial-time algorithms in temporal graphs, and not even standard results in combinatorics like Menger's theorem hold without adapting them adequately [6,8].

On the other hand, a growing number of positive results has been developed in recent years for various problems in temporal graphs [1,4,6,7,9,12].

R. Uehara et al. (Eds.): WALCOM 2021, LNCS 12635, pp. 40–51, 2021.
https://doi.org/10.1007/978-3-030-68211-8_4

In this paper, we focus on path problems in temporal graphs, for which, in contrast to static graphs, various notions of optimality exist [3,5,11]. For example, one may not only want to find the *fastest* paths mentioned above, but also *shortest* paths, which minimize the sum of durations of their edges (we give precise definitions in Sect. 1.2).

It was recently shown in [11] that, given a temporal graph G and two of its vertices s and z, both fastest and shortest paths from s to z can be computed efficiently in running times $O(n + m \log c_{min})$ and $O(n + m \log c_{in}(G))$, respectively, where $c_{in}(G)$ is the maximum number of ingoing edges over all vertices of G, S is the number of outgoing edges of s with distinct departure times, and $c_{min} = \min\{|S|, c_{in}(G)\}$.

A natural strengthening that we investigate here is to compute a fastest path from s to z that has minimal duration. To our surprise, no efficient algorithm seems to be known for this problem.

1.1 Temporal Graphs

A *temporal graph* G is a pair (V, E), where V is a finite set and $E := (e_1, e_2, \ldots, e_m)$ is a finite sequence such that $e_i := (v_i, w_i, t_i, d_i) \in V \times V \times \mathbb{N} \times \mathbb{N}^{>0}$ and $v_i \neq w_i$ for every $1 \leq i \leq m$. For every $1 \leq i \leq m$, we call e_i an *edge* of G, v_i and w_i the *source* and *target vertex* of e_i, t_i the *departure time* of e_i and d_i the *duration* of e_i. Hence, in the terminology of usual graphs, every edge e_i of G is directed (as e_i is ordered), not a self-loop (parallel edges may occur), and has positive duration. Every edge e_i is equipped with a departure time t_i and a duration d_i, where t_i is the point in time at which one may depart from v_i in order to arrive at w_i at time $t_i + d_i$; we call $arr(e_i) := t_i + d_i$ the *arrival time* of e_i.

This model generalizes the models of temporal graphs that were used in [3]. In the above definition, the edges $(e_i)_i$ are used in a stream representation: for temporal graphs, it is usually assumed that the edges in this stream $(e_i)_i$ are ordered with respect to some natural and easy-to-pick ordering such as their creation, collection or deletion [11, Section 4.1]. Here, we assume that the edges are ordered monotonically increasing according to their arrival times, so that we have $i < j$ if and only if $arr(e_i) \leq arr(e_j)$. If for some reason such an ordering cannot be expected in a particular use case, a sorting routine with additional running time $O(m \log m)$ has to be invoked in advance.

We inherit standard graph-theoretic notions like paths and cycles (both are always given as edge sequences) for temporal graphs G. A path from a vertex s to a vertex z ($s = z$ is possible) is called an *s-z-path*. For any $G = (V, E)$, we define $V(G) := V$, $E(G) := E$ and $n := |V(G)|$ (note that $m := |E(G)|$ by definition of E).

1.2 Our Result

A path $P := (e_{j_1}, \ldots, e_{j_k})$ of a temporal graph G is *temporal* if $t_{j_i} + d_{j_i} \leq t_{j_{i+1}}$ for every $1 \leq i < k$. We call $dep(P) := t_{j_1}$ the *departure time* of P and $arr(P) := t_{j_k} + d_{j_k}$ the *arrival time* of P if $k > 0$. The *journey time* of such a temporal path

P is $journey(P) := \text{arr}(P) - \text{dep}(P)$, and the *duration* of P is $dur(P) := \sum_{i=1}^{k} d_{j_i}$ (see Fig. 1).

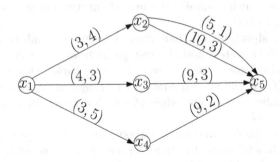

Fig. 1. A temporal graph G in which the departure time t_i and the duration d_i of every edge e_i is shown. The path $P := (x_1x_3, x_3x_5)$ is a fastest path that has journey time $8 = 9 + 3 - 4$, and $Q := (x_1x_4, x_4x_5)$ is another fastest path from x_1 to x_5 that has journey time $8 = 9 + 2 - 3$. However, only P is a short fastest path, as $dur(P) = 6 = 3 + 3 < dur(Q) = 7 = 5 + 2$.

Definition 1. *A temporal s-z-path P of a temporal graph G is called*

(i) fastest *if every temporal s-z-path Q of G satisfies journey$(P) \le$ journey(Q),*
(ii) shortest *if every temporal s-z-path Q of G satisfies dur$(P) \le$ dur(Q), and*
(iii) short fastest *if P is fastest and every fastest temporal s-z-path Q of G satisfies dur$(P) \le$ dur(Q).*

In other words, P is *short fastest* if P is fastest and has minimum duration among all fastest s-z-path of G. Note that all three notions fix the start- and end-vertex of the paths in question, while allowing an arbitrary departure time at vertex s. Short fastest paths arise naturally when we want to travel from s to z in the fastest journey time possible such that the total amount of time spent traveling is minimized (this maximizes the total amount of spare time at stops during the journey).

For an edge e_i, let $p(e_i) := |\{\text{arr}(e_j) : w_j = v_i \text{ and } t_i \le \text{arr}(e_j) \le \text{arr}(e_i)\}|$ be the number of integers in $[t_i, \text{arr}(e_i)]$ that are arrival times for at least one incoming edge to v_i. In particular, we have $p(e_i) \le d_i$ and $p(e_i)$ is at most the in-degree of v_i. Let $p(G) := \max\{p(e_i) : 1 \le i \le m\}$ and let $\delta^-(G)$ be the maximum in-degree of G. Given two vertices s and z of a temporal graph G, the problem SHORTFASTESTPATH(s, z, G) asks for a short fastest temporal s-z-path of G. We solve this problem as follows.

Theorem 1. *Given a source vertex s of a temporal graph G on n vertices and m edges, short fastest paths from s to every vertex $z \ne s$ can be computed in total time $O(n + m \log p(G))$, where $p(G) \le \min\{\delta^-(G), \max\{d_i : 1 \le i \le m\}\}$.*

As the duration in public-transport networks is often bounded by a constant, the factor $\log p(G)$ in our running time is typically insignificant for applications.

The algorithm of Theorem 1 may easily be adapted to compute short fastest paths in given time intervals, and to allow rational departure and duration times (e.g. by multiplying with the greatest common divisor in advance). Further, the algorithm may also be customized to solve related problems such as computing a fastest path with minimum waiting time, computing a fastest path with minimum fare, and computing a fastest path with minimum number of transfers at intermediate stations.

While our algorithm is inspired by the algorithm in [11] for fastest paths, it deviates from this algorithm by using a new dominance relation on paths and lexicographic orderings on the departure and arrival times of these paths. These two ideas allow us to perform various operations on dominating paths such as searching, insertion, and deletion efficiently. Another difference is that our algorithm processes the edges of G by increasing arrival time.

2 Dominating Paths

From now on, let a temporal graph G and a source vertex s of G for the problem SHORTFASTESTPATH be given. We first provide structural properties of temporal paths that are useful to reduce the search space.

Definition 2. *For temporal x-y-paths P and Q of G, P dominates Q if either*

(i) $dep(P) > dep(Q)$ and $arr(P) \leq arr(Q)$,
(ii) $dep(P) = dep(Q)$, $arr(P) < arr(Q)$ and $dur(P) \leq dur(Q)$, or
(iii) $dep(P) = dep(Q)$, $arr(P) = arr(Q)$ and $dur(P) < dur(Q)$.

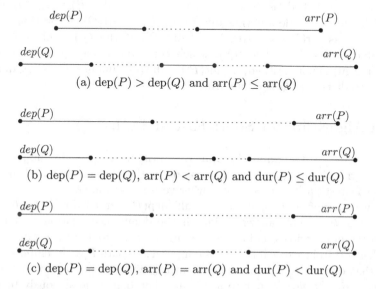

(a) $dep(P) > dep(Q)$ and $arr(P) \leq arr(Q)$

(b) $dep(P) = dep(Q)$, $arr(P) < arr(Q)$ and $dur(P) \leq dur(Q)$

(c) $dep(P) = dep(Q)$, $arr(P) = arr(Q)$ and $dur(P) < dur(Q)$

Fig. 2. Three instances of a path P that dominates Q. Solid and dotted lines depict the duration of edges and the waiting times for the next departure, respectively.

The three cases of Definition 2 are depicted in Fig. 2. A temporal x-y-path is *dominating* if it is not dominated by any other temporal x-y-path, and *non-dominating* otherwise. In order to motivate these definitions, observe that every short fastest path is dominating (Definition 2(i) and (ii) strictly decrease the journey time, while Definition 2(iii) strictly decreases the duration). We therefore are interested in computing all dominating paths. Definition 2 and the resulting properties of the dominance relation on paths will be crucial for establishing the structural properties that allow the algorithm to be efficient in the remainder of the paper.

For a path, a *prefix subpath* of this path is a subpath of this path that starts at the same start vertex. The correctness of shortest path algorithms in traditional graphs such as Dijkstra's algorithm rely heavily on the fact that subpaths of shortest paths are again shortest. For temporal graphs however, such properties are bound to fail (in fact, they fail for fastest as well as for shortest paths). For example, any prefix subpath of a fastest path in which the departure time of the last edge is sufficiently far away from the arrival time of the second last edge may not be fastest (as the second last vertex may be reached by much faster paths). The next lemma shows that paths that are dominating (as defined in the last paragraph) obey this property.

Lemma 1. *Every prefix subpath of every dominating path is dominating.*

Proof. Assume to the contrary that Q' is a (temporal) non-dominating prefix x-y-path of a dominating path Q. Then $Q' \neq Q$, since Q is dominating, and there is a temporal x-y-path P' that dominates Q'. Let P be the path obtained from Q by replacing the subpath Q' with P'; in particular, $\mathrm{arr}(P) = \mathrm{arr}(Q)$.

Since P' dominates Q', Q' satisfies exactly one of the Conditions 2(i)–(iii). The first is not satisfied, as $\mathrm{dep}(P') > \mathrm{dep}(Q')$ implies $\mathrm{dep}(P) > \mathrm{dep}(Q)$, which contradicts that Q is dominating due to $\mathrm{arr}(P) = \mathrm{arr}(Q)$. Condition 2(ii) is not satisfied, as $\mathrm{arr}(P') < \mathrm{arr}(Q')$ and $\mathrm{dur}(P') \leq \mathrm{dur}(Q')$ imply $\mathrm{dur}(P) < \mathrm{dur}(Q)$ due to $\mathrm{dep}(P') = \mathrm{dep}(Q')$, which contradicts that Q is dominating. Condition 2(iii) is not satisfied, as $\mathrm{dur}(P') < \mathrm{dur}(Q')$ implies $\mathrm{dur}(P) < \mathrm{dur}(Q)$, which contradicts that Q is dominating. This gives the claim.   □

3   An Algorithm for Short Fastest Paths

Given any vertex s of a temporal graph G, we describe an efficient one-pass algorithm by dynamic programming that computes the journey time and duration of a short fastest path from s to any other vertex $z \neq s$ in G.

For every temporal x-y-path P, we call $(\mathrm{dep}(P), \mathrm{arr}(P), \mathrm{dur}(P))$ a *temporal triple from x to y*. This allows to inherit the dominance relation of temporal paths to temporal triples as follows: a triple (t, a, d) from x to y *dominates* a triple (t', a', d') from x to y if there is a temporal x-y-path P with temporal triple (t, a, d) that dominates a temporal x-y-path Q with temporal triple (t', a', d'). The *dominating triples* from x to $y \neq x$ are then defined analogously to paths that are dominating; in addition, for every $1 \leq i \leq m$ such that $v_i = s$, let

$(t_i, t_i, 0)$ be an artificial *dominating triple* from s to s. These artificial dominating triples will later allow the algorithm to start at s using any outgoing edge e_j of s at its departure time t_j.

We will compute dominating triples of G by starting with an edge-less subgraph of G and updating these triples each time after the next edge of $E(G)$ in the given ordering of $E(G)$ is added. We therefore define a sequence of temporal graphs that adheres to this ordering (i.e. adds the edges of $E(G)$ one by one). For every $0 \leq i \leq m$, let G_i be the temporal graph $(V(G), \{e_1, \ldots, e_i\})$. Hence, G_0 has no edges at all and, for every $1 \leq i \leq m$, $V(G_i) = V(G)$ and e_i is the only edge of G_i that is not in G_{i-1}.

In every graph, we aim to maintain for every $y \neq s$ the list L_y of all dominating triples from s to y. The lists L_y are initially empty. We will store the final journey time and duration of a short fastest s-y-path for every $y \neq s$ in *journey(y)* and *dur(y)*, respectively.

Algorithm 1: Short Fastest Path

Input: A vertex s of a temporal graph $G = (V, E)$, where the sequence E is ordered increasingly according to arrival time.

Output: For every vertex $z \neq s$ in G, the journey time and duration of a short fastest path from s to z.

1 Initialize $L_y := \emptyset$ and journey$(y) :=$ dur$(y) := \infty$ for every vertex $y \neq s$;
2 **for** $i = 1$ *to* m **do**
3   **if** $w_i = s$ **then continue**;
4   **if** $v_i = s$ **then** Append artificial dominating triple $(t_i, t_i, 0)$ to L_{v_i};
5   **if** L_{v_i} *contains a predecessor triple of* e_i **then**
6   Choose a predecessor triple (t, a, d) of e_i in L_{v_i};
7            $T := (t, \mathrm{arr}(e_i), d + d_i)$;
8   **if** $T \notin L_{w_i}$ **then**
9   Append T to L_{w_i};
10   Delete all elements of L_{w_i} that are dominated by an element of L_{w_i};
11   **if** $\mathrm{arr}(e_i) - t < $ *journey(w_i) or* $(\mathrm{arr}(e_i) - t = $ *journey(w_i) and* $d + d_i < $ *dur(w_i))* **then**
12   journey$(w_i) := \mathrm{arr}(e_i) - t$;
13   dur$(w_i) := d + d_i$;

14 **return** *journey(z) and dur(z) for every* $z \neq s$;

Now we add the edges of $E(G)$ one by one, which effectively iterates through the sequence $G_0, \ldots, G_m$ (see Algorithm 1). After the edge e_i has been processed, we ensure that L_y stores the set of all dominating triples from s to $y \neq s$ in G_i. For an edge $e_i \in E(G)$, we say that (t, a, d) is a *predecessor triple of* e_i if

– (t, a, d) is a dominating triple from s to v_i,

– $a \leq t_i$, and
– a is maximal among all such triples.

A predecessor triple (t, a, d) thus allows to traverse e_i after taking its corresponding dominating s-v_i-path. Note that not every edge has a predecessor triple, and that the artificial dominating triples correspond to temporal paths having no edge (and thus duration 0) that allow to traverse any outgoing edge of s).

Without loss of generality, we may ignore all edges e_i whose target vertex is s, as s is the start vertex. In order to update L_y during the processing phase of e_i if $w_i \neq s$, we choose a predecessor triple of e_i in L_{v_i} (if exists) in Line 6 of Algorithm 1 and create from it a new triple T in Line 7. The newly created triple T is then appended to L_{w_i} in Line 9, followed by removing all triples of L_{w_i} that are dominated by an element of L_{w_i}. Finally, the journey time $journey(w_i)$ and duration $dur(w_i)$ that are attained by T are updated if they improve the solution.

4   Correctness

In order to show the correctness of Algorithm 1, we rely on the next three basic lemmas, which collect helpful properties of dominating paths with respect to the sequence $G_0, \ldots, G_m$.

Lemma 2. *For every* $1 \leq i \leq m$, *every temporal path of* G_i *that contains* e_i *has* e_i *as its last edge.*

Proof. Assume to the contrary that G_i has a temporal path P that contains e_i such that the last edge of P is $e_j \neq e_i$. Then $j < i$ by definition of G_i, which implies $\mathrm{arr}(e_j) \leq \mathrm{arr}(e_i)$ by the ordering assumed for E. This contradicts that P is temporal.   □

The next two lemmas explore whether dominating and non-dominating paths are preserved when going from G_{i-1} to G_i.

Lemma 3. *For every* $1 \leq i \leq m$, *every non-dominating path* P *of* G_{i-1} *is non-dominating in* G_i.

Proof. Since P is non-dominating, G_{i-1} contains a temporal path Q that dominates P. Since neither Q nor P contains e_i, Q dominates P also in G_i. Hence, P is non-dominating in G_i.   □

In contrast to Lemma 3, a dominating path of G_{i-1} may in general become non-dominating in G_i, for example by Definition 2(iii) if e_i and the path have the same source and target vertex and the duration of e_i is very small. The next lemma states a condition under which dominating paths stay dominating paths.

Lemma 4. *For every* $1 \leq i \leq m$, *every dominating* x-y-path P *of* G_{i-1} *that satisfies* $y \neq w_i$ *is dominating in* G_i.

Proof. Assume to the contrary that P is non-dominating in G_i. Then an x-y-path Q dominates P in G_i; in particular, Q is temporal. Since $y \neq w_i$, e_i is not the last edge of Q. By Lemma 2, e_i is not contained in Q at all, so that Q is a path of G_{i-1}. Since Q dominates P in G_i, Q does so in G_{i-1}, which contradicts that P is dominating in G_{i-1}. □

Let $L_y(i)$ be the list L_y in Algorithm 1 after the edge e_i has been processed. The correctness of Algorithm 1 is based on the invariant revealed in the next lemma.

Lemma 5. *For every $0 \leq i \leq m$ and every vertex $y \neq s$ of G_i, (t, a, d) is a dominating triple from s to y in G_i if and only if $(t, a, d) \in L_y(i)$.*

Proof. Due to space constraints, we defer this to the full version of this paper. □

For $i = m$, we conclude the following corollary.

Corollary 1. *At the end of Algorithm 1, L_y contains exactly the dominating triples from s to y in G for every $y \neq s$.*

Theorem 2. *Algorithm 1 computes the journey-time and duration of a short fastest path from s to every vertex $z \neq s$ in G.*

Proof. Every short fastest path of G from s to z is dominating. By Corollary 1, L_z therefore contains every short fastest path of G from s to z at the end of Algorithm 1. By comparing the journey-time and duration of every path that is added to L_z (this may be a superset of the short fastest paths), Algorithm 1 computes the journey-time and duration of a short fastest path from s to z in G. □

Now the correctness of Theorem 1 follows from Theorem 2 by tracing back the path from z to s. This may be done by storing an additional pointer to the last edge of the current short fastest path found for every vertex z. Since following this pointer is only a constant-time operation, we may trace back the short fastest path from z to s in time proportional to its length.

5 Running Time

We investigate the running time of Algorithm 1. If the edge e_i satisfies $v_i = s$, the artificial predecessor triple $(t_i, t_i, 0)$ of Line 4 will reside at the end of the ordered list $L_v(i)$, and therefore can be appended to and retrieved from $L_v(i)$ in constant time. It suffices to clarify how we implement Lines 5–6 and Line 10, since every other step is computable in constant time. In particular, we have to maintain dominating triples in L_y and be able to compute a predecessor triple of e_i in L_y efficiently.

For every L_y, we enforce a lexicographic order $<_{\text{lex}}$ on the first two elements of all dominating triples stored. The first two elements suffice, as two distinct dominating triples (we do not store duplicates) differ always in their first two elements by Definition 2(iii). The following basic lemma will be useful.

Lemma 6. *Let $T = (t, a, d)$ and $T' = (t', a', d')$ be two distinct dominating triples from x to y such that $T <_{lex} T'$. Then $a < a'$ and either $t < t'$ or ($t = t'$ and $d > d'$).*

Proof. First, assume $t < t'$. Then $a < a'$, as otherwise T' dominates T by Definition 2(ii), which contradicts that T is dominating. In the remaining case, we have $t = t'$ and $a < a'$ by the lexicographic order on the first two elements. Then $d > d'$, as otherwise T dominates T' by Definition 2(ii).   □

Let $T_1 <_{lex} T_2 <_{lex} \cdots <_{lex} T_r$ be the dominating triples of L_{w_i} in Line 10 and let $T_j = (t^j, a^j, d^j)$ for every $1 \leq j \leq r$. By Lemma 6, $a^1 < a^2 < \cdots < a^r$. Let $T = (t, \mathrm{arr}(e_i), d + d_i)$ be the new dominating triple in G_i that is created in Line 7. Since e_i is the currently processed edge of Algorithm 1 and $E(G)$ is ordered by increasing arrival times, we have $a^r \leq \mathrm{arr}(e_i)$. Thus, appending T to L_{w_i} in Line 9 preserves the lexicographic ordering of L_{w_i}. The next two lemmas determine which elements of L_{w_i} are dominated by an element of L_{w_i}.

Lemma 7. *(i) T does not dominate any element of $\{T_1, T_2, \ldots, T_{r-1}\}$.*
(ii) If an element of $\{T_1, T_2, \ldots, T_{r-1}\}$ dominates T, then T_r dominates T.

Proof. By Lemma 6 and since $E(G)$ is ordered by increasing arrival times, $a^1 < a^2 < \cdots < a^r \leq \mathrm{arr}(e_i)$. Then $a^j < \mathrm{arr}(e_i)$ for every $1 \leq j < r$, so that T does not dominate T_j by Definition 2. This gives the first claim.

For the second claim, let T_j dominate T for some $1 \leq j < r$. Then either Definition 2(i), (ii) or (iii) holds. In Case (i), we have $t^j > t$ and $a^j \leq \mathrm{arr}(e_i)$, which implies $t < t^r$ by the lexicographic ordering. Since $a^r \leq \mathrm{arr}(e_i)$, T_r dominates T. In both Cases (ii) and (iii), we have $t^j = t$, $a^j \leq \mathrm{arr}(e_i)$ and $d^j \leq d + d_i$. By Lemma 6, either $t^j < t^r$ or $t^j = t^r$. If $t^j < t^r$, we have $t < t^r$ and $a^r \leq \mathrm{arr}(e_i)$, so that T_r dominates T. If otherwise $t^j = t^r$, we have $d^j > d^r$ by Lemma 6. Then, since $d^j \leq d + d_i$, we have $t^r = t$, $a^r \leq \mathrm{arr}(e_i)$ and $d + d_i > d^r$, so that T_r dominates T.   □

Lemma 8. *There is no element of $\{T_1, \ldots, T_r\}$ that dominates another element of $\{T_1, \ldots, T_r\}$. After Line 10, L_{w_i} consists of*

(i) $(T_1, \ldots, T_{r-1}, T)$ if T dominates T_r (then Line 10 deletes only T_r from L_{w_i}),
(ii) $(T_1, \ldots, T_{r-1}, T_r, T)$ if no element of $\{T_r, T\}$ dominates the other element of $\{T_r, T\}$ (then Line 10 deletes nothing from L_{w_i}), and
(iii) $(T_1, \ldots, T_{r-1}, T_r)$ if T_r dominates T (then Line 10 deletes only T from L_{w_i}).

Proof. Since T is the only triple that was added to L_{w_i}, every element of $\{T_1, \ldots, T_r\}$ was dominating in G_{i-1}. Thus, no element of $\{T_1, \ldots, T_r\}$ dominates another element of this set in G_i. By Lemma 7(i), T does not dominate any element of $\{T_1, T_2, \ldots, T_{r-1}\}$.

Consider Claim (i). Since T dominates T_r, T_r does not dominate T. Then the contrapositive of Lemma 7(ii) implies that no element of $\{T_1, \ldots, T_r\}$ dominates T. Together with the first claim, this gives $L_{w_i} = (T_1, \ldots, T_{r-1}, T)$.

Consider Claim (ii). Since T_r does not dominate T, the same argument as before implies that no element of $\{T_1, \ldots, T_r\}$ dominates T. Since T does not dominate T_r, the first claim of this lemma implies $L_{w_i} = (T_1, \ldots, T_{r-1}, T_r, T)$. Claim (iii) follows directly from the first claim and the fact that T_r dominates T. □

Algorithm 2: Deleting Dominating Triples (Line 10 of Algorithm 1)

Input: A list L_{w_i} of dominating triples ordered by $<_{\text{lex}}$, and the new triple T of G_i from Line 7.

1 Retrieve the last element T_r of L_{w_i} if $L_{w_i} \neq \emptyset$;
2 **if** $L_{w_i} = \emptyset$ *or* $T \neq T_r$ **then** append T to L_{w_i};
3 **if** $|L_{w_i}| \geq 2$ **then**
4 | **if** T *dominates* T_r **then** delete T_r from L_{w_i};
5 | **if** T_r *dominates* T **then** delete T from L_{w_i};

Lemma 8 allows us to implement Line 10 of Algorithm 1 very efficiently by comparing just the last element T_r of L_{w_i} with T. This can be done in constant time by Algorithm 2.

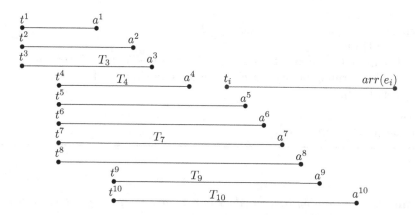

Fig. 3. A list $L_{v_i} = (T_1, \ldots, T_{10})$ containing dominating triples from s to v_i, ordered by $<_{\text{lex}}$ from top to bottom, and the departure and arrival times of e_i. The only predecessor triple of e_i is $T_4 = (t^4, a^4, d^4)$. In order to compute T_4, Algorithm 3 tests the arrival times of T_{10}, T_9 and T_7 until it stops at T_3 (because $a^3 \leq t_i$) and computes $T_4 \in \{T_3, \ldots, T_6\}$ by binary search.

It remains to show how a predecessor triple (t, a, d) of e_i in L_{v_i} in Lines 5+6 of Algorithm 1 can be computed efficiently. By Lemma 6 and its preceding remark, the arrival times of all triples in L_{v_i} are for every i distinct. Hence, for every i, the number of elements in L_{v_i} is at most the maximum in-degree $\delta^-(G)$ of G.

Since the triples in L_{v_i} are ordered by $<_{\text{lex}}$, the last triple in L_{v_i} (if exists) whose arrival time is at most t_i is a predecessor triple of e_i, so that we only have to compute this unique predecessor triple for every i. In order to do this, we might use binary search on the arrival times of triples of L_{v_i}. We achieve however the slightly better running time $O(\log p(G))$ when first using an exponential search [2] that starts with the triple having highest arrival time (which is at most $\text{arr}(e_i)$ due to the edge-ordering) until some triple with arrival time at most t_i is found (see Figure 3 for an example).

If such a triple exists, there is also a predecessor triple of e_i, which we then compute by binary search in the resulting range; see Algorithm 3 for a detailed description. This takes only time $O(\log p(G))$, which is upper bounded by $O(\min\{\delta^-(G), \max\{d_i : 1 \le i \le m\}\})$. If no such triple exists, there is no predecessor triple of e_i, and this information is given as output.

Algorithm 3: Computing a Predecessor Triple (Lines 5+6 of Algorithm 1)

Input: A list $L_{v_i} = (T_1, T_2, \ldots, T_r)$ of dominating triples in G_{i-1} ordered by $<_{\text{lex}}$ such that $T_j = (t^j, a^j, d^j)$ for every $1 \le j \le r$, and an edge e_i of G.

Output: A predecessor triple of e_i if exists, and otherwise the output "not existent"

1   $j := 1$;
2   **while** $j \le r$ and $a_{r+1-j} > t_i$ **do** $j := 2j$;
3   **if** $j \le r$ **then**
4   Compute the maximal $r + 1 - j \le l < r + 1 - \lfloor j/2 \rfloor$ such that $a_l \le t_i$ by performing binary search (then a_l is maximal by Lemma 6);
5   **return** T_l;
6   **else**
7   output "not existent";

Lemma 9. *The running time of Algorithm 1 for a temporal graph G on n vertices and m edges is $O(n + m \log p(G))$.*

Proof. Apart from Lines 5+6 and 10, every of the m edges can be processed in constant time. By Algorithms 2 and 3, the running times for Lines 5+6 and 10 amount to $O(\log p(G))$ and $O(1)$ time for every edge, which gives the claim. $\square$

This concludes the proof of Theorem 1.

References

1. Aaron, E., Krizanc, D., Meyerson, E.: DMVP: foremost waypoint coverage of time-varying graphs. In: Kratsch, D., Todinca, I. (eds.) WG 2014. LNCS, vol. 8747, pp. 29–41. Springer, Cham (2014). https://doi.org/10.1007/978-3-319-12340-0_3

2. Bentley, J.L., Yao, A.C.-C.: An almost optimal algorithm for unbounded searching. Inf. Process. Lett. **5**(3), 82–87 (1976)

3. Bui Xuan, B.-M., Ferreira, A., Jarry, A.: Computing shortest, fastest, and foremost journeys in dynamic networks. Int. J. Found. Comput. Sci. **14**(2), 267–285 (2003)

4. de Bernardo, G., Brisaboa, N.R., Caro, D., Rodríguez, M.A.: Compact data structures for temporal graphs. In: Data Compression Conference (DCC 2013), p. 477. IEEE (2013)

5. Dibbelt, J., Pajor, T., Strasser, B., Wagner, D.: Intriguingly simple and fast transit routing. In: Bonifaci, V., Demetrescu, C., Marchetti-Spaccamela, A. (eds.) SEA 2013. LNCS, vol. 7933, pp. 43–54. Springer, Heidelberg (2013). https://doi.org/10.1007/978-3-642-38527-8_6

6. Kempe, D., Kleinberg, J.M., Kumar, A.: Connectivity and inference problems for temporal networks. J. Comput. Syst. Sci. **64**(4), 820–842 (2002)

7. Liang, Q., Modiano, E.: Survivability in time-varying networks. IEEE Trans. Mob. Comput. **16**(9), 2668–2681 (2017)

8. Mertzios, G.B., Michail, O., Spirakis, P.G.: Temporal network optimization subject to connectivity constraints. Algorithmica **81**(4), 1416–1449 (2019)

9. Michail, O., Spirakis, P.G.: Traveling salesman problems in temporal graphs. Theoret. Comput. Sci. **634**, 1–23 (2016)

10. Tang, J.K., Musolesi, M., Mascolo, C., Latora, V.: Characterising temporal distance and reachability in mobile and online social networks. Comput. Commun. Rev. **40**(1), 118–124 (2010)

11. Huanhuan, W., Cheng, J., Ke, Y., Huang, S., Huang, Y., Hejun, W.: Efficient algorithms for temporal path computation. IEEE Trans. Knowl. Data Eng. **28**(11), 2927–2942 (2016)

12. Zschoche, P., Fluschnik, T., Molter, H., Niedermeier, R.: The complexity of finding small separators in temporal graphs. J. Comput. Syst. Sci. **107**, 72–92 (2020)

Minmax Regret 1-Sink Location Problems on Dynamic Flow Path Networks with Parametric Weights

Tetsuya Fujie[1], Yuya Higashikawa[1(✉)], Naoki Katoh[1], Junichi Teruyama[1], and Yuki Tokuni[2]

[1] University of Hyogo, Kobe, Japan
{fujie,higashikawa,naoki.katoh,junichi.teruyama}@sis.u-hyogo.ac.jp
[2] Kwansei Gakuin University, Sanda, Japan
enj32048@kwansei.ac.jp

Abstract. This paper addresses the minmax regret 1-sink location problem on dynamic flow path networks with parametric weights. We are given a *dynamic flow network* consisting of an undirected path with positive edge lengths, positive edge capacities, and nonnegative vertex weights. A path can be considered as a road, an edge length as the distance along the road and a vertex weight as the number of people at the site. An edge capacity limits the number of people that can enter the edge per unit time. We consider the problem of locating a *sink* in the network, to which all the people evacuate from the vertices as quickly as possible. In our model, each weight is represented by a linear function in a common parameter t, and the decision maker who determines the location of a sink does not know the value of t. We formulate the sink location problem under such uncertainty as the *minmax regret problem*. Given t and a sink location x, the cost of x under t is the sum of arrival times at x for all the people determined by t. The regret for x under t is the gap between the cost of x under t and the optimal cost under t. The task of the problem is formulated as the one to find a sink location that minimizes the maximum regret over all t. For the problem, we propose an $O(n^4 2^{\alpha(n)} \alpha(n) \log n)$ time algorithm where n is the number of vertices in the network and $\alpha(\cdot)$ is the inverse Ackermann function. Also for the special case in which every edge has the same capacity, we show that the complexity can be reduced to $O(n^3 2^{\alpha(n)} \alpha(n) \log n)$.

1 Introduction

Recently, many disasters, such as earthquakes, nuclear plant accidents, volcanic eruptions and flooding, have struck in many parts of the world, and it has been

A full version of the paper is available at [13]; https://arxiv.org/abs/2011.13569.
This work is supported by JSPS KAKENHI Grant-in-Aid for Early-Career Scientists (18K18003, 20K19746), JSPS KAKENHI Grant-in-Aid for Scientific Research (B) (19H04068), and JST CREST (JPMJCR1402).

R. Uehara et al. (Eds.): WALCOM 2021, LNCS 12635, pp. 52–64, 2021.
https://doi.org/10.1007/978-3-030-68211-8_5

recognized that orderly evacuation planning is urgently needed. A powerful tool for evacuation planning is the *dynamic flow model* introduced by Ford and Fulkerson [12], which represents movement of commodities over time in a network. In this model, we are given a graph with *source* vertices and *sink* vertices. Each source vertex is associated with a positive weight, called a *supply*, each sink vertex is associated with a positive weight, called a *demand*, and each edge is associated with positive length and capacity. An edge capacity limits the amount of supply that can enter the edge per unit time. One variant of the dynamic flow problem is the *quickest transshipment problem*, of which the objective is to send exactly the right amount of supply out of sources into sinks with satisfying the demand constraints in the minimum overall time. Hoppe and Tardos [22] provided a polynomial time algorithm for this problem in the case where the transit times are integral. However, the complexity of their algorithm is very high. Finding a practical polynomial time solution to this problem is still open. A reader is referred to a recent survey by Skutella [27] on dynamic flows.

This paper discusses a related problem, called the *sink location problem* [5–7, 10, 11, 20, 21, 26], of which the objective is to find a location of sinks in a given dynamic flow network so that all the supply is sent to the sinks as quickly as possible. For the optimality of location, the following two criteria can be naturally considered: the minimization of *evacuation completion time* and *aggregate evacuation time* (i.e., *sum of evacuation times*). We call the sink location problem that requires finding a location of sinks on a dynamic flow network that minimizes the evacuation completion time (resp. the aggregate evacuation time) the CTSL problem (resp. the ATSL problem). Several papers have studied the CTSL problems [7, 10, 11, 20, 21, 26]. On the other hand, for the ATSL problems, we have a few results only for path networks [5, 6, 21].

In order to model the evacuation behavior of people, it might be natural to treat each supply as a discrete quantity as in [22, 26]. Nevertheless, almost all the previous papers on sink location problems [7, 10, 11, 20, 21] treat each supply as a continuous quantity since it is easier for mathematically handling the problems and the effect of such treatment is small enough to ignore when the number of people is large. Throughout the paper, we adopt the model with continuous supplies.

Although the above two criteria are reasonable, they may not be practical since the population distribution is assumed to be fixed. In a real situation, the number of people in an area may vary depending on the time, e.g., in an office area in a big city, there are many people during the daytime on weekdays while there are much less people on weekends or during the night time. In order to take such the uncertainty into account, Kouvelis and Yu [23] introduced the *minmax regret model*. In the *minmax regret sink location problems*, we are given a finite or infinite set S of *scenarios*, where each scenario gives a particular assignment of weights on all the vertices. Here, for a sink location x and a scenario $s \in S$, we denote the evacuation completion time or aggregate evacuation time by $F(x, s)$. Then, the problem can be understood as a 2-person Stackelberg game as follows. The first player picks a sink location x and the second player chooses a scenario

$s \in S$ that maximizes the regret defined as $R(x, s) := F(x, s) - \min_x F(x, s)$. The objective of the first player is to choose x that minimizes the maximum regret. Throughout the paper, we call the minmax regret sink location problem, where the regret is defined with the evacuation completion time (resp. the aggregate evacuation time), the MMR-CTSL problem (resp. the MMR-ATSL problem). The MMR-CTSL problems have been studied so far [3, 9, 14, 18, 20, 24, 25]. On the other hand, for the MMR-ATSL problems, we have few results [8, 19] although the problems are also important theoretically and practically.

As for how to define a set of scenarios, all of the previous studies on the minmax regret sink location problems adopt the model with *interval weights*, in which each vertex is given the weight as a real interval, and a scenario is defined by choosing an element of the Cartesian product of all the weight intervals over the vertices. One drawback of the minmax regret model with interval weights is that each weight can take an independent value, thus we consider some extreme scenarios which may not happen in real situations, e.g, a scenario where all the vertices have maximum weights or minimum weights. To incorporate the dependency among weights of all the vertices into account, we adopt the model with *parametric weights* (first introduced by Vairaktarakis and Kouvelis [28] for the minmax regret median problem), in which each vertex is given the weight as a linear function in a common parameter t on a real interval, and a scenario is just determined by choosing t. Note that considering a real situation, each weight function should be more complex, however, such a function can be approximated by a piecewise linear function. Thus superimposing all such piecewise linear functions, it turns out that for a sufficiently small subinterval of t, every weight function can be regarded as linear, and by solving multiple subproblems with linear weight functions, we can obtain the solution.

In this paper, we study the MMR-ATSL problem on dynamic flow path networks with parametric weights. Our main theorem is below.

Theorem 1 (Main Results). *Suppose that we are given a dynamic flow path network of n vertices with parametric weights.*

(i) *The* MMR-ATSL *problem can be solved in time $O(n^4 2^{\alpha(n)} \alpha(n) \log n)$, where $\alpha(\cdot)$ is the inverse Ackermann function.*

(ii) *When all the edge capacities are uniform, the* MMR-ATSL *problem can be solved in time $O(n^3 2^{\alpha(n)} \alpha(n) \log n)$.*

Note that the MMR-ATSL problem with interval weights is studied by [8, 19], and only for the case with the uniform edge capacity, Higashikawa et al. [19] provide an $O(n^3)$ time algorithm, which is improved to one running in $O(n^2 \log^2 n)$ time by [8]. However, for the case with general edge capacities, no algorithm has been known so far. Therefore, our result implies that the problem becomes solvable in polynomial time by introducing parametric weights.

The rest of the paper is organized as follows. In Sect. 2, we give the notations and the fundamental properties that are used throughout the paper. In Sect. 3, we give the key lemmas and the algorithms that solves the problems, which concludes the paper.

2 Preliminaries

For two real values a, b with $a < b$, let $[a, b] = \{t \in \mathbb{R} \mid a \leq t \leq b\}$, $(a, b) = \{t \in \mathbb{R} \mid a < t < b\}$, and $(a, b] = \{t \in \mathbb{R} \mid a < t \leq b\}$.

In our problem, we are given a real interval $T = [t^-, t^+] \subset \mathbb{R}$ and a dynamic flow path network $\mathcal{P} = (P, \mathbf{w}(t), \mathbf{c}, \mathbf{l}, \tau)$, which consists of five elements: $P = (V, E)$ is a path with vertex set $V = \{v_i \mid 1 \leq i \leq n\}$ and edge set $E = \{e_i = (v_i, v_{i+1}) \mid 1 \leq i \leq n-1\}$, $\mathbf{w}(t)$ is a vector $\langle w_1(t), \ldots, w_n(t) \rangle$ of which component $w_i(t)$ is a *weight function* $w_i : T \to \mathbb{R}_{\geq 0}$ which is linear in a parameter t and nonnegative for any $t \in T$, a vector $\mathbf{c} = \langle c_1, \ldots, c_{n-1} \rangle$ consists of the *capacity* c_i of edge e_i, a vector $\mathbf{l} = \langle \ell_1, \ldots, \ell_{n-1} \rangle$ consists of the *length* ℓ_i of edge e_i, and τ is the *time* which the supply takes to move a unit distance on any edge. Let us explain how edge capacities and lengths affect the evacuation time. Consider an evacuation under fixed $t \in T$. Suppose that at time 0, the amount w of supply is at vertex v_{i+1} and going through edge e_i towards vertex v_i. The first fraction of supply from v_{i+1} can arrive at v_i at time $\tau \ell_i$. The edge capacity c_i represents the maximum amount of supply which can enter e_i in a unit time interval, so all the supply w can complete leaving v_{i+1} at time w/c_i. Therefore, all the supply w can complete arriving at v_i at time $\tau \ell_i + w/c_i$.

For any integers i, j with $1 \leq i \leq j \leq n$, we denote the sum of weights from v_i to v_j by $W_{i,j}(t) = \sum_{h=i}^{j} w_h(t)$. For the notation, we define $W_{i,j}(t) = 0$ for i, j with $i > j$. For a vertex $v_i \in V$, we abuse v_i to denote the distance between v_1 and v_i, i.e., $v_i = \sum_{j=1}^{i-1} \ell_j$. For an edge $e_i \in E$, we abuse e_i to denote a real open interval (v_i, v_{i+1}). We also abuse P to denote a real closed interval $[0, v_n]$. If a real value x satisfies $x \in (v_i, v_{i+1})$, x is said to be a point on edge e_i to which the distance from v_i is value $x - v_i$. Let $C_{i,j}$ be the minimum capacity for all the edges from e_i to e_j, i.e., $C_{i,j} = \min\{c_h \mid i \leq h \leq j\}$.

Note that we precompute values v_i and $W_{1,i}(t)$ for all i in $O(n)$ time, and then, $W_{i,j}(t)$ for any i, j can be obtained in $O(1)$ time as $W_{i,j}(t) = W_{1,j}(t) - W_{1,i-1}(t)$. In addition, $C_{i,j}$ for any i, j can be obtained in $O(1)$ time with $O(n)$ preprocessing time, which is known as the *range minimum query* [2,4].

2.1 Evacuation Completion Time on a Dynamic Flow Path Network

In this section, we see the details of evacuation phenomenon using a simple example, and eventually show the general formula of evacuation completion time on a path, first provided by Higashikawa [17]. W.l.o.g., an evacuation to a sink x follows the first-come first-served manner at each vertex, i.e., when a small fraction of supply arrives at a vertex v on its way to x, it has to wait for the departure if there already remains some supply waiting for leaving v.

Let us consider an example with $|V| = 3$ where $V = \{v_1, v_2, v_3\}, E = \{e_1 = (v_1, v_2), e_2 = (v_2, v_3)\}$. Assume that the sink x is located at v_1, and under a fixed parameter $t \in T$, the amount of supply at v_i is w_i for $i = 2, 3$.

All the supply w_1 at v_1 immediately completes its evacuation at time 0 and we send all the supply w_2 and w_3 to v_1 as quickly as possible. Let us focus on how the supply of v_3 moves to v_1. First, the foremost fraction of supply from

v_3 arrives at v_2 at time $\tau \ell_2$, and all the supply w_3 completes leaving v_3 at time w_3/c_2, i.e., it completes arriving at v_2 at time $\tau \ell_2 + w_3/c_2$. Suppose that at time $\tau \ell_2 + w_3/c_2$, the amount $w' (\geq 0)$ of supply remains at v_2. From then on, the time required to send all the supply w' to v_1 is $\tau \ell_1 + w'/c_1$. Thus, the evacuation completion time is expressed as

$$\tau(\ell_1 + \ell_2) + \frac{w_3}{c_2} + \frac{w'}{c_1}. \tag{1}$$

We observe what value w' takes in the following cases.

Case 1: It Holds $c_1 \geq c_2$. In this case, the amount of supply at v_2 should be non-increasing, because the amount c_1 of supply leaves v_2 and the amount at most c_2 of supply arrives at v_2 per unit time. Let us consider the following two situations at time $\tau \ell_2 + w_3/c_2$: When all the supply w_3 completes arriving at v_2, there remains no supply at v_2, that is, $w' = 0$ holds or not. If $w' = 0$ holds, then substituting it into (1), the evacuation completion time is expressed as

$$\tau(\ell_1 + \ell_2) + \frac{w_3}{c_2}. \tag{2}$$

Otherwise, that is $w' > 0$ holds, there remains a certain amount of supply at v_2 even at time $\tau \ell_2$ since the amount of supply at v_2 is non-increasing. Thus at time $\tau \ell_2$, the amount $w_2 - c_1 \tau \ell_2$ of supply remains at v_2. From time $\tau \ell_2$ to time $\tau \ell_2 + w_3/c_2$, the amount of supply waiting at v_2 decreases by $c_1 - c_2$ per unit time. Then, we have

$$w' = w_2 - c_1(\tau \ell_2) - (c_1 - c_2) \cdot \frac{w_3}{c_2} = w_2 + w_3 - c_1 \tau \ell_2 - \frac{c_1 w_3}{c_2}.$$

Thus, the evacuation completion time is expressed as

$$\tau(\ell_1 + \ell_2) + \frac{w_3}{c_2} + \frac{w_2 + w_3 - c_1 \tau \ell_2 - c_1 w_3/c_2}{c_1} = \tau \ell_1 + \frac{w_2 + w_3}{c_1}. \tag{3}$$

Case 2: It Holds $c_1 < c_2$. In this case, the amount of supply waiting at v_2 increases by $c_2 - c_1$ per unit time from time $\tau \ell_2$ (when the foremost supply from v_3 arrives at v_2) to time $\tau \ell_2 + w_3/c_2$ (when the supply from v_3 completes to arrive at v_2). Let us consider the following two situations at time $\tau \ell_2$. When the foremost supply from v_3 arrives at v_2, there remains no supply at v_2 or not.

If there remains no supply at v_2 at time $\tau \ell_2$, then it holds $w' = (c_2 - c_1)(w_3/c_2) = w_3 - c_1 w_3/c_2$ in (1). Thus, the evacuation completion time is expressed as

$$\tau(\ell_1 + \ell_2) + \frac{w_3}{c_2} + \frac{w_3 - c_1 w_3/c_2}{c_1} = \tau(\ell_1 + \ell_2) + \frac{w_3}{c_1}. \tag{4}$$

Otherwise, the situation is similar to the latter case of Case 1. The difference is that the amount of supply waiting at v_2 increases by $c_2 - c_1$ per unit time during from time $\tau\ell_2$ to time $\tau\ell_2 + w_3/c_2$, while in Case 1, it decreases by $c_1 - c_2$ per unit time. For this case, the evacuation completion time is given by formula (3).

In summary of formulae (2)–(4), the evacuation completion time for a dynamic flow path network with three vertices is given by the following formula:

$$\max\left\{\tau\ell_1 + \frac{w_2 + w_3}{c_1}, \tau(\ell_1 + \ell_2) + \frac{w_3}{\min\{c_1, c_2\}}\right\}. \tag{5}$$

Let us turn to the case with n vertices, that is, $V = \{v_i \mid 1 \leq i \leq n\}$. When the sink is located at v_1 and a parameter $t \in T$ is fixed, generalizing formula (5), the evacuation completion time is given by the following formula, which is provided by Higashikawa [17]:

$$\max_{2 \leq i \leq n}\left\{\tau\sum_{j=1}^{i-1}\ell_j + \frac{\sum_{j=i}^{n}w_j(t)}{\min_{1 \leq j \leq i-1} c_j}\right\} = \max_{2 \leq i \leq n}\left\{\tau v_i + \frac{W_{i,n}(t)}{C_{1,i}}\right\}. \tag{6}$$

An interesting observation is that each $\tau v_i + W_{i,n}(t)/C_{1,i}$ in (6) is equivalent to the evacuation completion time for the transformed input so that only v_i is given supply $W_{i,n}(t)$ and all the others are given zero supply.

Let us give explicit formula of the evacuation completion time for fixed $x \in P$ and parameter $t \in T$. Suppose that a sink x is on edge $e_i = (v_i, v_{i+1})$. In this case, all the supply on the right side (i.e., at $v_{i+1}, \ldots, v_n$) will flow left to sink x and all the supply on the left side (i.e., at $v_1, \ldots, v_i$) will flow right to sink x. First, we consider the evacuation for the supply on the right side of x. Supply on the path is viewed as a continuous value, and we regard that all the supply on the right side of x is mapped to the interval $(0, W_{i+1,n}(t)]$. The value z satisfying $z \in (W_{i+1,j-1}(t), W_{i+1,j}(t)]$ with $i + 1 \leq j \leq n$ represents all the supply at vertices $v_{i+1}, v_{i+2}, \ldots, v_{j-1}$ plus partial supply of $z - W_{i+1,j-1}(t)$ at v_j. Let $\theta_R^{e_i}(x, t, z)$ denote the time at which the first z amount of supply on the right side of x (i.e., $v_{i+1}, v_{i+2}, \ldots, v_n$) completes its evacuation to sink x. Modifying formula (6), $\theta_R^{e_i}(x, t, z)$ is given by the following formula: For $z \in (W_{i+1,j-1}(t), W_{i+1,j}(t)]$ with $i + 1 \leq j \leq n$,

$$\theta_R^{e_i}(x, t, z) = \max_{i+1 \leq h \leq j}\left\{\tau(v_h - x) + \frac{z - W_{i+1,h-1}(t)}{C_{i,h}}\right\}. \tag{7}$$

In a symmetric manner, we consider the evacuation for the supply on the left side of x (i.e., $v_1, \ldots, v_i$). The value z satisfying $z \in (W_{j+1,i}(t), W_{j,i}(t)]$ with $1 \leq j \leq i$ represents all the supply at vertices $v_i, v_{i-1}, \ldots, v_{j+1}$ plus partial supply of $z - W_{j+1,i}(t)$ at v_j. Let $\theta_L^{e_i}(x, t, z)$ denote the time at which the first z amount of supply on the left side of x completes its evacuation to sink x, which is given by the following formula: For $z \in (W_{j+1,i}(t), W_{j,i}(t)]$ with $1 \leq j \leq i$,

$$\theta_L^{e_i}(x, t, z) = \max_{j \leq h \leq i}\left\{\tau(x - v_h) + \frac{z - W_{h+1,i}(t)}{C_{h,i}}\right\}. \tag{8}$$

Let us turn to the case that sink x is at a vertex $v_i \in V$. We confirm that the evacuation times when the amount z of supply originating from the right side of and the left side of v_i to sink v_i are given by $\theta_R^{e_i}(v_i, t, z)$ and $\theta_L^{e_{i-1}}(v_i, t, z)$, respectively.

2.2   Aggregate Evacuation Time

Let $\Phi(x, t)$ be the aggregate evacuation time (i.e., sum of evacuation time) when a sink is at a point $x \in P$ and the weight functions are fixed by a parameter $t \in T$. For a point x on edge e_i and a parameter $t \in T$, the aggregate evacuation time $\Phi(x, t)$ is defined by the integrals of the evacuation completion times $\theta_L^{e_i}(x, t, z)$ over $z \in [0, W_{1,i}(t)]$ and $\theta_R^{e_i}(x, t, z)$ over $z \in [0, W_{i+1,n}(t)]$, i.e.,

$$\Phi(x, t) = \int_0^{W_{1,i}(t)} \theta_L^{e_i}(x, t, z)dz + \int_0^{W_{i+1,n}(t)} \theta_R^{e_i}(x, t, z)dz. \tag{9}$$

In a similar way, if a sink x is at vertex v_i, then $\Phi(v_i, t)$ is given by

$$\Phi(v_i, t) = \int_0^{W_{1,i-1}(t)} \theta_L^{e_{i-1}}(v_i, t, z)dz + \int_0^{W_{i+1,n}(t)} \theta_R^{e_i}(v_i, t, z)dz. \tag{10}$$

2.3   Minmax Regret Formulation

We denote by $\text{Opt}(t)$ the minimum aggregate evacuation time with respect to a parameter $t \in T$. Higashikawa et al. [21] and Benkoczi et al. [6] showed that for the minsum k-sink location problems, there exists an optimal k-sink such that all the k sinks are at vertices. This implies that we have

$$\text{Opt}(t) = \min_{x \in V} \Phi(x, t) \tag{11}$$

for any $t \in T$. For a point $x \in P$ and a value $t \in T$, a regret $R(x, t)$ with regard to x and t is a gap between $\Phi(x, t)$ and $\text{Opt}(t)$ that is defined as

$$R(x, t) = \Phi(x, t) - \text{Opt}(t). \tag{12}$$

The maximum regret for a sink $x \in P$, denoted by $MR(x)$, is the maximum value of $R(x, t)$ with respect to $t \in T$. Thus, $MR(x)$ is defined as

$$MR(x) = \max_{t \in T} R(x, t). \tag{13}$$

Given a dynamic flow path network $\mathcal{P}$ and a real interval T, the problem MMR-ATSL is defined as follows:

$$\text{minimize } MR(x) \text{ subject to } x \in P \tag{14}$$

Let x^* denote an optimal solution of (14).

2.4   Piecewise Functions and Upper/Lower Envelopes

A function $f : X(\subset \mathbb{R}) \to \mathbb{R}$ is called a *piecewise polynomial function* if and only if real interval X can be partitioned into subintervals $X_1, X_2, \ldots, X_m$ so that f forms as a polynomial f_i on each X_i. We denote such a piecewise polynomial function f by $f = \langle (f_1, X_1), \ldots, (f_m, X_m) \rangle$, or simply $f = \langle (f_i, X_i) \rangle$. We assume that such a partition into subintervals are maximal in the sense that $f_i \neq f_{i+1}$ for any i. We call each pair (f_i, X_i) a *piece* of f, and an endpoint of the closure of X_i a *breakpoint* of f. A piecewise polynomial function $f = \langle (f_i, X_i) \rangle$ is called a *piecewise polynomial function of degree at most two* if and only if each f_i is quadratic or linear. We confirm the following property about the sum of piecewise polynomial functions.

Proposition 1. *Let m and m' be positive integers, and $f, g : X(\subset \mathbb{R}) \to \mathbb{R}$ be piecewise polynomial functions of degree at most two with m and m' pieces, respectively. Then, a function $h = f + g$ is a piecewise polynomial function of degree at most two with at most $m + m'$ pieces. Moreover, given $f = \langle (f_i, X_i) \rangle$ and $g = \langle (g_j, X'_j) \rangle$, we can obtain $h = f + g = \langle (h_j, X''_j) \rangle$ in $O(m + m')$ time.*

Let $\mathcal{F} = \{f_1(y), \ldots, f_m(y)\}$ be a family of m polynomial functions where $f_i : Y_i(\subset \mathbb{R}) \to \mathbb{R}$ and Y denote the union of Y_i, that is, $Y = \cup_{i=1}^{m} Y_i$. An *upper envelope* $\mathcal{U}_{\mathcal{F}}(y)$ and a *lower envelope* $\mathcal{L}_{\mathcal{F}}(y)$ of $\mathcal{F}$ are functions from Y to $\mathbb{R}$ defined as follows:

$$\mathcal{U}_{\mathcal{F}}(y) = \max_{i=1,\ldots,m} f_i(y), \quad \mathcal{L}_{\mathcal{F}}(y) = \min_{i=1,\ldots,m} f_i(y), \tag{15}$$

where the maximum and the minimum are taken over those functions that are defined at y, respectively. For an upper envelope $\mathcal{U}_{\mathcal{F}}(y)$ of $\mathcal{F}$, there exist an integer sequence $U_{\mathcal{F}} = \langle u_1, \ldots, u_k \rangle$ and subintervals $I_1, \ldots, I_k$ of Y such that $\mathcal{U}_{\mathcal{F}}(y) = \langle (f_{u_1}(y), I_1), \ldots, (f_{u_k}(y), I_k) \rangle$ holds. That is, an upper envelope $\mathcal{U}_{\mathcal{F}}(y)$ can be represented as a piecewise polynomial function. We call the above sequence $U_{\mathcal{F}}$ the *upper-envelope sequence* of $\mathcal{U}_{\mathcal{F}}(y)$.

In our algorithm, we compute the upper/lower envelopes of partially defined, univariate polynomial functions. The following result is useful for this operation.

Theorem 2 ([1,15,16]). *Let $\mathcal{F}$ be a family of n partially defined, polynomial functions of degree at most two. Then, $\mathcal{U}_{\mathcal{F}}$ and $\mathcal{L}_{\mathcal{F}}$ consist of $O(n2^{\alpha}(n))$ pieces and one can obtain them in time $O(n\alpha(n) \log n)$, where $\alpha(n)$ is the inverse Ackermann function. Moreover, if $\mathcal{F}$ a family of n line segments, then $\mathcal{U}_{\mathcal{F}}$ and $\mathcal{L}_{\mathcal{F}}$ consist of $O(n)$ pieces and one can obtain them in time $O(n \log n)$.*

Note that the number of pieces and the computation time for the upper/lower envelopes are involved with the maximum length of Davenport–Schinzel sequences. See [15] for the details. For a family $\mathcal{F}$ of functions, if we say that we *obtain* envelopes $\mathcal{U}_{\mathcal{F}}(y)$ or $\mathcal{L}_{\mathcal{F}}(y)$, then we obtain the information of all pieces $(f_{u_i}(y), I_i)$.

3   Algorithms

The main task of the algorithm is to compute the following $O(n)$ values, $MR(v)$ for all $v \in V$ and $\min\{MR(x) \mid x \in e\}$ for all $e \in E$. Once we compute these values, we immediately obtain the solution of the problem by choosing the minimum one among them in $O(n)$ time.

Let us focus on computing $\min\{MR(x) \mid x \in e\}$ for each $e \in E$. (Note that we can compute $MR(v)$ for $v \in V$ in a similar manner.) Recall the definition of the maximum regret for x, $MR(x) = \max\{R(x,t) \mid t \in T\}$. A main difficulty lies in evaluating $R(x,t)$ over $t \in T$ even for a fixed x since interval T is infinite. Furthermore, we are also required to find an optimal location among an infinite set e. To tackle with this issue, our key idea is to partition the problem into a polynomial number of subproblems as follows: We partition interval T into a polynomial number of subintervals $T_1, \ldots, T_m$ so that $R(x,t)$ is represented as a (single) polynomial function in x and t on $\{x \in e\} \times T_j$ for each $j = 1, \ldots, m$. For each T_j, we compute the maximum regret for $x \in e$ over T_j denoted by $G_j(x) = \max\{R(x,t) \mid t \in T_j\}$. An explicit form of $G_j(x)$ is given in the full paper [13]. We then obtain $MR(x)$ for $x \in e$ as the upper envelope of functions $G_1(x), \ldots, G_m(x)$ and find the minimum value of $MR(x)$ for $x \in e$ by elementary calculation.

In the rest of the paper, we mainly show that for each e or v, there exists a partition of T with a polynomial number of subintervals such that the regret $R(x,t)$ is a polynomial function of degree at most two on each subinterval.

3.1   Key Lemmas

To understand $R(x,t)$, we observe function $\Phi(x,t)$. We give some other notations. Let $f_R^{e_i,j}(t,z)$ and $f_L^{e_i,j}(t,z)$ denote functions obtained by removing terms containing x from formulae (7) and (8). Formally, for $1 \leq i < j \leq n$, let function $f_R^{e_i,j}(t,z)$ be defined on $t \in T$ and $z \in (W_{i+1,j-1}(t), W_{i+1,n}(t)]$ as

$$f_R^{e_i,j}(t,z) = \tau v_j + \frac{z - W_{i+1,j-1}(t)}{C_{i,j}}, \tag{16}$$

and for $1 \leq j < i \leq n$, let function $f_L^{e_i,j}(t,z)$ be defined on $t \in T$ and $z \in (W_{j+1,i}(t), W_{1,i}(t)]$ as

$$f_L^{e_i,j}(t,z) = -\tau v_j + \frac{z - W_{j+1,i}(t)}{C_{j,i}}. \tag{17}$$

In addition, let $F_L^{e_i}(t)$ and $F_R^{e_i}(t)$ denote univariate functions defined as

$$F_L^{e_i}(t) = \int_0^{W_{1,i}(t)} f_L^{e_i}(t,z)dz, \quad F_R^{e_i}(t) = \int_0^{W_{i+1,n}(t)} f_R^{e_i}(t,z)dz, \tag{18}$$

where $f_L^{e_i}(t,z)$ and $f_R^{e_i}(t,z)$ denote functions defined as

$$f_L^{e_i}(t,z) = \max_{1 \leq j \leq i}\left\{f_L^{e_i,j}(t,z)\right\}, \quad f_R^{e_i}(t,z) = \max_{i+1 \leq j \leq n}\left\{f_R^{e_i,j}(t,z)\right\}.$$

Recall the definition of the aggregate evacuation time $\Phi(x, t)$ shown in (9). We observe that for $x \in e_i$, $\Phi(x, t)$ can be represented as

$$\Phi(x, t) = \big(W_{1,i}(t) - W_{i+1,n}(t)\big)\tau x + \int_0^{W_{1,i}(t)} f_L^{e_i}(t, z)dz + \int_0^{W_{i+1,n}(t)} f_R^{e_i}(t, z)dz$$

$$= \big(W_{1,i}(t) - W_{i+1,n}(t)\big)\tau x + F_L^{e_i}(t) + F_R^{e_i}(t). \tag{19}$$

In a similar manner, by the definition of (10) and formula (18), we have

$$\Phi(v_i, t) = \big(W_{1,i-1}(t) - W_{i+1,n}(t)\big)\tau v_i + F_L^{e_{i-1}}(t) + F_R^{e_i}(t). \tag{20}$$

Let us focus on function $F_R^e(t)$. As t increases, while the upper-envelope sequence of $f_R^e(t, z)$ w.r.t. z remains the same, function $F_R^e(t)$ is represented as the same polynomial, whose degree is at most two by formulae (16), (17) and (18). In other words, a breakpoint of $F_R^e(t)$ corresponds to the value t such that the upper-envelope sequence of $f_R^e(t, z)$ w.r.t. z changes. We notice that such a change happens only when three functions $f_R^{e,h}(t, z)$, $f_R^{e,i}(t, z)$ and $f_R^{e,j}(t, z)$ intersect each other, which can happen at most once. This implies that $F_R^e(t)$ consists of $O(n^3)$ breakpoints, that is, it is a piecewise polynomial function of degree at most two with $O(n^3)$ pieces. The following lemma shows that the number of pieces is actually $O(n^2)$. See [13] for details of the proof.

Lemma 1. *For each $e \in E$, $F_L^e(t)$ and $F_R^e(t)$ are piecewise polynomial functions of degree at most two with $O(n^2)$ pieces, and can be computed in $O(n^3 \log n)$ time. Especially, when all the edge capacities are uniform, the numbers of pieces of them are $O(n)$, and can be computed in $O(n^2 \log n)$ time.*

Let N_F denote the maximum number of pieces of $F_L^e(t)$ and $F_R^e(t)$ over $e \in E$. Then we have $N_F = O(n^2)$, and for the case with uniform edge capacity, $N_F = O(n)$. Next, we consider $\mathrm{Opt}(t) = \min\{\Phi(x, t) \mid x \in V\}$, which is the lower envelope of a family of n functions $\Phi(v_i, t)$ in t. Theorem 2 and Lemma 1 imply the following lemma. See [13] for the proof.

Lemma 2. *$\mathrm{Opt}(t)$ is a piecewise polynomial function of degree at most two with $O(nN_F 2^{\alpha(n)})$ pieces, and can be obtained in $O(nN_F \alpha(n) \log n)$ time if functions $F_L^e(t)$ and $F_R^e(t)$ for all $e \in E$ are available.*

Let N_{Opt} denote the number of pieces of $\mathrm{Opt}(t)$. Then we have $N_{\mathrm{Opt}} = O(nN_F 2^{\alpha(n)})$.

Let us consider $R(x, t)$ in the case that sink x is on an edge $e_i \in E$. Substituting formula (19) for (12), we have

$$R(x, t) = \Phi(x, t) - \mathrm{Opt}(t) = \big(W_{1,i}(t) - W_{i+1,n}(t)\big)\tau x + F_L^{e_i}(t) + F_R^{e_i}(t) - \mathrm{Opt}(t).$$

By Proposition 1, $F_L^{e_i}(t) + F_R^{e_i}(t) - \mathrm{Opt}(t)$ is a piecewise polynomial function of degree at most two with at most $2N_F + N_{\mathrm{Opt}} = O(N_{\mathrm{Opt}})$ pieces. Let N_{e_i} be the number of pieces of $F_L^{e_i}(t) + F_R^{e_i}(t) - \mathrm{Opt}(t)$ and $T_j^{e_i}$ be the interval of the j-th piece (from the left) of $F_L^{e_i}(t) + F_R^{e_i}(t) - \mathrm{Opt}(t)$. Thus, $R(x, t)$ is represented as a (single) polynomial function in x and t on $\{x \in e\} \times T_j$ for each T_j. For each integer j with $1 \leq j \leq N_{e_i}$, let $G_j^{e_i}(x)$ be a function defined as

$$G_j^{e_i}(x) = \max\{R(x,t) \mid t \in T_j^{e_i}\}. \tag{21}$$

We then have the following lemma. See [13] for the proof.

Lemma 3. *For each $e_i \in E$ and j with $1 \leq j \leq N_{e_i}$, $G_j^{e_i}(x)$ is a piecewise polynomial function of degree at most two with at most three pieces, and can be obtained in constant time if functions $F_{\mathrm{L}}^{e_i}(t)$, $F_{\mathrm{R}}^{e_i}(t)$ and $\mathrm{Opt}(t)$ are available.*

Recalling the definition of $MR(x)$, it holds that for $x \in e$,

$$MR(x) = \max\{R(x,t) \mid t \in T\} = \max\{G_j^e(x) \mid 1 \leq j \leq N_e\},$$

that is, $MR(x)$ is the upper envelope of functions $G_1^e(x), \ldots, G_{N_e}^e(x)$. Applying Theorem 2, we have the following lemma. See [13] for the proof.

Lemma 4. *For each $e \in E$, there exists an algorithm that finds a location that minimizes $MR(x)$ under the restriction with $x \in e$ in $O(N_{\mathrm{Opt}}\alpha(n)\log n)$ time if functions $F_{\mathrm{L}}^e(t)$, $F_{\mathrm{R}}^e(t)$ and $\mathrm{Opt}(t)$ are available.*

3.2   Algorithms and Time Analyses

Let us give an algorithm that finds a sink location that minimizes the maximal regret and the analysis of the running time of each step.

First, we obtain $F_{\mathrm{L}}^e(t)$ and $F_{\mathrm{R}}^e(t)$ for all $e \in E$, and function $\mathrm{Opt}(t)$ as a preprocess. Applying Lemmas 1 and 2, we take $O(n^2 N_F \log n)$ time for these operations. Next, we compute $x^{*,e} = \arg\min\{MR(x) \mid x \in e\}$ for all $e \in E$ in $O(n N_{\mathrm{Opt}}\alpha(n)\log n)$ time by applying Lemma 4. Note that the small modification for the algorithm of Lemma 4 leads that we can also compute $MR(v)$ for all $v \in V$ in $O(n N_{\mathrm{Opt}})$ time. (See Lemma 5 in [13].) Finally, we find an optimal sink location x^* in $O(n)$ time by evaluating the values $MR(x)$ for $x \in \{x^{*,e}\} \cup V$.

Since we have $N_{Opt} = O(n N_F 2^{\alpha(n)})$, the bottleneck of our algorithm is to compute $x^{*,e}$ for all $e \in E$. Thus, we see that the algorithm runs in $O(n^2 N_F 2^{\alpha(n)}\alpha(n)\log n)$ time, which completes the proof of our main theorem because $N_F = O(n^2)$, and for the case with uniform edge capacity, $N_F = O(n)$.

References

1. Agarwal, P.K., Sharir, M., Shor, P.: Sharp upper and lower bounds on the length of general Davenport-Schinzel sequences. J. Comb. Theory Ser. A **52**(2), 228–274 (1989)
2. Alstrup, S., Gavoille, C., Kaplan, H., Rauhe, T.: Nearest common ancestors: a survey and a new distributed algorithm. In: Proceedings of the 14th Annual ACM Symposium on Parallel Algorithms and Architectures (SPAA 2002), pp. 258–264 (2002)

3. Arumugam, G.P., Augustine, J., Golin, M.J., Srikanthan, P.: Minmax regret k-sink location on a dynamic path network with uniform capacities. Algorithmica **81**(9), 3534–3585 (2019)
4. Bender, M.A., Farach-Colton, M.: The LCA problem revisited. In: Gonnet, G.H., Viola, A. (eds.) LATIN 2000. LNCS, vol. 1776, pp. 88–94. Springer, Heidelberg (2000). https://doi.org/10.1007/10719839_9
5. Benkoczi, R., Bhattacharya, B., Higashikawa, Y., Kameda, T., Katoh, N.: Minsum k-sink problem on dynamic flow path networks. In: Iliopoulos, C., Leong, H.W., Sung, W.-K. (eds.) IWOCA 2018. LNCS, vol. 10979, pp. 78–89. Springer, Cham (2018). https://doi.org/10.1007/978-3-319-94667-2_7
6. Benkoczi, R., Bhattacharya, B., Higashikawa, Y., Kameda, T., Katoh, N.: Minsum k-sink problem on path networks. Theor. Comput. Sci. **806**, 388–401 (2020)
7. Bhattacharya, B., Golin, M.J., Higashikawa, Y., Kameda, T., Katoh, N.: Improved algorithms for computing k-sink on dynamic flow path networks. In: Ellen, F., Kolokolova, A., Sack, J.R. (eds.) WADS 2017. LNCS, vol. 10389, pp. 133–144. Springer, Cham (2017). https://doi.org/10.1007/978-3-319-62127-2_12
8. Bhattacharya, B., Higashikawa, Y., Kameda, T., Katoh, N.: An $O(n^2 \log^2 n)$ time algorithm for minmax regret minsum sink on path networks. In: Proceedings of the 29th International Symposium on Algorithms and Computation (2018)
9. Bhattacharya, B., Kameda, T.: Improved algorithms for computing minmax regret sinks on dynamic path and tree networks. Theor. Comput. Sci. **607**, 411–425 (2015)
10. Chen, D., Golin, M.J.: Sink evacuation on trees with dynamic confluent flows. In: 27th International Symposium on Algorithms and Computation (2016)
11. Chen, D., Golin, M.J.: Minmax centered k-partitioning of trees and applications to sink evacuation with dynamic confluent flows. CoRR abs/1803.09289 (2018)
12. Ford, L.R., Fulkerson, D.R.: Constructing maximal dynamic flows from static flows. Oper. Res. **6**(3), 419–433 (1958)
13. Fujie, T., Higashikawa, Y., Katoh, N., Teruyama, J., Tokuni, Y.: Minmax regret 1-sink location problems on dynamic flow path networks with parametric weights. CoRR abs/2011.13569 (2020)
14. Golin, M.J., Sandeep, S.: Minmax-regret k-sink location on a dynamic tree network with uniform capacities. CoRR abs/1806.03814 (2018)
15. Hart, S., Sharir, M.: Nonlinearity of Davenport-Schinzel sequences and of generalized path compression schemes. Combinatorica **6**(2), 151–177 (1986)
16. Hershberger, J.: Finding the upper envelope of n line segments in $O(n \log n)$ time. Inf. Process. Lett. **33**(4), 169–174 (1989)
17. Higashikawa, Y.: Studies on the space exploration and the sink location under incomplete information towards applications to evacuation planning. Ph.D. thesis, Kyoto University, Japan (2014)
18. Higashikawa, Y., et al.: Minimax regret 1-sink location problem in dynamic path networks. Theor. Comput. Sci. **588**, 24–36 (2015)
19. Higashikawa, Y., Cheng, S.W., Kameda, T., Katoh, N., Saburi, S.: Minimax regret 1-median problem in dynamic path networks. Theory Comput. Syst. **62**(6), 1392–1408 (2018)
20. Higashikawa, Y., Golin, M.J., Katoh, N.: Minimax regret sink location problem in dynamic tree networks with uniform capacity. J. Graph Algorithms Appl. **18**(4), 539–555 (2014)
21. Higashikawa, Y., Golin, M.J., Katoh, N.: Multiple sink location problems in dynamic path networks. Theor. Comput. Sci. **607**, 2–15 (2015)
22. Hoppe, B., Tardos, E.: The quickest transshipment problem. Math. Oper. Res. **25**(1), 36–62 (2000)

23. Kouvelis, P., Yu, G.: Robust Discrete Optimization and its Applications. Kluwer Academic Publishers, London (1997)
24. Li, H., Xu, Y.: Minimax regret 1-sink location problem with accessibility in dynamic general networks. Eur. J. Oper. Res. **250**(2), 360–366 (2016)
25. Li, H., Xu, Y., Ni, G.: Minimax regret vertex 2-sink location problem in dynamic path networks. J. Comb. Optim. **31**(1), 79–94 (2014). https://doi.org/10.1007/s10878-014-9716-2
26. Mamada, S., Uno, T., Makino, K., Fujishige, S.: An $O(n \log^2 n)$ algorithm for a sink location problem in dynamic tree networks. Discret. Appl. Math. **154**, 2387–2401 (2006)
27. Skutella, M.: An introduction to network flows over time. In: Cook, W., Lovász, L., Vygen, J. (eds.) Research Trends in Combinatorial Optimization, pp. 451–482. Springer, Heidelberg (2009). https://doi.org/10.1007/978-3-540-76796-1_21
28. Vairaktarakis, G.L., Kouvelis, P.: Incorporation dynamic aspects and uncertainty in 1-median location problems. Nav. Res. Logist. (NRL) **46**(2), 147–168 (1999)

The Bike Sharing Problem

Jurek Czyzowicz[1], Konstantinos Georgiou[2], Ryan Killick[3],
Evangelos Kranakis[3(✉)], Danny Krizanc[4], Lata Narayanan[5],
Jaroslav Opatrny[5], and Denis Pankratov[5]

[1] Dép. d'informatique, Université du Québec en Outaouais, Gatineau, Canada
[2] Department of Mathematics, Ryerson University, Toronto, Canada
[3] School of Computer Science, Carleton University, Ottawa, Canada
kranakis@scs.carleton.ca
[4] Department of Mathematics and Computer Science,
Wesleyan University, Middletown, USA
[5] Department of Computer Science and Software Engineering,
Concordia University, Montreal, Canada

Abstract. Assume that $m \geq 1$ autonomous mobile agents and $0 \leq b \leq m$ single-agent transportation devices (called *bikes*) are initially placed at the left endpoint 0 of the unit interval $[0, 1]$. The agents are identical in capability and can move at speed one. The bikes cannot move on their own, but any agent riding bike i can move at speed $v_i > 1$. An agent may ride at most one bike at a time. The agents can cooperate by sharing the bikes; an agent can ride a bike for a time, then drop it to be used by another agent, and possibly switch to a different bike.

We study two problems. In the Bike Sharing problem, we require all agents and bikes starting at the left endpoint of the interval to reach the end of the interval as soon as possible. In the Relaxed Bike Sharing problem, we aim to minimize the arrival time of the agents; the bikes can be used to increase the average speed of the agents, but are not required to reach the end of the interval.

Our main result is the construction of a polynomial time algorithm for the Bike Sharing problem that creates an arrival-time optimal schedule for travellers and bikes to travel across the interval. For the Relaxed Bike Sharing problem, we give an algorithm that gives an optimal solution for the case when at most one of the bikes can be abandoned.

Keywords: Transportation · Scheduling · Cooperation · Multi agent systems · Arrival time · Optimal schedules · Robots · Resource sharing

1 Introduction

Autonomous mobile robots are increasingly used in many manufacturing, warehousing, and logistics applications. A recent development is the increased interest in deployment of so-called *cobots* - collaborative robots - that are intended to

Research supported by Natural Sciences and Engineering Research Council of Canada.

R. Uehara et al. (Eds.): WALCOM 2021, LNCS 12635, pp. 65–77, 2021.
https://doi.org/10.1007/978-3-030-68211-8_6

work collaboratively and in close proximity with humans [1, 2, 13]. Such cobots are controlled by humans, but are intended to augment and enhance the capabilities of the humans with whom they work.

In this paper we study applications of cobots to transportation problems. We propose a new paradigm of cooperative transportation in which cooperating autonomous mobile agents (humans or robots) are aided by transportation cobots (called bikes) that increase the speed of the agents when they are used. The agents are autonomous, identical in capability, and can *walk* at maximum speed 1. The bikes are not autonomous and cannot move by themselves, but agents can move bikes by riding them. At any time, an agent can ride at most one bike, and a bike can be ridden by at most one agent. An agent riding bike i can move at speed $v_i > 1$; note that the bikes may have different speeds. The bikes play a dual role: on the one hand, they are *resources* that can be exploited by the agents to increase their own speed, but on the other hand, they are also *goods* that need to be transported. We assume that initially all agents and bikes are located at the start of the unit interval $[0, 1]$. The goal of the agents is to traverse the interval in such a way as to minimize the latest arrival time of agents and bikes at the endpoint 1.

If the number of the bikes is smaller than the number of agents, or if the speeds of the bikes are different, it is clear that agents can collectively benefit by *sharing* the bikes. This can be shown by a simple example of two agents with speed 1 and a bike with spedd v, where $v > 1$; It is easy to derive the optimal arrival time of $\frac{1+v}{2v}$ for the two agents and a bike.

From the above example, we can see that a solution to the cooperative transport problem described here consists of a *schedule:* a partition of the unit interval into sub-intervals, as well as a specification of the bike to be used by each agent in each sub-interval. It is easy to see that it is never helpful for agents to turn back and ride in the opposite direction. In [4] we show that it is also not helpful for agents to *wait* to switch bikes, or to ride/walk at less than the maximum speed possible. Thus we only consider *feasible* schedules wherein the bike to be used by an agent in a sub-interval is always available when the agent reaches the sub-interval. This will be defined formally in Sect. 2; for now it suffices to define the *Bike Sharing* problem (BS for short) informally. Given m agents and the speeds of b bikes $v_1 \geq v_2 \geq \cdots \geq v_b > 1$, find a feasible schedule specifying how the agents can travel and transport bikes from initial location 0 to location 1 so that the latest arrival time of agents and bikes is minimized.

We also consider a variation of the problem when there is no requirement for the bikes to be transported; bikes are simply resources used by the agents in order to minimize the agents' arrival time, and can be abandoned enroute if they are not useful towards achieving this objective. Thus the *Relaxed Bike Sharing* (RBS for short) problem is as follows. Given m agents and b bikes with speeds $v_1 \geq v_2 \geq \cdots \geq v_b > 1$, find a feasible schedule specifying how m agents can travel from initial location 0 to location 1 so that the latest arrival time of the agents is minimized. We also consider a version of the problem when the input also specifies an upper bound ℓ on the number of bikes that can be abandoned.

Notice that for the same input, the optimal arrival time for the RBS problem can be less than that for the BS problem. For example, suppose we have two agents and 2 bikes with speeds v_1, v_2 with $v_1 > v_2 > 1$. For the BS problem the best arrival time achievable is $1/v_2$, as both bikes have to make it to the end. But for the RBS problem, a better arrival time for the agents can be achieved by the following strategy (see Fig. 1): one agent rides the faster bike up to some point z and then walks the remaining distance taking time $z/v_1 + 1 - z$. The other agent rides the slower bike up to point z, abandons it, switches to the faster bike, and rides it to the end. This takes time $z/v_2 + (1 - z)/v_1$. By equating these two times, we get an overall arrival time of $\frac{v_1^2 - v_2}{v_2 v_1^2 + v_1^2 - 2v_1 v_2} \leq \frac{1}{v_2}$.

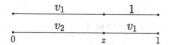

Fig. 1. Strategy for RBS for 2 agents and 2 bikes. The two rows correspond to the behaviour of two agents. Each line segment is labelled with a bike speed that corresponds to the agent travelling on that bike for the distance of the line segment. If an agent walks then the line segment is labelled with 1.

1.1 Related Work

The bike sharing problem involves a sort of cooperation between agents and bikes in order to optimize the arrival time. In this context, related studies concern cooperation between mobile agents and wireless sensors. For example, [5] and [12] use information obtained from wireless sensors for the problem of localization of a mobile robot while in [8], mobile robots and stationary sensors cooperate in a target tracking problem whereby stationary sensors track moving sensors in their sensor range and mobile agents explore regions not covered by the fixed sensors. There are also evaluation platforms for mixed-mode environments incorporating both mobile robots and wireless sensor networks, see e.g., [9].

Bike sharing systems have been installed in many cities as a low-cost and environmentally friendly transportation alternative; researchers have considered optimization problems such as placement of bike stations or rebalancing of inventory - see for example [3,11]. But this line of work is not concerned with the optimal usage of bikes by a particular set of users to travel between two points. The authors of [10] consider *trip planning* for a set of users in a bike sharing system, but the problem is quite different from ours: bikes are available at a predefined set of locations, each user has a different start and end location, and can use at most one bike in their trip, and the problem is concerned with optimizing the number of served users and minimizing their trip time. The authors show NP-hardness of their trip planning problem, and give approximation algorithms.

Our problem has similarities to job shop scheduling problems [6,7], where several jobs of the same processing time are to be scheduled on a collection of machines/servers with different processing power, and the objective is to minimize the *makespan*, the time when all jobs have finished. Indeed our agents can be seen as akin to jobs, we can assume that there is an unlimited number of basic servers of low processing power available (akin to agents walking), and bikes are similar to the fast servers. However, the constraints imposed by bikes being available only at the points they are dropped off do not seem to translate to any natural precedence constraints in the context of multiprocessor or job shop scheduling. To the best of our knowledge, a scheduling analog of our problem has not been studied.

1.2   Outline and Results of the Paper

We introduce and study variations of the Bike Sharing problem – a novel paradigm of mobile agent optimization problems that includes both active elements (agents) and passive elements (bikes). For ease of exposition, we assume that all agents always move at the maximum speed possible (walking or biking), and never wait. In [4] we show that this assumption does not lose any generality. For the BS problem, all bikes required to make it to the end and so the speed of the slowest bike imposes a lower bound on the arrival time. We prove another lower bound on the arrival time based on the "average" completion time of the agents. We design an algorithm that produces a schedule with completion time matching the maximum of the two lower bounds. Our main result is the following (see Theorems 7 and 8).

Theorem 1. *There is a polynomial time algorithm that constructs an optimal schedule for the BS problem.*

For the variation of the RBS problem, in which ℓ bikes are allowed to be abandoned, we demonstrate the following (see [4]).

Theorem 2. *There is a polynomial time algorithm that constructs an optimal schedule for the RBS problem when at most one bike can be abandoned.*

Finally, we show that the RBS problem can be solved efficiently and optimally for some special cases of bike speeds; we state the result informally here; see Sect. 4 for details.

Theorem 3 (informal). *There is a polynomial time algorithm that constructs an optimal schedule for the RBS problem when at most one bike is "slow".*

The outline of the paper is as follows. In Sect. 2 we introduce an intuitive matrix representation of bike sharing schedules and formally define the problems we are studying. In Sects. 3 and 4 we give our algorithms for the BS and RBS problems respectively. Finally, we conclude the paper in Sect. 5 with a discussion of open problems. Details of all missing proofs can be found in [4].

2   Definitions and Preliminary Observations

We assume there are m agents and b bikes with speeds >1 on a line. The bikes and agents are initially located at the origin 0. All agents have the same walking speed of 1 and will travel at speed $v_i > 1$ when using the bike i. An agent can ride at most one bike at a time, and a bike can be ridden by at most one agent at a time. An agent can use a bike for a portion of the trip and at any time drop the bike. If an agent comes across bike i while walking or riding a different bike j, it may choose to pick up bike i (and drop bike j), and continue the trip on bike i. We assume that pickups and dropoffs happen instantaneously, and that any number of agents and bikes can be present at the same point simultaneously. Thus, bikes can be dropped without blocking other bikes or agents, and agents and bikes can pass each other at will. We will assume that agents always move at the maximum speed allowed by walking/biking, and never stop and wait.

When necessary we use v_i to represent the speed of the ith bike, however, it is much more useful to work with the inverse speeds of the bikes – which we represent by $u_i \equiv 1/v_i$. Thus, we will assume that the (multi)set U lists the inverse speeds of the bikes, i.e. $U = \{u_1, u_2, \ldots, u_b\}$. An input to both BS and RBS problems is then represented by a pair (m, U) where m is the number of agents and the number of bikes is $b = |U|$. We assume without loss of generality that the bikes are labelled in increasing order of their inverse speeds. Thus, bike 1 is the fastest bike and bike b is the slowest bike. We reserve label 0 to denote absence of a bike, i.e., walking. The inverse speed associated with this label is 1.

We say that bike i is *dropped* at location $y \in [0, 1]$ if an agent leaves the bike at location y and it is picked up by another agent at a later time. We say that bike i is *abandoned* at location y if an agent leaves the bike at location y and it is never picked up by another agent. Note that bikes cannot be abandoned in the BS problem but may be abandoned in solutions to the RBS problem.

A solution to either problem is a schedule that should specify for each agent i a partition of the entire interval $[0, 1]$ and for each block of the partition whether the agent walks or uses a particular bike. By taking a common refinement of all these partitions we may assume that all partitions are exactly the same. The information about which bike gets used during which block of the partition and by which agent is collected in a single matrix. Thus, we define a *schedule* as a pair (X, M) where: $X = (x_1, x_2, \ldots, x_n)$ is a *partition vector* satisfying $\forall j \in [n]$ $x_j \geq 0$ and $\sum_{j=1}^{n} x_j = 1$; M is a *schedule matrix* of size $m \times n$ such that entry $M(i, j)$ indicates the label of the bike used by agent i during the interval x_j or 0 if agent i walked during the interval x_j. By a slight abuse of notation, we use x_j to refer to both the jth interval of the partition as well as its length.

We define the *size of a schedule* as n – the number of columns in the schedule matrix M (also the number of entries in the partition vector). While the number of rows m will be the same for any schedule, a priori it is not even clear that there has to be an optimal schedule with a finite number of columns. Clearly, it is desirable to minimize the size of a schedule.

The matrix M gives rise to an induced matrix of inverse speeds $\widetilde{M}$. More specifically, for $i \in [m]$ and $j \in [n]$ the entry $M(i,j) = k$ implies $\widetilde{M}(i,j) = u_k$ when $k \in [b]$, and $M(i,j) = 0$ implies $\widetilde{M}(i,j) = 1$. It will be convenient for us to treat these representations as interchangeable, which is easy to implement.

The main utility of the above definitions is that they allow one to easily compute the completion times of the agents. Indeed, the time at which agent i reaches the end of interval x_j – referred to as the jth completion time of agent i – is computed as follows: $t_{i,j}(X, M) = \sum_{k=1}^{j} \widetilde{M}(i,k)x_k$. Note that $t_{i,n}$ is the time at which agent i reaches the end of the entire interval. The overall completion time will be represented by τ and equals the maximum of the $t_{i,n}$, i.e. $\tau(X, M) \equiv \max_i t_{i,n}$. We may omit (X, M) when it is clear from the context and simply write $t_{i,j}$, and τ. We shall sometimes refer to the completion time as *arrival time*.

Not every schedule (X, M) is feasible. In particular, feasibility requires that (1) a bike cannot move on its own, (2) a bike is used by at most one agent during a particular interval of the partition, and (3) a bike is available when an agent is supposed to use that bike. In light of the introduced notation and terminology we can formalize the above three conditions as follows:

Definition 1. (X, M) *represents a feasible schedule if:*

1. $M(i,j) \neq 0$ *implies that* $\exists$ i' $M(i', j-1) = M(i,j)$,
2. $M(i,j) \neq 0$ *implies that* $\forall$ $i' \neq i$ $M(i',j) \neq M(i,j)$,
3. $M(i,j) = M(i', j-1) \neq 0$ *implies that* $t_{i',j-1} \leq t_{i,j-1}$.

In a feasible schedule (X, M) we say that *bike k makes it to the end* if this bike is used during the last interval, i.e., $\exists i$ such that $M(i,n) = k$. We are now ready to state the BS and RBS problems formally:

The Bike Sharing Problem
Input: $m \geq 1$ – number of agents; $b \geq 0$ – number of bikes; $0 < u_1 \leq u_2 \leq \cdots \leq u_b < 1$ – inverse speeds of bikes.
Output: A feasible schedule (X, M) such that all bikes reach the end.
Objective: Minimize $\tau(X, M)$.

The Relaxed Bike Sharing Problem
Input: $m \geq 1$ – number of agents; $b \geq 0$ – number of bikes; $0 < u_1 \leq u_2 \leq \cdots \leq u_b < 1$ – inverse speeds of bikes.
Output: A feasible schedule (X, M) such that all agents reach the end.
Objective: Minimize $\tau(X, M)$.

Notice that a solution to the BS problem has two components: the continuous component represented by the partition vector, and the discrete component represented by the schedule matrix. Intuitively, the core difficulty of the problem is in figuring out an optimal schedule matrix. Once we have a "good" schedule matrix, we can find an optimal partition that goes along with it via linear programming.

Lemma 1. *For any schedule matrix M satisfying parts 1 and 2 of Definition 1, we can in polynomial time find a partition vector X that makes schedule (X, M) feasible and achieves the smallest completion time among all schedules with the same schedule matrix M.*

We refer to the partition X guaranteed by the above lemma as *the induced partition vector* of M and denote it by X_M.

3 Optimal and Efficient Algorithm for the BS Problem

In this section we present a polynomial time algorithm that solves the BS problem optimally, i.e., our algorithm produces a feasible schedule that minimizes the latest arrival time.

High-level overview of the algorithm. The structure of our algorithm is quite involved and it relies on several subroutines, so we do not present the entire algorithm all at once. We build up towards the complete algorithm in stages:

1. First we derive a lower bound T on the minimum completion time of any feasible schedule and show that the relationship between u_b and T controls the structure of an optimal schedule matrix for the BS problem.
2. We then tackle the case when $u_b \leq T$ and describe a subroutine ALLMAKEIT that calls itself recursively to build up an optimal schedule matrix. We prove the optimality of this subroutine and compute the size of its output schedule.
3. Unfortunately, the ALLMAKEIT subroutine creates a schedule of size that is exponential in the number of bikes. We describe two more subroutines – STANDARDIZE and REDUCE that allow transforming any schedule of potentially large size n into an equally or more efficient schedule of size $n' \leq m$.
4. We finish the case $u_b \leq T$ by describing two modifications to ALLMAKEIT resulting in a polynomial time algorithm. First, we use the REDUCE subroutine to prevent the size of the schedule from growing exponentially. Second, we observe that recursive calls have overlapping substructures allowing us to rewrite the algorithm with the help of a dynamic programming table.
5. Finally, we tackle the case of $u_b > T$ by showing that it can be reduced, in a certain sense, to the case of $u_b \leq T$.

3.1 Lower Bound on the Arrival Time for the Bike Sharing Problem

Define $T(m, U)$ as follows: $T(m, U) := 1 - \frac{1}{m} \sum_{j=1}^{b}(1 - u_j)$. The following lemma shows that $T(m, U)$ is a lower bound on the optimum for the BS problem. For ease of reference, we also record another trivial lower bound in the statement of the lemma, observe a sufficient condition for the schedule to satisfy $\tau(X, M) = T(m, U)$, and give two useful properties that result from the definition of $T(m, U)$.

Lemma 2. *Let (X, M) be an arbitrary feasible schedule for the BS problem with inputs $m, b,$ and U. Then we have*

1. $\tau(X, M) \geq u_b$,
2. $\tau(X, M) \geq T(m, U)$,
3. $\tau(X, M) = T(m, U)$ if and only if all agents have the same arrival time in (X, M).
4. $u_1 \leq T(m, U)$.
5. $T(m, U) \leq T(m - 1, U \setminus \{u_k\})$ iff $u_k \leq T(m, U)$ with equality only when $u_k = T(m, U)$.

In spite of the simplicity of the above lower bounds, they turn out extremely important: in the following sections we show that if $u_b \leq T(m, U)$ then there is a feasible schedule (X, M) with completion time $\tau(X, M) = T(m, U)$, and otherwise there is a feasible schedule (X, M) with completion time $\tau(X, M) = u_b$. Putting it together the BS problem has an optimal schedule that has completion time $\max(u_b, T(m, U))$.

3.2 Finding an Optimal Schedule for the Case $u_b \leq T(m, U)$

The goal of this section is to present and analyze the algorithm ALLMAKEIT that on input (m, U) satisfying $u_b \leq T(m, U)$ produces a feasible schedule (X, M) with $\tau(X, M) = T(m, U)$.

In anticipation of the recursive nature of this algorithm, we introduce the following notation: for $k \in \{0, 1, 2, \ldots, b\}$ we define $m_k := m - b + k$, and $U_k := \{u_1, \ldots, u_k\}$. Note that we have $U_0 = \emptyset$ and $U_b = U$.

We say that a group of agents is *synchronized* at a particular time if they are found to be at the same location at that time. We say that agent i is a *walker* at time t in a given schedule if the agent does not use a bike at time t (see [4]).

At a high level, ALLMAKEIT has two phases. The goal of the first phase is to get a group of $m - b$ synchronized walkers ahead of all the bikes. At that time the bikes will be in use by the remaining b agents. Moreover, the tardiness of agents on bikes has a particular order: if we make no changes to the schedule after the first phase and let the walkers walk and biking agents bike, then the agent on bike 1 will catch up to the walkers first, followed by the agent on bike 2, and so on. This goal can be achieved by sharing bikes among m bikers as follows. At first, agents $1, \ldots, b$ use bikes while others walk. After travelling a certain distance, each agent on a bike drops their bike to be picked up by an agent immediately behind them. This way the set of agents that use bikes during consecutive intervals propagate from $1, \ldots, b$ to $2, \ldots, b+1$ to $3, \ldots, b+2$, and so on until all the bikes accumulate among the last b agents numbered $m - b + 1, \ldots, m$. The intervals of the partition vector during this phase can be arranged so that walkers $1, 2, \ldots, m - b$ are synchronized at the end of the first phase.

In the second phase, the agent using bike 1 is the first agent on a bike to catch up with the group of walkers. Rather than overtaking the group of walkers, the new agent and the bike "get absorbed" by the group. Intuitively, it means that now we have a group of $m - b + 1$ agents with 1 bike that can be shared in the entire group to increase the average speed with which this group can travel. It turns out that the best possible average speed of this group is exactly $1/T(m_1, U_1)$. Moreover, we can design a schedule which says how this 1 bike is shared with the group during any interval of length x by taking a schedule produced (recursively) by ALLMAKEIT on input (m_1, U_1) and scaling this schedule by x. This has the added benefit that if the group was synchronized at the beginning of the interval (which it was) the group remains synchronized at the end of the interval. The distance x is chosen so the agent using bike 2 catches up to the new group precisely after travelling distance x. When the agent using bike 2 catches up to the group, the new agent and the new bike again get absorbed by the group, further increasing the average speed. An optimal schedule for the new group is given by a scaled version of ALLMAKEIT on input (m_2, U_2). This process continues for bikes $3, 4, \ldots, b-1$. Since recursive calls have to have smaller inputs, we have to stop this process before bike b gets absorbed. What do we do with bike b then? The answer is simple: we arrange the entire schedule in a such a way that bike b catches up with the big group precisely at the end of the entire interval $[0, 1]$.

The main result of this section is that the expanded schedule produced by ALLMAKEIT is optimal under the condition $u_b \leq T(m, U)$.

Theorem 4. *Consider input* $(m, U = \{u_1 \leq u_2 \leq \cdots \leq u_b\})$ *to the* BS *problem s.t.* $u_b \leq T(m, U)$. *Let* (X, M) *be the schedule produced by* ALLMAKEIT(m, U). *The schedule* (X, M) *is feasible and has completion time* $\tau(X, M) = T(m, U)$.

The (rather technical) proof proceeds along the following lines: (1) we make sure that the recursive calls made in the ALLMAKEIT procedure satisfy the precondition $u_b \leq T(m, U)$, (2) we prove that there is a partition that makes schedule feasible and results in completion time $T(m, U)$. Thus, by Lemma 1 such a partition can be found by solving a related LP. The proof of step (2) is by induction on the number of bikes b and the difficulty is that we don't have a closed-form expression of the partition so we cannot easily verify property 3 of Definition 1. We sidestep this difficulty by working with so-called "unexpanded" schedule (unexpanded partition and matrix) that precisely captures the recursive structure of the solution (see [4]). We show how an unexpanded schedule can be transformed into a complete feasible schedule.

Unfortunately, the size of the schedule produced by ALLMAKEIT is exponential in the number of bikes.

Theorem 5. ALLMAKEIT(m, U) *has size* $(m-1)2^{b-1} + 1$ *for* $0 < b < m$.

The detailed proofs of the above theorems are given in [4].

3.3   STANDARDIZE and REDUCE Procedures

In this section, we present two procedures called STANDARDIZE and REDUCE that would allow us to overcome the main drawback of ALLMAKEIT algorithm – the exponential size of a schedule it outputs. We begin with the STANDARDIZE procedure.

Consider a feasible schedule (X, M) of size n. If there exists an entry $x_j = 0$ in X then this entry and the jth column of M do not contain any useful information. The *zero columns* only serve to increase the size of (X, M) and can be removed from the algorithm by deleting x_j and the corresponding column of M. Similarly, whenever we have two consecutive columns $j - 1$ and j of M that are identical (i.e. none of the agents switch bikes between the intervals x_{j-1} and x_j) then the jth column is a *redundant column* and can be removed from the algorithm by deleting the jth column of M and merging the intervals x_{j-1} and x_j in X.

In addition to removing these useless columns, we can also remove from the algorithm any useless switches that might occur. To be specific, if ever it happens during (X, M) that an agent i performs a pickup-switch with an agent i' at the same position and *time*, then this bike switch can be avoided. Indeed, assume that M instructs the agents i and i' to switch bikes at the end of some interval x_k and assume that this switch happens at the same time. Then we can remove this switch by swapping the schedules of agents i and i' in every interval succeeding the interval x_k. In other words, we need to swap $M(i, j)$ and $M(i', j)$ for each $j = k + 1, k + 2, \ldots, n$. We will call these types of switches *swap-switches*. Note that a swap-switch is also a pickup-switch. Clearly, the removal of swap-switches will not affect the feasibility nor the overall completion time of an algorithm.

We say that a schedule (X, M) is expressed in *standard form* if (X, M) does not contain any zero columns, redundant columns, or swap-switches.

In [4] we present and analyze the procedure STANDARDIZE(X, M) that takes as input a feasible schedule (X, M) and outputs a feasible schedule (X', M') in standard form.

Lemma 3. *A feasible schedule (X, M) of size n can be converted to standard form in $O(m^2 n)$ computational time using $O(mn)$ space.*

Next, we describe the REDUCE procedure that demonstrates that a feasible schedule is never required to have size larger than m. This procedure is presented in [4]. The idea is simple: we start with a feasible schedule (X_M, M) (X_M is the induced partition vector of M), apply the standardization procedure to get a new schedule (X', M') of reduced size, and then iterate on the schedule $(X_{M'}, M')$ ($X_{M'}$ is the induced partition vector of M'). The iteration stops once we reach a schedule $(X_{M''}, M'')$ that is already in standard form. The key thing is to demonstrate that the iteration always stops at a schedule with size $n \leq m$ (without increasing the completion time of the schedule). This is the subject of Theorem 6 (details of the proof can be found in [4]).

Theorem 6. *Consider a feasible schedule (X_M, M) of size $n > m$ that completes in time τ. Then REDUCE(M) produces a feasible schedule $(X_{M'}, M')$ with size $n' \leq m$ that completes in time $\tau' \leq \tau$. The computational complexity of REDUCE(U, M) is $O(\text{poly}(m, n))$ time and space.*

3.4   ALLMAKEIT*: Computationally Efficient Version of ALLMAKEIT

There are two obstacles we need to overcome in order to make ALLMAKEIT algorithm run in polynomial time. The first obstacle is information-theoretic: the size of the schedule produced by ALLMAKEIT algorithm is exponential in the number of bikes. The second obstacle is computational: the sheer number of recursive calls made by ALLMAKEIT algorithm is exponential in the number of bikes, so even if sizes of all matrices were 1 it still would not run in polynomial time. The REDUCE procedure allows us to overcome the first obstacle. We overcome the second obstacle by observing that the number of *distinct* sub-problems in ALLMAKEIT algorithm and all of its recursive calls is at most b. Thus, we can replace recursion with dynamic programming (DP) to turn ALLMAKEIT into a polynomial time algorithm ALLMAKEIT*. A more thorough explanation/analysis of ALLMAKEIT* algorithm is provided in [4].

Theorem 7. *Let (m, U) be the input to the BS problem such that $u_b \leq T(m, U)$. The algorithm ALLMAKEIT* runs in polynomial time on input (m, U) and returns an optimal schedule (X, M) with $\tau(X, M) = T(m, U)$.*

3.5   Finding an Optimal Schedule for the Case $u_b > T(m, U)$

In this section we solve the BS problem efficiently and optimally for the case when the slowest bike is the bottleneck. The idea is to reduce it to the case of Subsect. 3.2. We make the following observation which will allow us to prove the main result of this section.

Lemma 4. *Let $(m, U = \{u_1 \leq \ldots \leq u_b\})$ be input to the BS problem. If $u_b > T(m, U)$ then there exists $k \in \{1, 2, \ldots, b - 1\}$ such that $u_k \leq T(m_k, U_k) \leq u_b$.*

Theorem 8. *Let (m, U) be the input to the BS problem such that $u_b > T(m, U)$. There exists a polynomial time algorithm that constructs an optimal schedule (X, M) with $\tau(X, M) = u_b$.*

4   The RBS Problem

In Sect. 3 we presented an optimal algorithm for the BS problem. Recall that in the case $u_b > T(m, U)$ the minimum arrival time is u_b. Thus, the arrival time in that case is controlled by how much time it takes the slowest bike to travel from 0 to 1. This suggests that if we relax the requirement of all bikes making it to the end and allow agents to abandon the slowest bike, for example, the overall completion time for the agents and the remaining bikes might be improved. This naturally leads to the RBS problem. We begin by developing a polynomial time algorithm that solves the RBS problem optimally when the abandonment limit ℓ is 1, that is we allow at most one bike to be abandoned.

High-level overview of the algorithm. Our approach to the RBS problem with the abandonment limit $\ell = 1$ mimics to some extent our approach to the BS problem. The structure of the optimal schedule depends on the relationships between speeds of two slowest bikes and certain expressions lower bounding the optimal completion time. More specifically, in [4] we design an algorithm ALLBUTONE that solves the RBS problem with the abandonment limit $\ell = 1$ optimally in polynomial time. We follow the next steps:

1. We begin by generalizing the lower bound of Lemma 2 to the situation where multiple bikes may be abandoned.
2. Analyzing the lower bound from the first step we can immediately conclude that ALLMAKEIT* algorithm provides an optimal solution to the RBS problem (in fact for any value of ℓ) under the condition $u_b \leq T(m, U)$.
3. By the previous step, it remains to handle the case $u_b > T(m, U)$. We split this case into two sub-cases depending on the relationship between the inverse speed of the second slowest bike u_{b-1} and T_1, which is the lower bound from Step 1 specialized to the case $\ell = 1$ (the subscript in T_1 indicates the abandonment limit). We handle the case of $u_{b-1} \leq T_1(m, U)$ first.
4. Lastly, we show how to handle the remaining sub-case of $u_b > T(m, U)$, namely, when $u_{b-1} > T_1(m, U)$.

Carrying out the above steps results in the following two theorems that jointly provide a solution to the RBS problem with the abandonment limit $\ell = 1$.

Theorem 9. *Let* $(m, U = \{u_1 \leq u_2 \leq \cdots \leq u_b\})$ *be such that* $u_b > T(m, U)$ *and* $u_{b-1} \leq T_1(m, U)$. *The schedule* (X, M) *output by* ALLBUTONE(m, U) *is feasible and has completion time* $\tau(X, M) = T_1(m, U)$. *In particular,* (X, M) *is an optimal solution to the RBS problem with abandonment limit 1.*

Theorem 10. *Let* $(m, U = \{u_1 \leq u_2 \leq \cdots \leq u_b\})$ *be s.t.* $u_b > T(m, U)$ *and* $u_{b-1} > T_1(m, U)$. *There is a polynomial time computable feasible schedule* (X, M) *s.t. at most one bike is abandoned and* $\tau(X, M) = u_{b-1}$. *In particular,* (X, M) *is an optimal solution to the RBS problem with abandonment limit 1.*

A closer look at our results for the BS and RBS problems, indicates that we can solve the RBS problem optimally under some conditions. In particular, when all bikes are "fast enough", that is, when $u_b \leq T(m, U)$, we have:

Theorem 11. ALLMAKEIT* *solves the* RBS *problem optimally when* $u_b \leq T(m, U)$.

In fact, even if *all but* the slowest bike are fast enough, we can solve the RBS problem optimally. In particular, if $u_b > T(m, U)$ and $u_{b-1} \leq T_1(m, U)$ then ALLBUTONE produces a schedule where only a single bike is abandoned. This schedule is also a possible solution for the RBS problem for any $\ell \geq 1$. In the following theorem we demonstrate that there is no better schedule.

Theorem 12. ALLBUTONE *solves the* RBS *problem optimally when* $u_b > T(m, U)$ *and* $u_{b-1} \leq T_1(m, U)$.

5   Conclusion

There are many open questions that remain. First, the development of algorithms for the RBS problem when more than one bike can be abandoned is required. The techniques introduced in this paper can be extended further to cover the case that at most 2 bikes can be abandoned, however, this results in a messy case analysis that does not lend any intuition as to how the problem can be elegantly solved. Additionally, one can study more general versions of the problem where agents/bikes do not all begin at the same location, or even where the speed of a bike depends both on the bike and on the ID of the agent that is riding it.

References

1. Cobot. https://en.wikipedia.org/wiki/Cobot. Accessed 05 Feb 2020
2. Boy, E.S., Burdet, E., Teo, C.L., Colgate, J.E.: Investigation of motion guidance with scooter cobot and collaborative learning. IEEE Trans. Rob. **23**(2), 245–255 (2007)
3. Chen, L., et al.: Bike sharing station placement leveraging heterogeneous urban open data. In: Proceedings of the 2015 ACM International Joint Conference on Pervasive and Ubiquitous Computing, pp. 571–575 (2015)
4. Czyzowicz, J., et al.: The bike sharing problem. arXiv preprint arXiv:2006.13241 (2020)
5. Djugash, J., Singh, S., Kantor, G., Zhang, W.: Range-only SLAM for robots operating cooperatively with sensor networks. In: Proceedings of IEEE International Conference on Robotics and Automation, pp. 2078–2084, May 2006
6. Garey, M.R., Johnson, D.S., Sethi, R.: The complexity of flowshop and jobshop scheduling. Math. Oper. Res. **1**(2), 117–129 (1976)
7. Gonzalez, T., Sahni, S.: Preemptive scheduling of uniform processor systems. J. ACM (JACM) **25**(1), 92–101 (1978)
8. Jung, B., Sukhatme, G.: Cooperative tracking using mobile robots and environment-embedded networked sensors. In: International Symposium on Computational Intelligence in Robotics and Automation, pp. 206–211 (2001)
9. Kropff, M., et al.: MM-ulator: towards a common evaluation platform for mixed mode environments. In: Carpin, S., Noda, I., Pagello, E., Reggiani, M., von Stryk, O. (eds.) SIMPAR 2008. LNCS (LNAI), vol. 5325, pp. 41–52. Springer, Heidelberg (2008). https://doi.org/10.1007/978-3-540-89076-8_8
10. Li, Z., Zhang, J., Gan, J., Lu, P., Gao, Z., Kong, W.: Large-scale trip planning for bike-sharing systems. Pervasive Mob. Comput. **54**, 16–28 (2019)
11. Schuijbroek, J., Hampshire, R., van Hoeve, W.J.: Inventory rebalancing and vehicle routing in bike sharing systems. Eur. J. Oper. Res. **257**, 992–1004 (2017)
12. Seow, C.K., Seah, W.K.G., Liu, Z.: Hybrid mobile wireless sensor network cooperative localization. In: Proceedings IEEE 22nd International Symposium on Intelligent Control, pp. 29–34 (2007)
13. Veloso, M., et al.: Cobots: collaborative robots servicing multi-floor buildings. In: 2012 IEEE/RSJ International Conference on Intelligent Robots and Systems, pp. 5446–5447. IEEE (2012)

Efficient Generation of a Card-Based Uniformly Distributed Random Derangement

Soma Murata[1], Daiki Miyahara[1,3]([✉]), Takaaki Mizuki[2], and Hideaki Sone[2]

[1] Graduate School of Information Sciences, Tohoku University,
6–3–09 Aramaki-Aza-Aoba, Aoba-ku, Sendai 980–8578, Japan
soma.murata.p5@dc.tohoku.ac.jp, daiki.miyahara.q4@dc.tohoku.ac.jp
[2] Cyberscience Center, Tohoku University,
6–3 Aramaki-Aza-Aoba, Aoba-ku, Sendai 980–8578, Japan
mizuki+lncs@tohoku.ac.jp
[3] National Institute of Advanced Industrial Science and Technology,
2–3–26, Aomi, Koto-ku, Tokyo 135-0064, Japan

Abstract. Consider a situation, known as Secret Santa, where n players wish to exchange gifts such that each player receives exactly one gift and no one receives a gift from oneself. Each player only wants to know in advance for whom he/she should purchase a gift. That is, the players want to generate a hidden uniformly distributed random derangement. (Note that a permutation without any fixed points is called a derangement.) To solve this problem, in 2015, Ishikawa *et al.* proposed a simple protocol with a deck of physical cards. In their protocol, players first prepare n piles of cards, each of which corresponds to a player, and shuffle the piles. Subsequently, the players verify whether the resulting piles have fixed points somehow: If there is no fixed point, the piles serve as a hidden random derangement; otherwise, the players restart the shuffle process. Such a restart occurs with a probability of approximately 0.6. In this study, we consider how to decrease the probability of the need to restart the shuffle based on the aforementioned protocol. Specifically, we prepare more piles of cards than the number n of players. This potentially helps us avoid repeating the shuffle, because we can remove fixed points even if they arise (as long as the number of remaining piles is at least n). Accordingly, we propose an efficient protocol that generates a uniformly distributed random derangement. The probability of the need to restart the shuffle can be reduced to approximately 0.1.

Keywords: Card-based cryptography · Derangement (Permutation without fixed points) · Exchange of gifts · Secret Santa

1 Introduction

Let $n\ (\geq 3)$ be a natural number, and consider a situation, known as Secret Santa, where n players $P_1, P_2, \ldots, P_n$ wish to exchange gifts such that each

© Springer Nature Switzerland AG 2021
R. Uehara et al. (Eds.): WALCOM 2021, LNCS 12635, pp. 78–89, 2021.
https://doi.org/10.1007/978-3-030-68211-8_7

player receives exactly one gift and no one receives a gift from oneself. Every player wants to know in advance for whom he/she should purchase a gift. Mathematically, an assignment of a gift exchange can be regarded as a permutation, *i.e.*, an element in S_n, which is the symmetric group of degree n; in this context, a permutation $\pi \in S_n$ indicates that a player P_i for every i, $1 \leq i \leq n$, will purchase a gift for $P_{\pi(i)}$. Such a permutation $\pi \in S_n$ must not have any fixed points, *i.e.*, $\pi(i) \neq i$ for every i, $1 \leq i \leq n$, to prevent each player from receiving a gift from himself/herself. Note that a permutation is called a *derangement* if it has no fixed point. Therefore, the players want to generate a uniformly distributed random derangement. Furthermore, to make the exchange fun, it is necessary for each player P_i to know only the value of $\pi(i)$. Thus, we aim to generate a "hidden" uniformly distributed random derangement.

Physical cryptographic protocols are suitable for resolving this type of problem because they can be easily executed by using familiar physical tools without relying on complicated programs or computers.

1.1   Background

The problem of generating a hidden derangement was first studied by Crépeau and Kilian [2] in 1993. Since then, several solutions with physical tools have been proposed. (Refer to [17] for the non-physical solutions.) As practical protocols, Heather *et al.* [6] proposed a protocol with envelopes and fill-in-the-blank cards in 2014; Ibaraki *et al.* [7] proposed a protocol with two sequences of cards representing player IDs and gift IDs in 2016. The common feature of these two practical protocols is that the generated derangement is not uniformly distributed; it always includes a cycle of a specific length.

Let us focus on generating a *uniformly distributed* random derangement. A protocol that generates a uniformly distributed random derangement was proposed by Crépeau and Kilian [2] with a four-colored deck of $4n^2$ cards. Ishikawa, Chida, and Mizuki [8] subsequently improved the aforementioned protocol by introducing a *pile-scramble shuffle* that "scrambles" piles of cards. Their improved protocol, which we refer to as the *ICM protocol* hereinafter, uses a two-colored deck of n^2 cards. It is described briefly as follows. (Further details will be presented in Sect. 2.5).

1. Prepare n piles of cards, each of which corresponds to a player.
2. Apply a pile-scramble shuffle to the n piles to permute them randomly.
3. Check whether there are fixed points in the n piles somehow.
   - If there is at least one fixed point, restart the shuffle process, *i.e.*, go back to Step 2.
   - If there is no fixed point, the piles serve as a hidden random derangement.

Thus, the ICM protocol is not guaranteed to terminate within a finite runtime, because it restarts the shuffle process whenever a fixed point arises. The probability that at least one fixed point appears in Step 3 is $1 - \sum_{k=0}^{n} (-1)^k/k! \approx 1 - 1/e \approx 0.63$ (where e is the base of the natural logarithm), which will be described later in Sect. 2.5.

In 2018, Hashimoto *et al.* [4] proposed the first finite-runtime protocol for generating a uniformly distributed random derangement by using the properties of the types of permutations. While their proposed protocol is innovative, its feasibility to be performed by humans has not been studied, as it requires a shuffle operation with a nonuniform probability distribution.

1.2 Contributions

In this study, we also deal with generating a uniformly distributed random derangement and propose a new card-based protocol by improving the ICM protocol. Specifically, we devise a method to reduce the probability of the need to restart in the ICM protocol. Recall that, after one shuffle is applied in Step 2, the ICM protocol returns to Step 2 with a probability of approximately 0.6. In card-based protocols, it is preferable to avoid repeating shuffle operations because players manipulate the deck of physical cards by hand. Here, we prepare more piles of cards than the number n of players, *i.e.*, we prepare $n + t$ piles for some $t \geq 1$. This potentially helps us remove fixed points (if they arise); hence, we can reduce the probability of the need to restart the shuffle. In the same manner as the ICM protocol, the proposed protocol generates a hidden uniformly distributed random derangement. The probability of the need to restart the shuffle is reduced by increasing the number t of additional piles. Specifically, the probability of the need for such a restart can be reduced to approximately 0.1 by setting $t = 3$.

The remainder of this paper is organized as follows. In Sect. 2, we introduce the notions of card-based cryptography, the properties of permutations, and the ICM protocol. In Sect. 3, we present our protocol. In Sect. 4, we demonstrate the relationship between the number t of additional piles and the probability of the need to restart the shuffle; we illustrate how the probability can be reduced by increasing the number t.

1.3 Related Works

Card-based cryptography involves performing cryptographic tasks, such as secure multi-party computations, using a deck of physical cards; since den Boer [1] first proposed a protocol for a secure computation of the AND function with five cards, many elementary computations have been devised (e.g., [11, 14]). For more complex tasks, millionaire protocols [9, 12, 13] that securely compare the properties of two players, a secure grouping protocol [5] that securely divides players into groups, and zero-knowledge proof protocols for pencil puzzles [3, 10, 15, 16, 18] were also proposed.

2 Preliminaries

In this section, we introduce the notions of cards and the pile-scramble shuffle used in this study, and the properties of permutations. Furthermore, we introduce the ICM protocol proposed by Ishikawa *et al.* [8].

2.1 Cards

In this study, we use a two-colored (black ♣ and red ♡) deck of cards. The rear sides of the cards have the same pattern ?. The cards of the same color are indistinguishable. Using n cards consisting of $n-1$ black cards and one red card, we represent a natural number i, $1 \leq i \leq n$, using a sequence such that the i-th card is red and the remaining cards are black:

$$\overset{1}{♣}\;\overset{2}{♣}\;\cdots\;\overset{i}{♡}\;\cdots\;\overset{n-1}{♣}\;\overset{n}{♣}.$$

If a sequence of face-down cards represents a natural number i according to the above encoding rule, we refer to it as a *commitment to i* and express it as follows:

$$\underbrace{\overset{1}{?}\;\overset{2}{?}\;\cdots\;\overset{n}{?}}_{i}.$$

2.2 Pile-scramble Shuffle

A *pile-scramble shuffle* is a shuffle operation proposed by Ishikawa *et al.* [8]. Let $(\mathsf{pile}_1, \mathsf{pile}_2, \ldots, \mathsf{pile}_n)$ be a sequence of n piles, each consisting of the same number of cards. By applying a pile-scramble shuffle to the sequence, we obtain a sequence of piles $(\mathsf{pile}_{\pi^{-1}(1)}, \mathsf{pile}_{\pi^{-1}(2)}, \ldots, \mathsf{pile}_{\pi^{-1}(n)})$ where $\pi \in S_n$ is a uniformly distributed random permutation. Humans can easily implement a pile-scramble shuffle by using rubber bands or envelopes.

2.3 Properties of Permutations

An arbitrary permutation can be expressed as a product of disjoint cyclic permutations. For example, the permutation

$$\tau = \begin{pmatrix} 1\,2\,3\,4\,5\,6\,7 \\ 3\,5\,6\,4\,2\,7\,1 \end{pmatrix}$$

can be expressed as the product of three disjoint cyclic permutations $\tau_1 = (4), \tau_2 = (25), \tau_3 = (1367)$: $\tau = \tau_1\tau_2\tau_3 = (4)(25)(1367)$. The lengths of the cyclic permutations τ_1, τ_2, and τ_3 are 1, 2, and 4, respectively. A cycle of length one is a fixed point.

Let d_n denote the number of all derangements in S_n; then, d_n can be expressed as follows:

$$d_n = n! \sum_{k=0}^{n} \frac{(-1)^k}{k!}$$

for $n \geq 2$, and $d_1 = 0$. The number of permutations (in S_n) having exactly f fixed points is ${}_nC_f \cdot d_{n-f}$, where we define $d_0 = 1$.

2.4 Expression of Permutation Using Cards

Hereinafter, we use the expression $[1 : m]$ to represent the set $\{1, 2, \ldots, m\}$ for a positive integer m. Remember that a commitment to $i \in [1 : n]$ consists of one red card at the i-th position and $n - 1$ black cards at the remaining positions. In this paper, we represent a *hidden permutation* $\pi \in S_n$ using a sequence of n distinct commitments $(X_1, \ldots, X_n)$ such that

$$X_1 : \underbrace{\boxed{?}\,\boxed{?}\cdots\boxed{?}}_{\pi(1)}$$

$$\vdots$$

$$X_n : \underbrace{\boxed{?}\,\boxed{?}\cdots\boxed{?}}_{\pi(n)}. \tag{1}$$

Given a hidden permutation $\pi \in S_n$ in the above form (1), to check whether an element $i \in \pi$ is a fixed point, it suffices to reveal the i-th card of the i-th commitment: if the revealed card is red, the element is a fixed point, *i.e.*, $\pi(i) = i$.

2.5 The Existing Protocol

We introduce the ICM protocol [8], which generates a uniformly distributed random derangement using n^2 cards with the pile-scramble shuffle, as follows.

1. Arrange n distinct commitments corresponding to the identity permutation (in S_n) according to the form (1). That is, all the cards on the diagonal are red ♡ and the remaining cards are black ♣.
2. Apply a pile-scramble shuffle to the sequence of n commitments. Note that the resulting n commitments correspond to a certain permutation $\pi \in S_n$; moreover, π is uniformly randomly distributed.
3. Turn over the n cards on the diagonal to check whether there are fixed points in the permutation π.
   - If at least one red card appears, return to Step 2.
   - If all the revealed cards are black, π has no fixed point; hence, π is a uniformly distributed random derangement.

After n players $P_1, P_2, \ldots, P_n$ obtain a hidden derangement (consisting of n commitments) through this protocol, Secret Santa can be implemented by P_i receiving the i-th commitment; he/she reveals the commitment privately to confirm the value of $\pi(i)$, and then purchases a gift for $P_{\pi(i)}$.

Whenever a generated permutation π is not a derangement, the protocol returns to Step 2. The probability that a generated permutation uniformly randomly chosen from S_n is a derangement is $d_n/n! = \sum_{k=0}^{n} (-1)^k/k!$. As $\lim_{n\to\infty} \sum_{k=0}^{n} (-1)^k/k! = 1/e$, the probability of the need to restart the shuffle in the ICM protocol is approximately $1 - 1/e \approx 0.63$.

Note that Ishikawa *et.al.* [8] also showed that the number of required cards can be reduced from n^2 to $2n\lceil \log_2 n \rceil + 6$ by arranging each pile of cards corresponding to a player based on a binary number.

3 Proposed Protocol for Generating a Derangement

In this section, we improve the ICM protocol [8] described in Sect. 2.5 so that the probability of the need to restart the shuffle is decreased. Here, we prepare more piles of cards than the number n of players.

3.1 Overview of the Proposed Protocol

Let us provide an overview of the proposed protocol.

We first prepare $n + t$ commitments instead of n commitments, and apply a pile-scramble shuffle to them. These t additional commitments provide a buffer that absorbs any fixed points that may arise. By revealing the cards on the diagonal, we determine all the fixed points; let f be their number. If the fixed points are too many to be absorbed, i.e., $f > t$, restart the shuffle. If $f \leq t$, we apply the "fixed-point removal" operation (described in Sect. 3.2), resulting in $n + t - f$ commitments. Subsequently, we apply the "reduction" operation (described in Sect. 3.2) to eliminate the $t - f$ extra commitments.

We explain both the fixed-point removal and reduction operations in the following subsection.

3.2 Definitions of the Two Operations

Suppose that we execute Steps 1 and 2 in the ICM protocol (shown in Sect. 2.5), starting with the identity permutation of degree $n + t$ (instead of degree n). Then, we obtain a sequence of $n + t$ commitments corresponding to a uniformly distributed random permutation in S_{n+t}: we refer to such a sequence of commitments as a *committed permutation on* $[1 : n + t]$.

Fixed-point Removal Operation. For the above committed permutation on $[1 : n + t]$, let us reveal all the $n + t$ cards on the diagonal as in Step 3 of the ICM protocol. Subsequently, we determine all the fixed points in the permutation. Let I_{FP} be the set of indices of these fixed points. Ignoring the commitments corresponding to the fixed points, namely, the commitments whose positions are in I_{FP}, the sequence of the remaining commitments corresponds to a derangement uniformly distributed on $[1 : n + t] \backslash I_{\mathrm{FP}}$: we refer to this sequence as a *committed derangement on* $[1 : n + t] \backslash I_{\mathrm{FP}}$.

Through the fixed-point removal operation, a committed permutation of degree $n + t$ is transformed into a committed derangement on $[1 : n + t] \backslash I_{\mathrm{FP}}$ of degree $n + t - |I_{\mathrm{FP}}|$.

Consider the case of $(n, t) = (4, 3)$ as an example. Let us transform a committed permutation shown in Fig. 1a. Then, after the fixed-point removal operation is applied to the committed permutation of degree seven, all the seven cards on the diagonal are revealed as shown in Fig. 1b. In this example, the commitment X_2 is a fixed point; hence, we have $I_{\mathrm{FP}} = \{2\}$ and the sequence of the remaining six commitments $(X_1, X_3, X_4, X_5, X_6, X_7)$ is a committed derangement on $[1 : 7] \backslash \{2\} = \{1, 3, 4, 5, 6, 7\}$ of degree six.

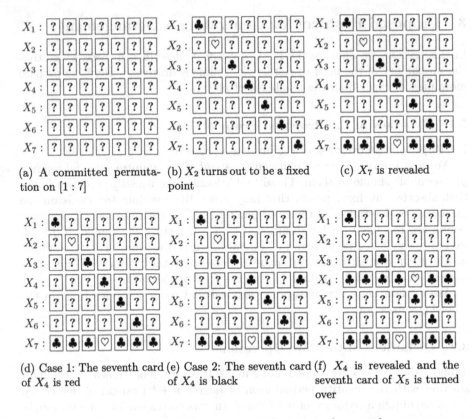

(a) A committed permutation on [1 : 7]

(b) X_2 turns out to be a fixed point

(c) X_7 is revealed

(d) Case 1: The seventh card of X_4 is red

(e) Case 2: The seventh card of X_4 is black

(f) X_4 is revealed and the seventh card of X_5 is turned over

Fig. 1. Example of execution of the proposed protocol

As $n = 4$, the current committed derangement (depicted in Fig. 1b) has two "extra" commitments, $i.e.$, we aim to reduce the degree by two. To this end, we turn over the last commitment, $i.e.$, the seventh commitment X_7. Assume that the revealed value of X_7 is 4 as illustrated in Fig. 1c, indicating a mapping $7 \mapsto 4$, which we refer to as a *bypass*. That is, let us ignore the seventh revealed commitment and regard mapping to 7 as virtually mapping to 4 (via the bypass $7 \mapsto 4$). Consequently, we obtain a committed permutation on $\{1, 3, 4, 5, 6\}$ of degree five, which has been reduced by one.

We now have the committed permutation of degree five (as in Fig. 1c). It may not be a derangement because if $4 \mapsto 7$, it (virtually) has a fixed point (due to the cycle $7 \mapsto 4 \mapsto 7$). Therefore, we turn over the seventh card of the fourth commitment X_4 to check whether it is a fixed point.

- If a red card appears as shown in Fig. 1d, we have the cycle $7 \mapsto 4 \mapsto 7$, indicating a fixed point. Let I_{cycle} denote the set of all the indices of the cycle, $i.e.$, $I_{\text{cycle}} = \{4, 7\}$. Ignoring this cycle, namely, the commitments X_4 and X_7, the sequence of the remaining commitments (X_1, X_3, X_5, X_6) becomes a committed derangement uniformly distributed on $[1 : 7] \backslash (I_{\text{FP}} \cup I_{\text{cycle}}) = \{1, 3, 5, 6\}$.

Thus, we obtain a committed derangement of degree four, as desired. Note that, in this case, the degree decreases by two.

- If a black card appears as shown in Fig. 1e, there is no fixed point; hence, this committed permutation $(X_1, X_3, X_4, X_5, X_6)$ is a uniformly distributed derangement of degree five under the bypass $7 \mapsto 4$. In this case, the degree decreases by one.

In the above example, if a black card appears, we obtain a derangement of degree five; hence, we need to reduce the degree further (because $n = 4$). Therefore, we are expected to reveal another commitment (in this case, we reveal the fourth commitment X_4 because of the bypass $7 \mapsto 4$, as illustrated in Fig. 1f; we will revisit it later).

In general, we define the reduction operation for a committed derangement on $[1 : n + t] \backslash (I_{FP} \cup I_{cycle} \cup I_{BP})$ as follows, where I_{FP} is the set of fixed points, I_{cycle} is the set of indices in cycles, there is a bypass $i_1 \mapsto i_2 \mapsto \cdots \mapsto i_{\ell-1} \mapsto i_\ell$, and $I_{BP} = \{i_1, i_2, \ldots, i_{\ell-1}\}$.

Reduction Operation. If $I_{BP} \neq \phi$, turn over the i_ℓ-th commitment X_{i_ℓ} (which is the end of the bypass). If $I_{BP} = \phi$, turn over the last of the remaining commitments, i.e., the $(\max([1 : n + t] \backslash (I_{FP} \cup I_{cycle})))$-th commitment; in this case, we set $i_\ell = i_1 = \max([1 : n + t] \backslash (I_{FP} \cup I_{cycle}))$ for the sake of convenience. In either case, let $i_{\ell+1}$ be the value of the turned over commitment. Then, turn over the i_1-th card of the $i_{\ell+1}$-th commitment $X_{i_{\ell+1}}$.

- If a red card appears, this committed permutation has the cycle $i_1 \mapsto \cdots \mapsto i_{\ell+1} \mapsto i_1$. The indices $i_1, \ldots, i_{\ell+1}$ of this cycle, namely, all the elements in set $I_{BP} \cup \{i_\ell, i_{\ell+1}\}$, are added to the set I_{cycle}, and the bypass disappears; hence, we set $I_{BP} = \phi$. Ignoring all the commitments whose positions are in $I_{FP} \cup I_{cycle}$, the sequence of the remaining commitments becomes a committed derangement on $[1 : n + t] \backslash (I_{FP} \cup I_{cycle})$. Note that the degree of the committed derangement has been reduced by two (because of ignoring X_{i_ℓ} and $X_{i_{\ell+1}}$).
- If a black card appears, the sequence of the remaining commitments is a committed derangement uniformly distributed on $[1 : n + t] \backslash (I_{FP} \cup I_{cycle} \cup I_{BP})$ under the bypass $i_1 \mapsto \cdots \mapsto i_\ell \mapsto i_{\ell+1}$ (where $I_{BP} = \{i_1, i_2, \ldots, i_\ell\}$). In this case, the degree of the committed derangement has been reduced by one.

3.3 Description of the Proposed Protocol

We describe the proposed protocol using the two aforementioned operations. This protocol uses $n + t$ piles (whereas the ICM protocol [8] uses n piles), as follows.

1. Arrange $n+t$ distinct commitments corresponding to the identity permutation (in S_{n+t}) according to the form (1). That is, all the $n+t$ cards on the diagonal are red $\heartsuit$ and the remaining cards are black $\clubsuit$.

2. Apply a pile-scramble shuffle to the sequence of $n + t$ commitments, and the resulting $n + t$ commitments become a committed permutation on $[1 : n + t]$.

3. Apply the fixed-point removal operation described in Sect. 3.2 to the committed permutation obtained in Step 2. Let I_{FP} be the set of fixed points and $f = |I_{\text{FP}}|$. We obtain a committed derangement on $[1 : n + t] \backslash I_{\text{FP}}$ of degree $n + t - f$.

   - In the case of $f > t$, the committed derangement is insufficient because its degree is less than the number n of players. Therefore, turn all the cards face-down, and return to Step 2.
   - In the case of $f = t$, the degree of the committed derangement is n, as desired. Therefore, proceed to Step 5.
   - In the case of $f < t$, proceed to Step 4.

4. Repeatedly apply the reduction operation described in Sect. 3.2 to the committed derangement on $[1 : n + t] \backslash I_{\text{FP}}$ obtained in Step 3, until its degree becomes n or less. Recall that each application of the reduction operation reduces the degree by one or two.

   - If a committed derangement of degree $n - 1$ is obtained, turn all the cards face-down, and return to Step 2.
   - If a committed derangement of degree n is obtained, proceed to Step 5.

5. We have a committed derangement of degree n on $[1 : n + t] \backslash (I_{\text{FP}} \cup I_{\text{cycle}} \cup I_{\text{BP}})$, as desired.

After we obtain a committed derangement in Step 5, we renumber the players based on the remaining n commitments. If there is a bypass $i_1 \mapsto i_2 \mapsto \cdots \mapsto i_\ell$, a player who turns over the commitment to i_1 should purchase a gift for the player corresponding to i_ℓ. For example, consider the committed derangement illustrated in Fig. 1f, which is obtained from Fig. 1e by revealing X_4 and the seventh card of X_5. We renumber the four players such that $P_1 = P_1'$, $P_2 = P_3'$, $P_3 = P_5'$, and $P_4 = P_6'$, and make P_1', P_3', P_5', and P_6' receive commitments X_1, X_3, X_5, and X_6, respectively. Each player secretly turns over the assigned commitment to know for whom he/she should purchase a gift. Because of the bypass $7 \mapsto 4 \mapsto 5$, the player who reveals the commitment to 7 should purchase a gift for P_5'.

Thus, the proposed protocol generates a committed derangement. As there is a trade-off between the number t of additional piles and the probability of returning to Step 2, we comprehensively analyze the probability of the need to restart the shuffle in the following section.

4 Probability of the Need to Restart the Shuffle

In the proposed protocol presented in the previous section, the probability of the need to restart the shuffle depends on the number $t \, (\geq 1)$ of additional piles. In this section, we analyze this probability. Recall that the restart occurs when either more than t fixed points appear in Step 3 or a derangement of degree $n - 1$ is obtained in Step 4.

Let f be the number of fixed points determined in Step 2. A restart from Step 3 occurs if $t + 1 \leq f \leq n + t$. As the probability that a uniformly distributed random permutation in S_{n+t} has exactly f fixed points is $_{n+t}C_f \cdot d_{n+t-f}/(n+t)!$, the following equation holds:

$$\Pr[\text{Restart from Step 3}] = \sum_{f=t+1}^{n+t} \frac{_{n+t}C_f \cdot d_{n+t-f}}{(n+t)!}. \tag{2}$$

Next, we consider a restart from Step 4. Suppose that we have a committed derangement of degree $n + x$ for a non-negative integer x. Let $\epsilon(n, x)$ be the probability that repeated applications of the reduction operation result in a committed derangement of degree $n - 1$ and we return to Step 2. Each application of the reduction operation to the committed derangement of degree $n + x$ produces a committed derangement of degree either $n + x - 2$ or $n + x - 1$. The former occurs when the commitment to be revealed is included in a cycle of length two; therefore, its occurrence probability is $(n + x - 1)d_{n+x-2}/d_{n+x}$. The latter occurs with a probability of $1 - (n + x - 1)d_{n+x-2}/d_{n+x}$. Thus, $\epsilon(n, x)$ can be expressed recursively as follows:

$$\epsilon(n, x) = \left(1 - \frac{(n + x - 1)d_{n+x-2}}{d_{n+x}}\right) \cdot \epsilon(n, x - 1)$$
$$+ \frac{(n + x - 1)d_{n+x-2}}{d_{n+x}} \cdot \epsilon(n, x - 2),$$

where $\epsilon(n, 0) = 0$ and $\epsilon(n, 1) = n \cdot d_{n-1}/d_{n+1}$. As Step 4 occurs when f lies between 0 and $t - 1$ with a probability of $_{n+t}C_f \cdot d_{n+t-f}/(n+t)!$, the following equation holds:

$$\Pr[\text{Restart from Step 4}] = \sum_{f=0}^{t-1} \frac{_{n+t}C_f \cdot d_{n+t-f}}{(n+t)!} \cdot \epsilon(n, t - f). \tag{3}$$

The probability of the need to return to Step 2 for the entire proposed protocol, denoted by $\Pr[\text{Restart}^{(n,t)}]$, is the sum of Eqs. (2) and (3). Figure 2 shows the relationship between $t (\leq 10)$ and the probability $\Pr[\text{Restart}^{(n,t)}]$ for the number of players from $n = 3$ to $n = 20$. $\Pr[\text{Restart}^{(n,0)}]$ is the same as the probability for the ICM protocol. The proposed protocol improves significantly as t increases. If we prepare t additional piles, the number of required cards increases by $t(t + 2n)$; considering an unnecessarily large t is not realistic. Even if we set t to a small value such as 2 or 3, the probability can be reduced to approximately 0.1 compared with that of the ICM protocol (approximately 0.6).

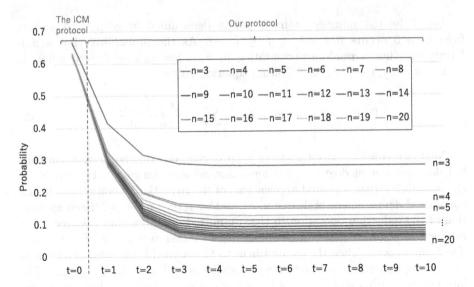

Fig. 2. Relationship between the number t of additional piles and the probability of the need to restart $\Pr[\text{Restart}^{(n,t)}]$

5   Conclusion

In this paper, we proposed a new efficient protocol that generates a uniformly distributed random derangement. We prepared more piles of cards than the number n of players to suppress the need to restart the shuffle process. There is a trade-off between the number t of additional piles and the probability of the need to restart the shuffle. When executing the proposed protocol, it is better to set $t = 2$ or $t = 3$, as shown in Fig. 2.

The proposed technique can also be applied to the existing protocol based on the binary expression of the indices of players [8].

Acknowledgement. We thank the anonymous referees, whose comments have helped us to improve the presentation of the paper. This work was supported in part by JSPS KAKENHI Grant Number JP19J21153.

References

1. Boer, B.: More efficient match-making and satisfiability *the five card trick*. In: Quisquater, J.-J., Vandewalle, J. (eds.) EUROCRYPT 1989. LNCS, vol. 434, pp. 208–217. Springer, Heidelberg (1990). https://doi.org/10.1007/3-540-46885-4_23

2. Crépeau, C., Kilian, J.: Discreet solitary games. In: Stinson, D.R. (ed.) CRYPTO 1993. LNCS, vol. 773, pp. 319–330. Springer, Heidelberg (1994). https://doi.org/10.1007/3-540-48329-2_27

3. Gradwohl, R., Naor, M., Pinkas, B., Rothblum, G.N.: Cryptographic and physical zero-knowledge proof systems for solutions of Sudoku puzzles. Theory Comput. Syst. **44**(2), 245–268 (2009)

4. Hashimoto, Y., Nuida, K., Shinagawa, K., Inamura, M., Hanaoka, G.: Toward finite-runtime card-based protocol for generating a hidden random permutation without fixed points. IEICE Trans. Fundam. Electron. Commun. Comput. Sci. **E101.A**, 1503–1511 (2018)
5. Hashimoto, Y., Shinagawa, K., Nuida, K., Inamura, M., Hanaoka, G.: Secure grouping protocol using a deck of cards. In: Shikata, J. (ed.) ICITS 2017. LNCS, vol. 10681, pp. 135–152. Springer, Cham (2017). https://doi.org/10.1007/978-3-319-72089-0_8
6. Heather, J., Schneider, S., Teague, V.: Cryptographic protocols with everyday objects. Formal Aspects Comput. **26**, 37–62 (2014)
7. Ibaraki, T., Manabe, Y.: A more efficient card-based protocol for generating a random permutation without fixed points. In: 2016 Third International Conference on Mathematics and Computers in Sciences and in Industry (MCSI), pp. 252–257, August 2016
8. Ishikawa, R., Chida, E., Mizuki, T.: Efficient card-based protocols for generating a hidden random permutation without fixed points. In: Calude, C.S., Dinneen, M.J. (eds.) UCNC 2015. LNCS, vol. 9252, pp. 215–226. Springer, Cham (2015). https://doi.org/10.1007/978-3-319-21819-9_16
9. Miyahara, D., Hayashi, Y., Mizuki, T., Sone, H.: Practical card-based implementations of Yao's millionaire protocol. Theor. Comput. Sci. **803**, 207–221 (2020)
10. Miyahara, D., et al.: Card-based ZKP protocols for Takuzu and Juosan. In: Farach-Colton, M., Prencipe, G., Uehara, R. (eds.) Fun with Algorithms. Leibniz International Proceedings in Informatics, LIPIcs, Schloss Dagstuhl-Leibniz-Zentrum fur Informatik GmbH, Dagstuhl Publishing, September 2020
11. Mizuki, T., Sone, H.: Six-card secure AND and four-card secure XOR. In: Deng, X., Hopcroft, J.E., Xue, J. (eds.) FAW 2009. LNCS, vol. 5598, pp. 358–369. Springer, Heidelberg (2009). https://doi.org/10.1007/978-3-642-02270-8_36
12. Nakai, T., Tokushige, Y., Misawa, Y., Iwamoto, M., Ohta, K.: Efficient card-based cryptographic protocols for millionaires' problem utilizing private permutations. In: Foresti, S., Persiano, G. (eds.) CANS 2016. LNCS, vol. 10052, pp. 500–517. Springer, Cham (2016). https://doi.org/10.1007/978-3-319-48965-0_30
13. Ono, H., Manabe, Y.: Efficient card-based cryptographic protocols for the millionaires' problem using private input operations. In: 2018 13th Asia Joint Conference on Information Security (AsiaJCIS), pp. 23–28, August 2018
14. Ono, H., Manabe, Y.: Card-based cryptographic logical computations using private operations. New Gener. Comput. (2020, in press). https://doi.org/10.1007/s00354-020-00113-z
15. Robert, L., Miyahara, D., Lafourcade, P., Mizuki, T.: Physical zero-knowledge proof for Suguru puzzle. In: Devismes, S., Mittal, N. (eds.) SSS 2020. LNCS, vol. 12514, pp. 235–247. Springer, Cham (2020). https://doi.org/10.1007/978-3-030-64348-5_19
16. Ruangwises, S., Itoh, T.: Physical zero-knowledge proof for Numberlink puzzle and k vertex-disjoint paths problem. New Gener. Comput. (2020, in press). https://doi.org/10.1007/s00354-020-00114-y
17. Ryan, P.Y.A.: Crypto Santa. In: Ryan, P.Y.A., Naccache, D., Quisquater, J.-J. (eds.) The New Codebreakers. LNCS, vol. 9100, pp. 543–549. Springer, Heidelberg (2016). https://doi.org/10.1007/978-3-662-49301-4_33
18. Sasaki, T., Miyahara, D., Mizuki, T., Sone, H.: Efficient card-based zero-knowledge proof for Sudoku. Theor. Comput. Sci. **839**, 135–142 (2020)

Compact Data Structures for Dedekind Groups and Finite Rings

Bireswar Das and Shivdutt Sharma[⊠]

IIT Gandhinagar, Gandhinagar, India
`{bireswar,shiv.sharma}@iitgn.ac.in`

Abstract. A group with n elements can be stored using $\mathcal{O}(n^2)$ space via its Cayley table which can answer a group multiplication query in $\mathcal{O}(1)$ time. Information theoretically it needs $\Omega(n \log n)$ bits or $\Omega(n)$ words in word-RAM model just to store a group (Farzan and Munro, ISSAC 2006).

For functions $s, t : \mathbb{N} \longrightarrow \mathbb{R}_{\geq 0}$, we say that a data structure is *an* $(\mathcal{O}(s), \mathcal{O}(t))$-*data structure* if it uses $\mathcal{O}(s)$ space and answers a query in $\mathcal{O}(t)$ time. Except for cyclic groups it was not known if we can design $(\mathcal{O}(n), \mathcal{O}(1))$-data structure for interesting classes of groups.

In this paper, we show that there exist $(\mathcal{O}(n), \mathcal{O}(1))$-data structures for several classes of groups and for *any* ring and thus achieve information theoretic lower bound asymptotically. More precisely, we show that there exist $(\mathcal{O}(n), \mathcal{O}(1))$-data structures for the following algebraic structures with n elements:

- Dedekind groups: This class contains abelian groups, Hamiltonian groups.
- Groups whose indecomposable factors admit $(\mathcal{O}(n), \mathcal{O}(1))$-data structures.
- Groups whose indecomposable factors are strongly indecomposable.
- Groups defined as a semidirect product of groups that admit $(\mathcal{O}(n), \mathcal{O}(1))$-data structures.
- Finite rings.

Keywords: Theoretical computer science · Abelian groups · Dedekind groups · Finite rings · Linear space representations · Compact data structures · Strongly indecomposable groups

1 Introduction

A group can be represented as the input to an algorithm in various formats. Some of these representations include the Cayley table representation, the permutation group representation, the polycyclic representation and the generator-relator representation.

Several computational group theoretic problems such as various property testing problems, group factoring problem, minimum generating set problem, computing basis for an abelian group etc., where the input groups are represented

© Springer Nature Switzerland AG 2021
R. Uehara et al. (Eds.): WALCOM 2021, LNCS 12635, pp. 90–102, 2021.
https://doi.org/10.1007/978-3-030-68211-8_8

by their Cayley tables (or the multiplication tables), have been studied in the past [2,4,11,13,16,21]. Perhaps the most important problem when the input is represented by Cayley tables is the group isomorphism problem (GrISO). This problem has been studied extensively in the field of computational complexity theory and algorithms because of its unresolved complexity status.

The Cayley table for a group with elements $\{1, 2, \ldots, n\}$ is a two dimensional table indexed by a pair of group elements (i, j). The (i, j)-th entry of the table is the product of i and j. A *multiplication query* where the user asks "what is the product of element i and j?" can thus be answered in constant time. However, it needs $\mathcal{O}(n^2 \log n)$ bits or $\mathcal{O}(n^2)$ words in the word-RAM model to store a Cayley table for a group of size n. Cayley tables can also be used to store quasigroups (latin squares) and semigroups. However the number of quasigroups or semigroups of size n are so large (see [15,18]) that the information theoretic lower bound to store quasigroups or semigroups is $\Omega(n^2 \log n)$ bits or $\Omega(n^2)$ words. In this sense, the Cayley group representations for quasigroups or semigroup is optimal.

It is natural to ask if a group can be represented using subquadratic space while still being able to answer a multiplication query in constant time. Farzan and Munro [9] gave an information theoretic lower bound of $n \log n$ bits to store a group of size[1] n. This lower bound implies an $\Omega(n)$ lower bound on the number of words required to store a group in word-RAM model. A data structure that achieves the optimum information theoretic lower bound asymptotically is known as a *compact data structure*.

A recent result [7] shows that a finite group G with n elements can be stored in a data structure that uses $\mathcal{O}(n^{1+\delta}/\delta)$ space and answers a multiplication query in time $\mathcal{O}(1/\delta)$ for any δ such that $1/\log n \le \delta \le 1$. This result shows, for example, that a finite group can be stored in a data structure that uses $\mathcal{O}(n^{1.01})$ words and answers a query in constant time. However, it is not clear if this data structure is compact as the lower bound given by Farzan and Munro may not be the optimal lower bound for the class of all groups. For general group class it is not clear if we can design a data structure that uses $\mathcal{O}(n \log^{\mathcal{O}(1)} n)$ space and answers a query in $\mathcal{O}(1)$ time. Or if we are willing to spend more time, say $\mathcal{O}(\log n)$, to answer a query can we design a data structure that takes $\mathcal{O}(n)$ space? Of course the best scenario would be to obtain a data structure for groups that uses $\mathcal{O}(n)$ space and answers queries in $\mathcal{O}(1)$ time.

Except for cyclic groups, it was unknown if we could design compact data structures for other classes of groups while still retaining the capability to answer queries in constant time. As an application of this result we show that groups

---

[1] Without loss of generality, the group elements are assumed to be $\{1, \ldots, n\}$, and the task is to store the information about the multiplication of any two elements. Here the user *knows* the labels or the names of each element and has a direct and explicit access to each element. We can compare the situation with permutation group representation where a group is represented as a subgroup of a symmetric group. The group is given by a set of generators. Here the user does not have an explicit representation for each group element.

whose indecomposable direct factors are *strongly indecomposable* have compact data structure with constant query time. We also show that if a group can be written as the semidirect product of two groups which admit linear space data structures with constant query time then the original group also admits a linear space data structure with constant query time.

A common theme in most of our results is to judiciously use the structure of the group to obtain *a constant number* of instances of the problem where the group sizes are small, and the data structures used in [7] can be used. We also use simple data structures to 'combine' the data structures for the smaller instances to get a data structure for the original group.

One natural way to decompose a group into smaller groups is to use the Remak-Krull-Schmidt theorem. The Remak-Krull-Schmidt theorem says that any finite group can be factored as a direct product of indecomposable groups. We prove that if the indecomposable direct factors of a group have a linear space data structure with constant query time then the group also has a linear space data structure with constant query time. As an application of this result we show that groups whose indecomposable direct factors are *strongly indecomposable* have compact data structure with constant query time. We also show that if a group can be written as the semidirect product of two groups which admit linear space data structures with constant query time then the original group also admits linear space data structure with constant query time.

We also study space efficient representations of finite rings. A simple idea to store a finite ring is to store its addition and multiplication tables using quadratic space. These two tables can answer an addition or a multiplication query in constant time. Kayal and Saxena [14] used a representation for finite rings which takes polylogarithmic time in the size of the given ring to answer an addition or a multiplication query. However, this representation do not label or name each element and the user needs to specify an element as a $\mathbb{Z}$-linear sum of the basis element of the additive structure for the group. We show that for *any* finite ring there is a data structure that uses linear space and answers an addition query or a multiplication query in constant time.

Related Work: Farzan and Munro [9] studied space efficient representations of finite abelian groups in a specific model of computation that has a query processing unit that performs the group operation. The query processing unit accepts the elements associated with a query in a certain format and the user is responsible to supply the elements in that specific format. The query processing unit also supports some extra operations such as bit reversals. They show that the query processing unit needs to store just a constant number of words and can answer a query in constant time. The result of Farzan and Munro has been extended [7] to some non-abelian group classes such as Dedekind groups. We note that in this paper we do not assume that the user converts the group elements into any specific format. The user's query will involve the elements in their original format only.

2 Preliminaries

In this paper all groups considered are finite. Let G be a group and S be a subset of group G then $\langle S \rangle$ denotes the subgroup generated S. The *order* of a G, denoted by $|G|$, is the cardinality of G. The order of an element $x \in G$ is denoted by $\mathrm{ord}_G(x)$ and it is the minimum positive integer m such that x^m is identity.

A group is said to be *indecomposable* if cannot be decomposed as a direct product of nontrival subgroups. The Remak-Krull-Schmidt theorem says that any finite group can be decomposed "uniquely" as the direct product of indecomposable subgroups.

Theorem 1 (Remak-Krull-Schmidt, see e.g., [10]). *Let G be a group. Let $G = G_1 \times G_2 \times \ldots \times G_t$ where G_i for $i \in [t]$ is indecomposable. If $G = H_1 \times H_2 \times \ldots \times H_s$ such that H_i for all $i \in [s]$ is indecomposable then $s = t$ and after reindexing $G_i \cong H_i$ for every i, and for any $r < t$, $G = G_1 \times \ldots \times G_r \times H_{r+1} \times \ldots \times H_t$.*

The *structure theorem for finite abelian groups* states that any finite abelian group G can be decomposed as $G = \langle g_1 \rangle \times \langle g_2 \rangle \times \ldots \times \langle g_t \rangle$, where each $\langle g_i \rangle$, $i \in [t]$ is a cyclic group of prime power order generated by the element g_i. The set $B = \{g_1, g_2, \ldots, g_t\}$ is called as a *basis* of the abelian group G.

Let A and B are two groups, let $\varphi : B \longrightarrow \mathrm{Aut}(A)$ be a homomorphism. The *semidirect product* of A and B with respect to φ, denoted $A \rtimes_\varphi B$, is the group with elements $A \times B$ along with the following binary operation: The product of two elements (a_1, b_1) and (a_2, b_2) is the element $(a_1(\varphi(b_1))(a_2), b_1 b_2)$.

The *generalized quaternion* group (see [5]) is defined as follows. For $n \geq 3$, set $Q_{2^n} = (\mathbb{Z}_{2^{n-1}} \rtimes \mathbb{Z}_4)/\langle(2^{n-2}, 2)\rangle$ where, semidirect product obeys the following product rule $(a, b)(c, d) = (a + (-1)^b c, b + d)$. The group Q_{2^n} is called the *generalized quaternion group*. For $n = 3$, Q_8 is called as *quaternion group*.

A ring is a set R with two binary operations $+$ and $\times$ (generally called as addition and multiplication, respectively) satisfying the following conditions: 1) $(R, +)$ is an abelian group, 2) For all $r, s, t \in R$, $r \times (s + t) = (r \times s) + (r \times t)$ and $(s + t) \times r = (s \times r) + (t \times r)$, 3) For all $r, s, t \in R$, $r \times (s \times t) = (r \times s) \times t$.

For integers a and $b > 0$, $a\%b$ denotes the positive remainder when a is divided by b.

Model of Computation: In this paper we use word-RAM (Random Access Machine) as the model of computation. A RAM is an abstract machine in the general class of register machines. In this model each register and memory unit can store $\mathcal{O}(\log n)$ bits where n is the input size. The unit of storage is called a *word*. In a word-RAM machine, word operations can be performed in constant time. In this paper, n denotes the order of the input group or ring. All arithmetic, logical, and comparison operations on words take constant time. Each memory access also takes constant time. Unless stated otherwise we assume without loss of generality that elements of the input group are encoded as $1, 2, \ldots, n$.

There will be two phases: *preprocessing phase* and *multiplication query phase* or simply *query phase*. In the preprocessing phase, if we are given a group G

by its Cayley table, then at the end of the preprocessing phase, we output a few data structures needed for multiplication query. In the multiplication query phase, we are given two group elements, g_1 and g_2 of the group G, and we want to know the result of the $g_1 g_2$. A multiplication query is answered using the data structures computed in the preprocessing phase. The situations is similar for rings except for the fact that there are two types of queries instead of one: an addition query or a multiplication query. The time and space used in the preprocessing phase are *not* counted towards time and space complexity. The space complexity refers only to the space used by the data structure in terms of number of words.

Suppose that B is the information theoretical optimal number of bits needed to store some data. A representation of this data is called *compact* if it uses $\mathcal{O}(B)$ bits of space.

Definition 1. Let $s : \mathbb{N} \longrightarrow \mathbb{R}_{\geq 0}$ and $t : \mathbb{N} \longrightarrow \mathbb{R}_{\geq 0}$ be two functions. We say that a group G has an (s,t)-*data structure* if we can construct a data structure for representing the group that uses at most $s(n)$ space to store G and can answer a multiplication query in time at most $t(n)$ where n is the size of the group G.

Definition 2. Let $\mathcal{G}$ be a class of groups and let $s : \mathbb{N} \longrightarrow \mathbb{R}_{\geq 0}$ and $t : \mathbb{N} \longrightarrow \mathbb{R}_{\geq 0}$ be two functions. We say that $\mathcal{G}$ has an $(\mathcal{O}(s(n)), \mathcal{O}(t(n)))$-data structure if for every $G \in \mathcal{G}$ with $|G| = n$ we can construct a data structure that uses $\mathcal{O}(s(n))$ space to store G and answers a multiplication query in time $\mathcal{O}(t(n))$.

We would say, with slight abuse of notation, that a group G has an $(\mathcal{O}(s(n)), \mathcal{O}(t(n)))$-data structure when the group class $\mathcal{G}$ containing G is clear from the context and $\mathcal{G}$ has an $(\mathcal{O}(s(n)), \mathcal{O}(t(n)))$-data structure. Note that the Cayley table representation is an $(\mathcal{O}(n^2), \mathcal{O}(1))$-data structure for the class of all groups.

Our Results: There is an $(\mathcal{O}(n), \mathcal{O}(1))$-data structure for algebraic structures Abelian (**AB**) (Theorem 4), Dedekind (**DDK**) (Theorem 7), groups whose indecomposable factors are strongly indecomposable ($\mathcal{G}_{strong}$) (Theorem 10), semidirect product groups whose factors have linear space constant query data structures (Theorem 8) and rings (**RING**) (Theorem 11).

3   Meta Theorems

In this section we present some results that will be used in the later section to give succinct representations of abelian, Hamiltonian, Dedekind groups etc. One of the ingredients for proving these results is the following theorem.

Theorem 2 ([7]). *Let G be a group of order n. Then for any δ such that $\frac{1}{\log n} \leq \delta \leq 1$, there is a data structure D that uses $\mathcal{O}(\frac{n^{1+\delta}}{\delta})$ space and answers multiplication queries in $\mathcal{O}(\frac{1}{\delta})$ time.*

The proof of the above result, which could be found in [7], uses cube generating sequences of size $\mathcal{O}(\log n)$, where n is the size of the group. Erdös and Renyi [8] proved that such sequences always exist. We note that the data structure D mentioned in the above theorem can actually be computed efficiently in polynomial time [7].

By setting, for example $\delta = \frac{1}{20}$ in the above theorem, the class of all groups has an $(\mathcal{O}(n^{1.05}), \mathcal{O}(1))$-data structure.

Lemma 1. *Let $\delta > 0$ be a fixed constant. Let $\mathcal{G}_\delta$ be the class of groups G such that G factors as $G_1 \times G_2$ with $|G_1|, |G_2| \leq n^{\frac{1}{1+\delta}}$ where $n = |G|$. Then $\mathcal{G}_\delta$ has an $(\mathcal{O}(n), \mathcal{O}(1/\delta))$-data structure.*

Proof. Let $G = G_1 \times G_2$ such that $|G_1|, |G_2| \leq n^{1/(1+\delta)}$. We use data structures D_1 and D_2 from Theorem 2 to store G_1 and G_2 in $\mathcal{O}(n)$ space that can support multiplication queries in $\mathcal{O}(1/\delta)$ time. Let g be an element of G suh that $g = (g_1, g_2)$ where $g_1 \in G_1$ and $g_2 \in G_2$. We use a forward-map array $\mathcal{F}$ indexed by the group elements of G such that $\mathcal{F}[g] = (g_1, g_2)$ and a backward-map array $\mathcal{B}$ such that $\mathcal{B}[g_1][g_2] = g$. In the query processing phase, given two elements g and h in G, access forward map array $\mathcal{F}$ at index g and h to obtain the decompositions $g = (g_1, g_2)$ and $h = (h_1, h_2)$. Using the data structures D_1 and D_2 compute $g_1 h_1$ and $g_2 h_2$ in time $\mathcal{O}(1/\delta)$. Finally we use $\mathcal{B}[g_1 h_1][g_2 h_2]$ to obtain the final result $g_1 h_1 g_2 h_2$. ∎

The next lemma states that if we have compact representations for the indecomposable factors then we can design compact representations of any group.

Lemma 2. *Let $\mathcal{G}$ be a group class. Let*

$$Fact(\mathcal{G}) = \{H | H \text{ is an indecomposable factor of } G \in \mathcal{G}\}.$$

Then if $Fact(\mathcal{G})$ has an $(\mathcal{O}(n), \mathcal{O}(1))$-data structure then $\mathcal{G}$ also has an $(\mathcal{O}(n), \mathcal{O}(1))$-data structure.

Proof. Let $G \in \mathcal{G}$ and $n = |G|$. If G has an indecomposable factor G_1 such that $|G_1| \geq n^{1/3}$ then the other factor G_2 (may not be indecomposable) has size at most $n^{2/3}$. Thus G_2 has a data structure D_2 that uses $\mathcal{O}(n)$ space and answers a multiplication query in time $\mathcal{O}(1)$ (we use $\delta = 1/2$ in Theorem 2). Since G_1 is indecomposable it has a data structure D_1 that uses at most $\mathcal{O}(|G_1|)$ space and answer queries in time $\mathcal{O}(1)$. We can use techniques similar to the proof of Lemma 1 to obtain the desired result.

On the other hand if all the indecomposable factors of G are of size at most $n^{1/3}$ then it is easy to see that G could be factored as $H_1 \times H_2$ where $|H_1|, |H_2| \leq n^{2/3}$. We are back again to a situation where we can use ideas similar to the proof of Lemma 1. ∎

Remark 1. The preprocessing time in Lemma 1 and Lemma 2 depends on how efficiently we can factor groups given by their Cayley tables. Kayal and Nezhmetdinov [13] showed that, factoring a group and more generally computing a Remak-Krull-Schmidt decomposition[2] is polynomial-time.

4   Compact Data Structures for Special Group Classes

In this section we show that several group classes such as Dedekind groups and some groups which can be written as a semidirect product have compact data structures which can answer multiplication queries in constant time.

Abelian Groups: In the following part we show that abelian groups have $(\mathcal{O}(n), \mathcal{O}(1))$-data structure. First we state the following well known lemma.

Lemma 3. *The indecomposable factors* $Fact(\mathbf{AB})$ *of the class of finite abelian groups* $\mathbf{AB}$ *are cyclic of prime power order. In other words*

$$Fact(\mathbf{AB}) = \{\mathbb{Z}_{p^r} | p \text{ is prime}\}.$$

Theorem 3. *Let* $\mathbf{CYC}$ *be the class of finite cyclic groups. Then* $\mathbf{CYC}$ *has an* $(\mathcal{O}(n), \mathcal{O}(1))$-*data structure. Moreover the data structure can be computed in time linear in the size of the group.*

Proof. Let C be a cyclic group of order n. By repeated powering of all the elements in C we can find a generator g of C in time $\mathcal{O}(n^2)$. However if we use the results by Kavitha [12] we can find a generator of C in $\mathcal{O}(n)$ time.

For the data structure we use a *forward-map array* $\mathcal{F}$ indexed by the group elements. For an element g^i we set $\mathcal{F}[g^i] = i$. We use a *backward-map array* $\mathcal{B}$ indexed by $\{0, 1, \cdots, |C| - 1\}$. We set $\mathcal{B}[i] = g^i$. Given two elements $a, b \in C$ we can compute ab by finding $k = (\mathcal{F}[a] + \mathcal{F}[b])\%|C|$ and then finding $\mathcal{B}[k]$. ∎

As a consequence of Theorem 3 and Lemma 2 we can conclude that the class of abelian groups $\mathbf{AB}$ has an $(\mathcal{O}(n), \mathcal{O}(1))$-data structure. Lemma 2 uses Theorem 2 and algorithms to factor groups. Group factoring is linear time for abelian groups [4]. The preprocessing used in Theorem 2 involves computing a cube generating sequence of logarithmic length which can be done in linear time. Moreover, if we have the cube generating sequence, the rest of the preprocessing can be done in linear time [7].

Theorem 4. *The class of abelian groups* $\mathbf{AB}$ *has an* $(\mathcal{O}(n), \mathcal{O}(1))$-*data structure. Moreover, the data structures supporting the representation can be computed in time linear in the size of the input group.*

---

[2] A decomposition of group G is said to be *Remak-Krull-Schmidt* if all the factors in the decomposition are indecomposable.

Dedekind Groups: A group G is called a Dedekind group if every subgroup of G is normal. The nonabelian Dedekind groups are called Hamiltonian groups. Since we already know how to succinctly represent abelian groups in this section we focus mainly on Hamiltonian groups. The Hamiltonian groups are characterized by the following interesting theorem.

Theorem 5 ([3]). *A nonabelian group G is Hamiltonian if and only if G is the direct product of Q_8 and B and A, where A is an abelian group of odd order and B is an elementary abelian 2-group*[3].

Theorem 6. *Let* **HAM** *be the class of Hamiltonian groups. Then,* **HAM** *has an $(\mathcal{O}(n), \mathcal{O}(1))$-data structure. The data structures supporting the representations can be computed in time $\mathcal{O}(n)$ where n is the size of the group to be succinctly represented.*

Proof Idea. Let G be a Hamiltonian group and $G = Q_8 \times A \times B$. Let g be an element of G such that $g = (q, a, b)$ where $q \in Q_8, a \in A, b \in B$. We use a forward-map array $\mathcal{F}$ such that $\mathcal{F}[g] = (q, a, b)$ and a backward-map array $\mathcal{B}$ such that $\mathcal{B}[q][a][b] = g$. The indecomposable factors of Hamiltonian groups are Q_8 and cyclic groups. So as a consequence of Theorem 3 and Lemma 2 we can conclude that the class of Hamiltonian groups **HAM** has an $(\mathcal{O}(n), \mathcal{O}(1))$-data structure. The preprocessing step involves decomposition of the Hamiltonian group G as $Q_8 \times A \times B$ which can be performed in $\mathcal{O}(n)$ time by a recent result of Das and Sharma [6].

Remark 2. We note that a group class in which every group can be factored as the direct product of a nonabelian group of bounded size and an abelian group, has an $(\mathcal{O}(n), \mathcal{O}(1))$-data structure. The preprocessing can be done in time $n(\log n)^{\mathcal{O}(1)}$ using a factoring algorithm for such groups described in [6].

Theorem 7. *Let* **DDK** *be the class of Dedekind groups. The group class* **DDK** *has an $(\mathcal{O}(n), \mathcal{O}(1))$-data structure.*

Semidirect Product Classes: There many groups which can be constructed as a semidirect product of two groups. One important example is the class of Z-groups. These are groups that can be written as the semidirect product of two of its cyclic subgroups. Our result for semidirect product classes is as follows.

Theorem 8. *Let c be a constant. Let $\mathcal{G}_c$ be the class of groups that can be written as a semidirect product $A \rtimes_\varphi B$ for some $\varphi : B \longrightarrow Aut(A)$ where A and B have (cn, c)-data structures. Then $\mathcal{G}_c$ has an $(\mathcal{O}(n), \mathcal{O}(1))$-data structure.*

Proof Idea. Let $G = A \rtimes_\varphi B$. Let g be an element of G such that $g = (a, b)$ where $a \in A$ and $b \in B$. We use a forward-map array $\mathcal{F}$ such that $\mathcal{F}[g] = (a, b)$ and a backward-map array $\mathcal{B}$ such that $\mathcal{B}[a][b] = g$. Both A and B have data structures

---

[3] An elementary abelian 2-group is an abelian group in which every nontrivial element has order 2. The groups A and B in the theorem can be the trivial group.

that use linear space to answer a multiplication query in constant time. We use an array $\mathcal{M}$ of size $|A||B|$ to store $\varphi(b) \in \text{Aut}(A)$ for each $b \in B$. In the query phase given two elements g and h of the group G, access $\mathcal{F}$ at index g and h. Let $\mathcal{F}[g] = (a_1, b_1)$ where $a_1 \in A$ and $b_1 \in B$ and $\mathcal{F}[h] = (a_2, b_2)$ where $a_2 \in A$ and $b_2 \in B$. The result of the multiplication of g and h is $(a_1(\varphi(b_1)(a_2)),\ b_1 b_2)$ which can be performed in constant time by using the data structures.

We can apply Theorem 8 in conjunction with Theorem 7 to obtain the result that if a group can be written as the semidirect product of two Dedekind groups then it has $(\mathcal{O}(n), \mathcal{O}(1))$-data structure. The class of groups that can be written as the semidirect product of Dedekind groups is fairly large class of groups as it contains the Z-groups, and some generalization of Dedekind groups [17].

Strongly Indecomposable Groups: Remak-Krull-Schmidt states that any group G can be decomposed as a direct product of indecomposable groups. If each indecomposable factor has a linear space constant query data structure then by Lemma 2, G has an $(\mathcal{O}(n), \mathcal{O}(1))$-data structure. We do not know how to design constant query compact data structures for indecomposable groups but for a class of indecomposable groups known as strongly indecomposable we have designed such data structures. An indecomposable group G is *strongly indecomposable* if each subgroups H of G is also indecomposable. We denote the class of strongly indecomposable groups by **STR**. Our results in this part are based on the characterization of strongly indecomposable groups given below.

Theorem 9 ([19]). *A group G is a strongly indecomposable group if and only if G is one of following types:* 1) G *is isomorphic to* $\mathbb{Z}_{p^n}$ *where p is some prime,* 2) G *is generalized quaternion, isomorphic to* Q_{2^n} *for $n \geq 3$,* 3) $G = \mathbb{Z}_{p^\alpha} \rtimes \mathbb{Z}_{p^\beta}$ *with p and q are different primes, p is odd such that q^β divides $p - 1$ and the image of $\mathbb{Z}_{q^\beta}$ in $\mathbb{Z}_{p^\alpha}^\times$ has order q^β.*

Lemma 4. *The class of finite strongly indecomposable groups* **STR** *has an $(\mathcal{O}(n), \mathcal{O}(1))$-data structure.*

Proof. If the structure of the input groups is as described in the case 1 or case 3 of above theorem, then Theorem 3 or Theorem 8 (respectively) proves the result. Now let G be a generalized quaternion group. By the structure of generalized quaternion groups (see Sect. 2) it will be isomorphic to $(\mathbb{Z}_{2^{m-1}} \rtimes \mathbb{Z}_4)/\langle (2^{m-2}, 2) \rangle$. Let $N = \langle (2^{m-2}, 2) \rangle$.[4] Notice that $|N| = 2$. Thus for any $g_1 \in \mathbb{Z}_{2^{m-1}} \rtimes \mathbb{Z}_4$ there is exactly one element g_2 in $\mathbb{Z}_{2^{m-1}} \rtimes \mathbb{Z}_4$ such that $g_1 N = g_2 N$ and $g_1 \neq g_2$. In other words, exactly two elements of $\mathbb{Z}_{2^{m-1}} \rtimes \mathbb{Z}_4$ corresponds to an element of G. This correspondence can be stored in an array of size $2|G|$. To multiply two elements of $x, y \in G$ we can view these elements as gN and $g'N$ and perform the multiplication of g and g' using Theorem 8. The final result can then be obtained using the $2|G|$ sized array. ∎

---

[4] This isomorphism can be computed in the preprocessing phase.

Theorem 10. *Let $\mathcal{G}_{Strong}$ be a class of groups whose indecomposable factors are strongly indecomposable. The group class $\mathcal{G}_{Strong}$ has an $(\mathcal{O}(n), \mathcal{O}(1))$-data structure.*

Proof. This follows as an easy consequence of Lemma 4 and Lemma 2. ∎

5 Linear Space Representation of Finite Rings

A table representation (see e.g., [1,14,20]) of rings requires two tables; one stores the result of the multiplication operation and the second stores the result of the addition operation. There are two types of queries: given two elements $x, y \in R$ an addition query return $x + y$ whereas a multiplication query returns $x \times y$. In table representation these queries can be answered in constant time by accessing the addition table and the multiplication table[5]. In this section we show that rings has an $(\mathcal{O}(n), \mathcal{O}(1))$-data structure[6]. Since $(R, +)$ is an abelian group, Theorem 4 implies that there is an $(\mathcal{O}(n), \mathcal{O}(1))$-data structure for $(R, +)$. The table which stores the multiplication of the elements of the ring R does not follow the group axioms. Thus the techniques used for groups can not be applied to the multiplicative structure.

Theorem 11. *Let* **RING** *be a class of finite rings. Then* **RING** *has an $(\mathcal{O}(n), \mathcal{O}(1))$-data structure.*

Proof. Let $(R, +, \times)$ be a ring of size n. The additive structure $(R, +)$ is an abelian group. Let $R = R_1 \oplus R_2 \oplus \ldots \oplus R_k$ be a decomposition of R by using the structure theorem of abelian groups. The proof of the theorem is divided into two cases on the basis of the size of the factors R_i, $i \in [k]$.

Case 1: There exists an indecomposable factor R_1 of order n^{α_1} where $\alpha_1 \geq \frac{1}{2}$ for any $i \in [k]$. In this case the ring R can be decomposed as a direct sum of two factors R_1 and R_2 i.e. $R = R_1 \oplus R_2$ where the sizes of R_1 and R_2 are n^{α_1} and n^{α_2} for some $\alpha_1 \geq \frac{1}{2}$ and $\alpha_2 \leq \frac{1}{2}$. It is important to note that $R_1 = \langle g \rangle$ is a *cyclic* group generated by an element g (see Lemma 3). Also note that any element r of R_1 can be written as a $n_r \times g$ where n_r is a non-negative integer less than the additive order $ord(g)$ of g.

Data Structures: We store a forward-map array $\mathcal{F}$ indexed by the elements of R. A table $\mathcal{T}_{cross}$ of size $\mathcal{O}(|R_1||R_2|)$ which stores the multiplication of R_1 and R_2. We store a table $\mathcal{T}_2$ for additive group R_2 which stores the multiplication of the elements of R_2 with itself. Note that $R_1 = \langle g \rangle$. Let $C = \langle g \rangle$. We store a forward-map array $\mathcal{F}_1$ indexed by the elements of R_1 such that $\mathcal{F}_1[r_1] = n_{r_1}$ where for n_{r_1} is the unique integer satisfying the conditions $r_1 = n_{r_1} g$ and $0 \leq n_{r_1} < ord(g)$. Finally we need an array $\mathcal{A}$ indexed by integers $\{0, \ldots, m-1\}$ such that $\mathcal{A}[i] = ig^2$ for $0 \leq i \leq m - 1$.

---

[5] The class of finite rings **RING** has $(\mathcal{O}(n^2), \mathcal{O}(1))$-data structures.

[6] The notion of (s, t)-data structure can be easily generalized for any finite algebraic structure.

Query Processing: Let r and s are two elements of ring R. Using the forward-map array $\mathcal{F}$ at index r and s we obtain $\mathcal{F}[r] = (r_1, r_2)$ and $\mathcal{F}[s] = (s_1, s_2)$ where $r_1, s_1 \in R_1$ and $r_2, s_2 \in R_2$. This gives us sums $r = r_1 + r_2$ and $s = s_1 + s_2$ according to the decomposition $R = R_1 \oplus R_2$. The result of the multiplication of the r and s is $(r_1 + r_2) \times (s_1 + s_2) = r_1 \times s_1 + r_1 \times s_2 + r_2 \times s_1 + r_2 \times s_2$. To multiply r_1 and s_1, access array $\mathcal{F}_1$ at index r_1 and at index s_1. Let $\mathcal{F}_1[r_1] = n_{r_1}$ and $\mathcal{F}_1[s_1] = n_{s_1}$. The result of $r_1 \times s_1$ is $(n_{r_1} g) \times (n_{s_1} g) = n_{r_1} n_{s_1} g^2$. The multiplication $n_{r_1} n_{s_1}$ is a simple integer multiplication. Let $n_3 = n_{r_1} n_{s_1}$. We compute $n_4 = n_3 \% m$ and access array $\mathcal{A}$ at index n_4 to obtain the result of $n_4 g^2$. Access $\mathcal{T}_{cross}[r_2][s_1]$ and $\mathcal{T}_{cross}[r_1][s_2]$ to obtain the result of $r_2 \times s_1$ and $r_1 \times s_2$ respectively. To obtain the result of $r_2 \times s_2$ access $\mathcal{T}_2[r_2][s_2]$. Once we have the result of all four multiplications, additions operations can be done in $\mathcal{O}(1)$ time using Theorem 4.

Space Complexity: The forward-map arrays $\mathcal{F}$ and $\mathcal{F}_1$ takes $\mathcal{O}(n)$ space. The space used by the tables $\mathcal{T}_{cross}$, $\mathcal{T}_2$ and $\mathcal{A}$ is easily seen to be $\mathcal{O}(n)$. Thus overall space complexity is $\mathcal{O}(n)$.

Case 2: Each indecomposable factor R' of R has size n^α where $\alpha < \frac{1}{2}$. Such a ring R can be decomposed as a direct sum of three factors[7], $R = R_1 \oplus R_2 \oplus R_3$ such that size of each factor R_i is n^{α_i} where each $\alpha_i < \frac{1}{2}$. To see this, we imagine three buckets B_1, B_2, B_3 each of capacity $n^{1/2}$ and we try to pack these buckets with the indecomposable factors as tightly as possible starting with bucket B_1. Once we cannot accommodate any more indecomposable factor in the bucket B_1 we go to bucket B_2 and so on. Note that if $i_1, i_2, \ldots, i_r$ are the sizes of the indecomposable factors in a bucket, then these factors occupy $i_1 i_2 \ldots i_r$ space out of the total capacity of the bucket. Let R_i be the direct sum of all the indecomposable factors in the bucket B_i, $i \in \{1, 2, 3\}$. Since each bucket has capacity $n^{1/2}$, $|R_i| \leq n^{1/2}$. Recalling the fact that each indecomposable factor is of size at most $n^{1/2}$ we can see that $|R_1 \oplus R_2| \geq n^{1/2}$. Thus, the direct sum $R_1 \oplus R_2 \oplus R_3$ covers each indecomposable factor. We note that $R_3 = \{0\}$ is a valid possibility but this case can be handled easily.

Data Structures: Let $\mathcal{F}$ be a forward-map array indexed by the ring elements r of R such that $\mathcal{F}(r) = (r_1, r_2, r_3)$ where $r = r_1 + r_2 + r_3$ is a decomposition of r as described in Case 2, where $r_i \in R_i$ for $i \in [3]$. We store the data structure $\mathcal{T} = \{\mathcal{T}_{(i,j)} \mid \mathcal{T}_{(i,j)}[r][s] = r \times s, \forall r \in R_i, \forall s \in R_j$ and $\forall i, j \in [3]\}$. The data structure $\mathcal{T}$ is a collection of 9 tables which stores the result of the multiplication of the elements of R_i with R_j for all $i, j \in [3]$.

Query Processing: Given two elements r and s of a ring R. First we access the forward-map array $\mathcal{F}$ at index r and s. Let $\mathcal{F}[r] = (r_1, r_2, r_3)$ and $\mathcal{F}[s] = (s_1, s_2, s_3)$, where $r_i, s_i \in R_i$ for all $i \in [3]$. The multiplication of r and s is $(r_1 + r_2 + r_3) \times (s_1 + s_2 + s_3)$ which can be written as $\sum_{i,j \in [3]} r_i \times s_j$. We need to perform 9 multiplication operations of the form $r_i \times s_j$ where $r_i \in R_i$ and $s_j \in R_j$ and $i, j \in [3]$, which can be performed by accessing data structure $\mathcal{T}$

---

[7] These factors may or may not be indecomposable.

followed by 8 addition operations which can be performed by maintaining the data structures discussed in the Theorem 4 for the additive structure. Thus the overall time to process a multiplication query is $\mathcal{O}(1)$.

Space Complexity: The forward-map array $\mathcal{F}$ takes $\mathcal{O}(n)$ space. The data structure $\mathcal{T}$ is a collection of 9 tables each of them takes $\mathcal{O}(n)$ space as each R_i has size at most $n^{1/2}$. Thus, the space required by $\mathcal{T}$ is $\mathcal{O}(n)$. Thus the overall space required is $\mathcal{O}(n)$.

This completes the proof that **RING** has an $(\mathcal{O}(n), \mathcal{O}(1))$-data structure. ∎

References

1. Agrawal, M., Saxena, N.: Automorphisms of finite rings and applications to complexity of problems. In: Diekert, V., Durand, B. (eds.) STACS 2005. LNCS, vol. 3404, pp. 1–17. Springer, Heidelberg (2005). https://doi.org/10.1007/978-3-540-31856-9_1

2. Arvind, V., Torán, J.: The complexity of quasigroup isomorphism and the minimum generating set problem. In: Asano, T. (ed.) ISAAC 2006. LNCS, vol. 4288, pp. 233–242. Springer, Heidelberg (2006). https://doi.org/10.1007/11940128_25

3. Carmichael, R.D.: Introduction to the Theory of Groups of Finite Order. GINN and Company (1937)

4. Chen, L., Fu, B.: Linear and sublinear time algorithms for the basis of abelian groups. Theor. Comput. Sci. **412**, 4110–4122 (2011)

5. Conrad, K.: Generalized quaternions (2013). https://kconrad.math.uconn.edu/blurbs/grouptheory/genquat.pdf

6. Das, B., Sharma, S.: Nearly linear time isomorphism algorithms for some non-abelian group classes. In: van Bevern, R., Kucherov, G. (eds.) CSR 2019. LNCS, vol. 11532, pp. 80–92. Springer, Cham (2019). https://doi.org/10.1007/978-3-030-19955-5_8

7. Das, B., Sharma, S., Vaidyanathan, P.: Space efficient representations of finite groups. In: Journal of Computer and System Sciences (Special Issue on Fundamentals of Computation Theory FCT 2019), pp. 137–146 (2020)

8. Erdös, P., Rényi, A.: Probabilistic methods in group theory. J. d'Analyse Mathématique **14**(1), 127–138 (1965). https://doi.org/10.1007/BF02806383

9. Farzan, A., Munro, J.I.: Succinct representation of finite abelian groups. In: Proceedings of the 2006 International Symposium on Symbolic and Algebraic Computation, pp. 87–92. ACM (2006)

10. Hungerford, T.W.: Abstract Algebra: An Introduction. Cengage Learning (2012)

11. Karagiorgos, G., Poulakis, D.: Efficient algorithms for the basis of finite abelian groups. Discret. Math. Algorithms Appl. **3**(4), 537–552 (2011)

12. Kavitha, T.: Linear time algorithms for abelian group isomorphism and related problems. J. Comput. Syst. Sci. **73**, 986–996 (2007)

13. Kayal, N., Nezhmetdinov, T.: Factoring groups efficiently. In: Albers, S., Marchetti-Spaccamela, A., Matias, Y., Nikoletseas, S., Thomas, W. (eds.) ICALP 2009. LNCS, vol. 5555, pp. 585–596. Springer, Heidelberg (2009). https://doi.org/10.1007/978-3-642-02927-1_49

14. Kayal, N., Saxena, N.: Complexity of ring morphism problems. Comput. Complex. **15**(4), 342–390 (2006)

15. Kleitman, D.J., Rothschild, B.R., Spencer, J.H.: The number of semigroups of order n. Proc. Am. Math. Soc. **55**(1), 227–232 (1976)
16. Kumar, S.R., Rubinfeld, R.: Property testing of abelian group operations (1998)
17. Li, S., Liu, J.: On hall subnormally embedded and generalized nilpotent groups. J. Algebra **388**, 1–9 (2013)
18. van Lint, J.H., Wilson, R.M.: A Course in Combinatorics. Cambridge University Press (1992)
19. Marin, I.: Strongly indecomposable finite groups. Expositiones Mathematicae **26**(3), 261–267 (2008)
20. Saxena, N.: Morphisms of Rings and Applications to Complexity. Indian Institute of Technology Kanpur (2006)
21. Wilson, J.B.: Existence, algorithms, and asymptotics of direct product decompositions, I. Groups Complex. Cryptol. **4**, 33–72 (2012)

Competitive Location Problems: Balanced Facility Location and the One-Round Manhattan Voronoi Game

Thomas Byrne[1] , Sándor P. Fekete[2] , Jörg Kalcsics[1] ,
and Linda Kleist[2(✉)]

[1] School of Mathematics,
University of Edinburgh, Edinburgh, UK
{tbyrne,joerg.kalcsics}@ed.ac.uk
[2] Department of Computer Science,
TU Braunschweig, Braunschweig, Germany
{s.fekete,l.kleist}@tu-bs.de

Abstract. We study competitive location problems in a continuous setting, in which facilities have to be placed in a rectangular domain R of normalized dimensions of 1 and $\rho \geq 1$, and distances are measured according to the Manhattan metric. We show that the family of *balanced* configurations (in which the Voronoi cells of individual facilities are equalized with respect to geometric properties) is richer in this metric than for Euclidean distances. Our main result considers the *One-Round Voronoi Game* with Manhattan distances, in which first player White and then player Black each place n points in R; each player scores the area for which one of its facilities is closer than the facilities of the opponent. We give a tight characterization: White has a winning strategy if and only if $\rho \geq n$; for all other cases, we present a winning strategy for Black.

Keywords: Facility location · Competitive location · Manhattan distances · Voronoi game · Geometric optimization

1 Introduction

Problems of optimal location are arguably among the most important in a wide range of areas, such as economics, engineering, and biology, as well as in mathematics and computer science. In recent years, they have gained importance through clustering problems in artificial intelligence. In all scenarios, the task is to choose a set of positions from a given domain, such that some optimality criteria for the resulting distances to a set of demand points are satisfied; in a geometric setting, Euclidean or Manhattan distances are natural choices. Another challenge is that facility location problems often happen in a *competitive* setting, in which two or more players contend for the best locations.

A full version can be found at arXiv:2011.13275 [6].

R. Uehara et al. (Eds.): WALCOM 2021, LNCS 12635, pp. 103–115, 2021.
https://doi.org/10.1007/978-3-030-68211-8_9

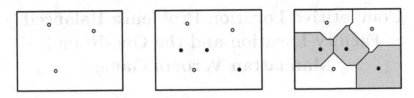

Fig. 1. Example of a one-round Manhattan Voronoi game: (Left) White places 3 points; (Middle) Black places 3 points; (Right) the dominated areas.

This change to competitive, multi-player versions can have a serious impact on the algorithmic difficulty of optimization problems: e.g., the classic Travelling Salesman Problem is NP-hard, while the competitive two-player variant is even PSPACE-complete [10].

In this paper, we consider problems of facility location under Manhattan distances; while frequently studied in location theory and applications (e.g., see [15,16,19]), they have received limited attention in a setting in which facilities compete for customers. We study a natural scenario in which facilities have to be chosen in a rectangle R of normalized dimensions with height 1 and width $\rho \geq 1$. A facility dominates the set of points for which it is strictly closer than any other facility, i.e., the respective (open) Voronoi cell, subject to the applicable metric. While for Euclidean distances a bisector (the set of points that are of equal distance from two facilities) is the boundary of the open Voronoi cells, so its area is zero, Manhattan bisectors may have positive area, as shown in Fig. 2. As we show below, accounting for fairness and local optimality, we consider *balanced* configurations for which the respective Voronoi cells are equalized.

Exploiting the geometric nature of Voronoi cells, we completely resolve a classic problem of competitive location theory for the previously open case of Manhattan distances. In the *One-Round Voronoi Game*, first player *White* and then player *Black* each place n points in R. Each player scores the area consisting of the points that are closer to one of their facilities than to any one of the opponent's; see Fig. 1 for an example. The goal for each player is to obtain the higher score. Owing to the different nature of the Manhattan metric, both players may dominate strictly less than $\rho/2$, the remaining area belonging to *neutral zones*.

1.1 Related Work

Problems of location are crucial in economics, optimization, and geometry; see the classic book of Drezner [8] with over 1200 citations, or the more recent book by Laporte et al. [17]. Many applications arise from multi-dimensional data sets with heterogeneous dimensions, so the Manhattan metric (which compares coordinate distances separately) is a compelling choice. The ensuing problems have also received algorithmic attention. Fekete et al. [12] provide several algorithmic results, including an NP-hardness proof for the k-median problem of minimizing the average distance. Based on finding an optimal location for an additional facility in a convex region with n existing facilities, Averbakh et al. [3] derive exact algorithms for a variety of conditional facility location problems.

An important scenario for competitive facility location is the *Voronoi game*, first introduced by Ahn et al. [1], in which two players take turns placing one facility a time. In the end, each player scores the total area of all of their Voronoi regions. As Teramoto et al. [18] showed, the problem is PSPACE-complete, even in a discrete graph setting.

Special attention has been paid to the *One-Round Voronoi Game*, in which each player places their n facilities at once. Cheong et al. [7] showed that for Euclidean distances in the plane, White can always win for a one-dimensional region, while Black has a winning strategy if the region is a square and n is sufficiently large. Fekete and Meijer [11] refined this by showing that in a rectangle of dimensions $1 \times \rho$ with $\rho \geq 1$, Black has a winning strategy for $n \geq 3$ and $\rho < n/\sqrt{2}$, and for $n = 2$ and $\rho < 2/\sqrt{3}$; White wins in all other cases. In this paper, we give a complementary characterization for the case of Manhattan distances; because of the different geometry, this requires several additional tools.

There is a considerable amount of other work on variants of the Voronoi game. Bandyapadhyay et al. [4] consider the one-round game in trees, providing a polynomial-time algorithm for the second player. As Fekete and Meijer [11] have shown, the problem is NP-hard for polygons with holes, corresponding to a planar graph with cycles. For a spectrum of other variants and results, see [5,9,13,14]. For an overview of work on Voronoi diagrams, we refer to the surveys by Aurenhammer and Klein [2].

1.2   Main Results

Our main results are twofold. Firstly, we show that for location problems with Manhattan distances in the plane, the properties of *fairness* and *local optimality* lead to a geometric condition called *balancedness*. While the analogue concept for Euclidean distances in a rectangle implies grid configurations [11], we demonstrate that there are balanced configurations of much greater variety.

Secondly, we give a full characterization of the One-Round Manhattan Voronoi Game in a rectangle R with aspect ratio $\rho \geq 1$. We show that White has a winning strategy if and only if $\rho \geq n$; for all other cases, Black has a winning strategy.

2   Preliminaries

Let P denote a finite set of points in a rectangle R. For two points $p_1 = (x_1, y_1)$ and $p_2 = (x_2, y_2)$, we define $\Delta_x(p_1, p_2) := |x_1 - x_2|$ and $\Delta_y(p_1, p_2) := |y_1 - y_2|$. Then their Manhattan distance is given by $d_M(p_1, p_2) := \Delta_x(p_1, p_2) + \Delta_y(p_1, p_2)$. Defining $D(p_1, p_2) := \{p \in R \mid d_M(p, p_1) < d_M(p, p_2)\}$ as a set of points that are closer to p_1 than to p_2, the *Voronoi cell* of p in P is $V^P(p) := \bigcap_{q \in P \setminus \{p\}} D(p, q)$. The *Manhattan Voronoi diagram* $\mathcal{V}(P)$ is the complement of the union of all

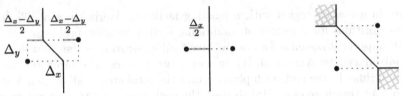

(a) General vertical bisector.   (b) Case: $\Delta_y(p_1, p_2) = 0$.   (c) Degenerate bisector.

Fig. 2. Illustration of the three types of bisectors.

Voronoi cells of P. The *bisector* of p_1 and p_2 is the set of all points that are of equal distance to p_1 and p_2, i.e., $\mathcal{B}(p_1, p_2) := \{q \in R \mid d_M(q, p_1) = d_M(q, p_2)\}$. There are three types of bisectors, as shown in Fig. 2. Typically, a bisector consists of three one-dimensional parts, namely two (vertical or horizontal) segments that are connected by a segment of slope ± 1. If $\Delta_x(p_1, p_2) = 0$ or $\Delta_y(p_1, p_2) = 0$, then the diagonal segment shrinks to a point and the bisector consists of a (vertical or horizontal) segment. However, when $\Delta_x(p_1, p_2) = \Delta_y(p_1, p_2)$, then the bisector $\mathcal{B}(p_1, p_2)$ contains two regions and is called *degenerate*. Further, a non-degenerate bisector is *vertical (horizontal)* if it contains vertical (horizontal) segments.

For $p = (x_p, y_p) \in P$, both the vertical line $\ell_v(p)$ and the horizontal line $\ell_h(p)$ through p split the Voronoi cell $V^P(p)$ into two pieces, which we call *half cells*. We denote the set of all half cells of P obtained by vertical lines by $\mathcal{H}^|$ and those obtained by horizontal lines by $\mathcal{H}^-$. Furthermore, we define $\mathcal{H} := \mathcal{H}^| \cup \mathcal{H}^-$ as the set of all half cells of P. Applying both $\ell_v(p)$ and $\ell_h(p)$ to p yields a subdivision into four quadrants, which we denote by $Q_i(p)$, $i \in \{1, \ldots, 4\}$; see Fig. 3(a). Moreover, $C_i(p) := V^P(p) \cap Q_i(p)$ is called the ith *quarter cell* of p. We also consider the eight regions of every $p \in P$ obtained by cutting R along the lines $\ell_v(p)$, $\ell_h(p)$, and the two diagonal lines of slope ± 1 through p. We refer to each such (open) region as an *octant* of p denoted by $O_i(p)$ for $i \in \{1, \ldots, 8\}$ as illustrated in Fig. 3(b); a closed octant is denoted by $\overline{O}_i(p)$. The area of a subset S of R is denoted by $\mathcal{A}(S)$.

For a point $p \in P$, we call the four horizontal and vertical rays rooted at p, contained with $V^P(p)$, the four *arms* of $V^P(p)$ (or of p). Two arms are *neighbouring* if they appear consecutively in the cyclic order; otherwise they are *opposite*. Moreover, we say an arm is a *boundary arm* if its end point touches the boundary of R; otherwise it is *inner*. For later reference, we note the following.

Observation 1. *The following properties hold:*

(i) *If the bisector $\mathcal{B}(p, q)$ is non-degenerate and vertical (horizontal), then it does not intersect both the left and right (top and bottom) half cells of p.*

(ii) *For every i and every $q_1, q_2 \in O_i(p)$, the bisectors $\mathcal{B}(p, q_1)$ and $\mathcal{B}(p, q_2)$ have the same type (vertical/horizontal).*

(iii) *A Voronoi cell is contained in the axis-aligned rectangle spanned by its arms.*

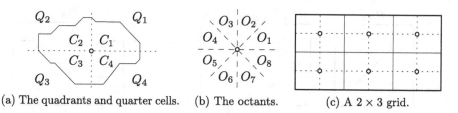

(a) The quadrants and quarter cells.   (b) The octants.   (c) A 2 × 3 grid.

Fig. 3. Illustration of crucial definitions.

3 Balanced Point Sets

In a competitive setting for facility location, it is a natural *fairness property* to allocate the same amount of influence to each facility. A second *local optimality property* arises from choosing an efficient location for a facility within its individual Voronoi cell. Combining both properties, we say a point set P in a rectangle R is *balanced* if the following two conditions are satisfied:

- **Fairness:** for all $p_1, p_2 \in P$, $V^P(p_1)$ and $V^P(p_2)$ have the same area.
- **Local optimality:** for all $p \in P$, p minimizes the average distance to the points in $V^P(p)$.

For Manhattan distances, there is a simple geometric characterization for the local optimality depending on the area of the half and quarter cells; see Fig. 3(a).

Lemma 2. *A point p minimizes the average Manhattan distance to the points in $V^P(p)$ if and only if either of the following properties holds:*

(i) p is a Manhattan median of $V^P(p)$: all four half cells of $V^P(p)$ have the same area.

(ii) p satisfies the quarter-cell property: diagonally opposite quarter cells of $V^P(p)$ have the same area.

The proof uses straightforward local optimality considerations; see full version for details. Lemma 2 immediately implies the following characterization.

Corollary 3. *A point set P in a rectangle R is balanced if and only if all half cells of P have the same area.*

A simple family of balanced sets arise from regular, $a \times b$ grids; see Fig. 3(c). In contrast to the Euclidean case, there exist a large variety of other balanced sets: Fig. 4 depicts balanced point sets for which *no* cell is a rectangle.

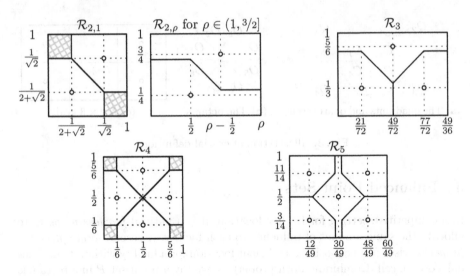

Fig. 4. Non-grid examples of balanced point sets of cardinality 2, 3, 4, and 5.

Lemma 4. *The configurations $\mathcal{R}_{2,\rho}, \mathcal{R}_3, \mathcal{R}_4, \mathcal{R}_5$, depicted in Fig. 4, are balanced. Moreover, $\mathcal{R}_{2,\rho}, \rho \in [1, 3/2]$, and $\mathcal{R}_3$ are the only balanced non-grid point sets with two and three points, respectively.*

Simple calculations show that the configurations are balanced. In order to prove the uniqueness, we make use of Lemma 2. While the analysis for $n = 2$ can be easily conducted manually, for $n = 3$, the relative point positions lead to about 20 cases of structurally different Voronoi diagrams, which were checked using MATLAB®; for more details see full version [6].

Observe that $\mathcal{R}_{2,\rho}, \mathcal{R}_3, \mathcal{R}_4$, and $\mathcal{R}_5$ are *atomic*, i.e., they cannot be decomposed into subconfigurations whose union of Voronoi cells is a rectangle. We show how they serve as building blocks to induce large families of balanced configurations.

Theorem 5. *For every n, $n \neq 7$, there exists a rectangle R and a set P of n points such that P is balanced and no Voronoi cell is a rectangle.*

Proof. For every $n = 3k + 5\ell$ with $k, \ell \in \{0, 1, \ldots\}$, we construct a configuration by combining k blocks of $\mathcal{R}_3$ and ℓ blocks of $\mathcal{R}_5$, as shown in Fig. 5. This yields configurations with $n = 3k$ for $k \geq 1$, $n = 3k + 2 = 3(k-1) + 5$ for $k \geq 1$, or $n = 3k+1 = 3(k-3)+10$ for $k \geq 3$, so we obtain configurations for all $n \geq 8$ and $n = 3, 5, 6$. Configurations with $n = 2k$ ($k \in \mathbb{N}$) points are obtained by combining k blocks of $\mathcal{R}_2$ as shown in Fig. 5; alternatively, recall the configurations in Fig. 4.

While none of the configurations in Theorem 5 contains a rectangular Voronoi cell, they contain many immediate repetitions of the same atomic components. In fact, there are arbitrarily large *non-repetitive* balanced configurations without directly adjacent congruent atomic subconfigurations.

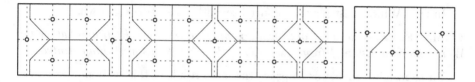

Fig. 5. Illustration of the proof of Theorem 5. (Left) Combining k blocks of $\mathcal{R}_3$ and l blocks of $\mathcal{R}_5$ for a configuration with $n = 3k + 5l$ points in a rectangle R with $\rho(R) = 1/49(36k + 60l)$. (Right) Combining k blocks of $\mathcal{R}_2$ for $n = 2k$ points.

Theorem 6. *There is a injection between the family of 0–1 strings and a family of non-repetitive balanced configurations without any rectangular Voronoi cells.*

Proof. For a given 0–1 string $\mathcal{S}$ of length s, we use s pairs of blocks $\mathcal{R}_3$ and its reflected version $\mathcal{R}'_3$ to build a sequence of $2s$ blocks. We insert a block $\mathcal{R}_5$ after the ith pair if $\mathcal{S}$ has a 1 in position i; otherwise the block sequence remains.

4   The Manhattan Voronoi Game

An instance of the One-Round Manhattan Voronoi Game consists of a rectangle R and the number n of points to be played by each player. Without loss of generality, R has height 1 and width $\rho \geq 1$. Player White chooses a set W of n white points in R, followed by the player Black selecting a set B of n black points, with $W \cap B = \varnothing$. Each player scores the area consisting of the points that are closer to one of their facilities than to any one of the opponent's. Hence, if two points of one player share a degenerate bisector, the possible neutral regions are assigned to this player. Therefore, by replacing each degenerate bisector between points of one player by a (w.l.o.g. horizontal) non-degenerate bisector, each player scores the area of its (horizontally enlarged) Manhattan Voronoi cells. With slight abuse of notation, we denote the resulting (horizontally enlarged) Voronoi cells of colored point sets by $V^{W \cup B}(p)$ similar as before. The player with the higher score wins, or the game ends in a tie.

For an instance (R, n) and a set W of white points, a set B of n black points is a *winning set* for Black if Black wins the game by playing B; likewise, B is a *tie set* if the game ends in a ti.e. A black point b is a *winning point* if its cell area $\mathcal{A}(V^{W \cup B}(b))$ exceeds $1/2n \cdot \mathcal{A}(R)$. A white point set W is *unbeatable* if it does not admit a winning set for Black, and W is a *winning set* if there exists neither a tie nor a winning set for Black. If Black or White can always identify a winning set, we say they have a *winning strategy*.

Despite the possible existence of degenerate bisectors for Manhattan distances, we show that Black has a winning strategy if and only if Black has a winning point. We make use of the following two lemmas, proved in full version [6].

Lemma 7. *Consider a rectangle R with a set W of white points. Then for every $\varepsilon > 0$ and every half cell H of W, Black can place a point b such that the area of $V^{W \cup \{b\}}(b) \cap H$ is at least $(\mathcal{A}(H) - \varepsilon)$.*

In fact, White must play a balanced set; otherwise Black can win.

Lemma 8. *Let W be a set of n white points in a rectangle R. If any half cell of W has an area different from $1/2n \cdot \mathcal{A}(R)$, then Black has a winning strategy.*

These insights enable us to prove the main result of this section.

Theorem 9. *Black has a winning strategy for a set W of n white points in a rectangle R if and only if Black has a winning point.*

Proof. If Black wins and their score exceeds $1/2 \cdot \mathcal{A}(R)$ then, by the pigeonhole principle, a black cell area exceeds $1/2n \cdot \mathcal{A}(R)$, confirming a winning point. Otherwise, Black wins by forcing neutral zones. By allowing a net loss of an arbitrarily small $\varepsilon > 0$, Black can slightly perturb their point set to obtain a winning set without neutral zones; for details consider full version [6].

Now suppose that there exists a winning point b, i.e., $\mathcal{A}(V^{W \cup \{b\}}(b)) = 1/2n \cdot \mathcal{A}(R) + \delta$ for some $\delta > 0$. If $n = 1$, Black clearly wins with b. If $n \geq 2$, Black places $n - 1$ further black points: consider $w_i \in W$. By Lemma 8, we may assume that each half cell of W has area $1/2n \cdot \mathcal{A}(R)$. By Observation 1(i), w_i has a half cell H_i that is disjoint from $V^{W \cup \{b\}}(b)$. By Lemma 7, Black can place a point b_i to capture the area of H_i up to every $\varepsilon > 0$. Choosing $\varepsilon < \delta/n-1$ and placing one black point for $n - 1$ distinct white points with Lemma 7, Black achieves a score of $\sum_{p \in B} \mathcal{A}(V^{W \cup B}(p)) = (1/2n \cdot \mathcal{A}(R) + \delta) + (n - 1)(1/2n \cdot \mathcal{A}(R) - \varepsilon) > 1/2\mathcal{A}(R)$.

5   Properties of Unbeatable Winning Sets

In this section, we identify necessary properties of *unbeatable* white sets, for which the game ends in a tie or White wins. We call a cell a *bridge* if it has two opposite boundary arms.

Theorem 10. *If W is an unbeatable white point set in a rectangle R, then it fulfills the following properties:*

(P1) The area of every half cell of W is $1/2n \cdot \mathcal{A}(R)$.
(P2) The arms of a non-bridge cell are equally long; the opposite boundary arms of a bridge cell are of equal length and, if $|W| > 1$, they are shortest among all arms.

Proof. Because W is unbeatable, property (P1) follows immediately from Lemma 8. Moreover, in case $|W| = 1$, (P1) implies that opposite arms of the unique (bridge) cell have equal length, i.e., (P2) holds for $|W| = 1$.

It remains to prove property (P2) for $|W| \geq 2$. By Theorem 9, it suffices to identify a black winning point if (P2) is violated. We start with an observation.

Claim 11. *Let P be a point set containing $p = (0,0)$ and let P' be obtained from P by adding $p' = (\delta, \delta)$ where $\delta > 0$ such that p' lies within $V^P(p)$. Restricted to $Q := Q_1(p')$, the cell $V^{P'}(p')$ contains all points that are obtained when the boundary of $V^P(p) \cap Q$ is shifted upwards (rightwards) by δ (if it does not intersect the boundary of R).*

To prove this claim, it suffices to consider the individual bisectors of p' and any other point $q \in P'$. Note that all points shaping the first quarter cell of p' are contained in an octant $O_i(p')$ with $i \in \{1, 2, 3, 8\}$. Observe that vertical bisectors move rightwards and horizontal bisectors move upwards by δ; see also Fig. 6.

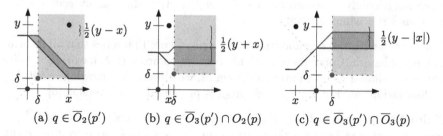

(a) $q \in \overline{O}_2(p')$ (b) $q \in \overline{O}_3(p') \cap O_2(p)$ (c) $q \in \overline{O}_3(p') \cap \overline{O}_3(p)$

Fig. 6. Illustration of Claim 11. If $q = (x, y)$ lies in $\overline{O}_2(p') \cup \overline{O}_3(p')$, the part of the bisector $\mathcal{B}(q, p')$ within the first quadrant Q of p' coincides with $\mathcal{B}(q, p) \cap Q$ shifted upwards by δ.

We use our insight of Claim 11 to show property (P2) in two steps.

Claim 12. *Let $w \in W$ be a point such that an arm A_1 of $V^W(w)$ is shorter than a neighbouring arm A_2 and the arm A_3 opposite to A_1 is inner. Then Black has a winning point.*

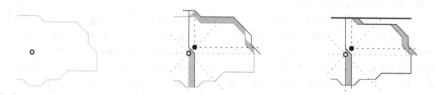

Fig. 7. Illustration of Claim 12 and Claim 13: the gain and loss of $V(b)$ compared to H. (Left) The right half cell H. (Middle) Case: The bottom arm A_1 is shorter than the right arm A_2 and the top arm A_3 is inner. (Right) Case: The bottom arm A_1 is shorter than the top arm A_3 and the right arm A_2 is inner.

Proof. Without loss of generality, we consider the case that A_1 is the bottom arm of $V^W(w)$, A_2 its right arm, and $w = (0, 0)$; see Fig. 7 (Middle). We denote the length of A_i by $|A_i|$. Now we consider Black placing a point b within $V^W(w)$ at (δ, δ) for some $\delta > 0$. To ensure that the cell of b contains almost all of the right half cell of $V^W(w)$, we infinitesimally perturb b rightwards; for ease of notation in the following analysis, we omit the corresponding infinitesimal terms and assume that the bisector of b and w is vertical. We compare the area of $V(b) := V^{W+b}(b)$ with the right half cell H of w. In particular, we show that there exists $\delta > 0$ such that the area of $V(b)$ exceeds the area of H. Because $\mathcal{A}(H) = 1/2n \cdot \mathcal{A}(R)$ by (P1), b is a winning point.

Clearly, all points in H to the right of the (vertical) bisector of b and w are closer to b. Consequently, when compared to H, the loss of $V(b)$ is upper bounded by $\delta|A_1| + 1/2\delta^2$; see also Fig. 7 (middle). By Claim 11 and the fact that A_3 is inner, $V(b) \cap Q_1(p)$ gains at least $\delta(|A_2| - \delta)$ when compared to $H \cap Q_1(p)$. When additionally guaranteeing $\delta < 2/3(|A_2| - |A_1|)$, the gain exceeds the loss and thus b is a winning point.

For a cell with two neighbouring inner arms, Claim 12 implies that all its arms have equal length. Consequently, it only remains to prove (P2) for bridges. With arguments similar to those proving Claim 12, we obtain the following result. For an illustration, see Fig. 7 (right). A proof is presented in full version [6].

Claim 13. *If there exists a point $w \in W$ such that two opposite arms of $V^W(w)$ have different lengths and a third arm is inner, then Black has a winning point.*

If $|W| > 1$, every cell has at least one inner arm. Therefore Claim 13 yields that opposite boundary arms of a bridge cell have equal length. Moreover, Claim 12 implies that the remaining arms are not shorter. This proves (P2) for bridges.

We now show that unbeatable white sets are grids; in some cases they are even *square grids*, i.e., every cell is a square.

Lemma 14. *Let P be a set of n points in a $(1 \times \rho)$ rectangle R with $\rho \geq 1$ fulfilling properties (P1) and (P2). Then P is a grid. More precisely, if $\rho \geq n$, then P is a $1 \times n$ grid; otherwise, P is a square grid.*

Proof. We distinguish two cases.

Case: $\rho \geq n$. By (P1), every half cell has area $\frac{1}{2n}\mathcal{A}(R) = \frac{1}{2n}\rho \geq \frac{1}{2}$. Since the height of every half cell is bounded by 1, every left and right arm has a length of at least $1/2$. Then, property (P2) implies that each top and bottom arm has length $1/2$, i.e., every $p \in P$ is placed on the horizontal centre line of R. Finally, again by (P1), the points must be evenly spread. Hence, P is a $1 \times n$ grid.

Case: $\rho < n$. We consider the point p whose cell $V^P(p)$ contains the top left corner of R and denote its quarter cells by C_i. Then, C_2 is a rectangle. Moreover, $V^P(p)$ is not a bridge; otherwise its left half cell has area $\geq \frac{1}{2} > \frac{1}{2n}\rho = \frac{1}{2n}\mathcal{A}(R)$. Therefore, by (P2), all arms of $V^P(p)$ have the same length; we denote this length by d. Together with the fact that C_2 and C_4 have the same area by (P1) and Lemma 2, it follows that C_2 and C_4 are squares of side length d.

We consider the right boundary of C_4. Since the right arm of $V^P(p)$ has length d (and the boundary continues vertically below), some point q has distance $2d$ to p and lies in $Q_1(p)$. The set of all these possible point locations forms a segment, which is highlighted in red in Fig. 8(a). Consequently, the left arm of q has length d. By (P2), the top arm of q must also have length d. Hence, q lies at the grid location illustrated in Fig. 8(b). Moreover, it follows that q is the unique point whose cell shares part of the boundary with C_1; otherwise the top arm of q does not have length d.

By symmetry, a point q' lies at a distance $2d$ below p and distance d to the boundary. Thus, every quarter cell of $V^P(p)$ is a square of size d; thus, the arms of all cells have length at least d. Moreover, the top left quarter cells of $V^P(q)$ and $V^P(q')$ are squares, so their bottom right quadrants must be as well. Using this argument iteratively along the boundary implies that boundary cells are squares. Applying it to the remaining rectangular hole shows that P is a square grid.

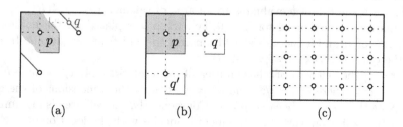

(a)   (b)   (c)

Fig. 8. Illustration of the proof of Lemma 14. (Color figure online)

We now come to our main result.

Theorem 15. *White has a winning strategy for placing n points in a $(1 \times \rho)$ rectangle with $\rho \geq 1$ if and only if $\rho \geq n$; otherwise Black has a winning strategy. Moreover, if $\rho \geq n$, the unique winning strategy for White is to place a $1 \times n$ grid.*

Proof. First we show that Black has a winning strategy if $\rho < n$. Suppose that Black cannot win. Note that $\rho < n$ implies $n \geq 2$. Consequently, by Theorem 10 and Lemma 14, the white point set W is a square $a \times b$ grid with $a, b \geq 2$, and thus the four cells in the top left corner induce a 2×2 grid. By Theorem 9, it suffices to identify a winning point for Black. Thus, we show that Black has a winning point in a square 2×2 grid: suppose the arms of all cells have length d. Then a black point p is a winning point if its cell has an area exceeding $2d^2$. With p placed at a distance $3d/2$ from the top and left boundary as depicted in Fig. 9, the cell of p has an area of $2d^2 + d^2/4$.

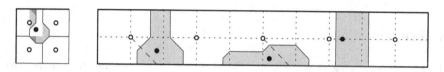

Fig. 9. Illustration of the proof of Theorem 15. (Left) A black winning point in a 2×2 grid. (Right) Every black cell has an area $\leq 1/2n \cdot \mathcal{A}(R)$. Moreover, only $n - 1$ locations result in cells of that size.

Secondly, we consider the case $\rho \geq n$ and show that White has a winning strategy. Theorem 10 and Theorem 14 imply that White must place its points in a $1 \times n$ grid; otherwise Black can win. We show that Black has no option to beat this placement; i.e., if $\rho \geq n$, then Black has no winning point and cannot force a tie in a $1 \times n$ grid.

By symmetry, there essentially exist two different placements of a black point b with respect to a closest white point w_b. Without loss of generality, we assume that w_b is to the left and not below b. Let x and y denote the horizontal and vertical distance of b to w_b, respectively. For a unified presentation, we add half of potential neutral zones in case $x = y$ to the area of the black cell. As a consequence, Black loses if its cells have an area of less than $1/2 \cdot \mathcal{A}(R)$.

If $x > y$, the cell of b evaluates to an area of (at most) $1/2n \cdot \mathcal{A}(R) - y^2$. In particular, it is maximized for $y = 0$, i.e., when b is placed on the horizontal centre line of R and if there exist white points to the left and right of b. In this case the cell area is exactly $1/2n \cdot \mathcal{A}(R)$.

If $x \leq y$, the cell area of b has an area of (at most) $1/2n \cdot \mathcal{A}(R) - y(w' - h') - 1/4(3y^2 + x^2)$, where $w' := w/2n$ and $h' := h/2$ denote the dimensions of the grid cells. Note that $w' \geq h'$ because $\rho \geq n$. Consequently, the cell area is maximized for $x = 0, y = 0$. However, this placement coincides with the location of a white point and is thus forbidden. Therefore every valid placement results in a cell area strictly smaller than $1/2n \cdot \mathcal{A}(R)$. Consequently, Black has no winning point.

Note that the cell area is indeed strictly smaller than the abovementioned maximum values if the black point does not have white points on both sides. Therefore the (unique) best placement of a black point is on the centre line between two white points, as illustrated by the rightmost black point in Fig. 9. However, there exist only $n-1$ distinct positions of this type; all other placements result in strictly smaller cells. Consequently, Black cannot force a tie and so loses. This completes the proof.

6 Open Problems

There exist many directions for future work. Are there further atomic configurations? Is it possible to combine them into intricate two-dimensional patterns? The biggest challenge is a full characterization of balanced configurations, with further generalizations to other metrics and dimensions. Moreover, the multi-round variant of the Voronoi game is wide open, as are higher-dimensional variants.

References

1. Ahn, H.K., Cheng, S.W., Cheong, O., Golin, M., van Oostrum, R.: Competitive facility location: the Voronoi game. Theoret. Comput. Sci. **310**, 357–372 (2004)
2. Aurenhammer, F., Klein, R.: Voronoi diagrams. In: Handbook of Computational Geometry, vol. 5, no. 10, pp. 201–290 (2000)
3. Averbakh, I., Berman, O., Kalcsics, J., Krass, D.: Structural properties of Voronoi diagrams in facility location problems with continuous demand. Oper. Res. **62**(2), 394–411 (2015)

4. Bandyapadhyay, S., Banik, A., Das, S., Sarkar, H.: Voronoi game on graphs. Theor. Comput. Sci. **562**, 270–282 (2015)
5. Banik, A., Bhattacharya, B.B., Das, S., Mukherjee, S.: One-round discrete Voronoi game in $\mathbb{R}^2$ in presence of existing facilities. In: CCCG (2013)
6. Byrne, T., Fekete, S.P., Kalcsics, J., Kleist, L.: Competitive facility location: balanced facility location and the one-round Manhattan Voronoi game. arXiv:2011.13275 (2020)
7. Cheong, O., Har-Peled, S., Linial, N., Matousek, J.: The one-round Voronoi game. Discrete Comput. Geom. **31**(1), 125–138 (2004). https://doi.org/10.1007/s00454-003-2951-4
8. Drezner, Z.: Facility Location: A Survey of Applications and Methods. Springer Series in Operations Research. Springer, New York (1995)
9. Dürr, C., Thang, N.K.: Nash equilibria in Voronoi games on graphs. In: Arge, L., Hoffmann, M., Welzl, E. (eds.) ESA 2007. LNCS, vol. 4698, pp. 17–28. Springer, Heidelberg (2007). https://doi.org/10.1007/978-3-540-75520-3_4
10. Fekete, S.P., Fleischer, R., Fraenkel, A.S., Schmitt, M.: Traveling salesmen in the presence of competition. Theoret. Comput. Sci. **313**(3), 377–392 (2004)
11. Fekete, S.P., Meijer, H.: The one-round Voronoi game replayed. Comput. Geom. Theory Appl. **30**(2), 81–94 (2005)
12. Fekete, S.P., Mitchell, J.S.B., Beurer, K.: On the continuous Fermat-Weber problems. Oper. Res. **53**, 61–76 (2005)
13. Gerbner, D., Mészáros, V., Pálvölgyi, D., Pokrovskiy, A., Rote, G.: Advantage in the discrete Voronoi game. arXiv preprint:1303.0523 (2013)
14. Kiyomi, M., Saitoh, T., Uehara, R.: Voronoi game on a path. IEICE Trans. Inf. Syst. **94**(6), 1185–1189 (2011)
15. Kolen, A.: Equivalence between the direct search approach and the cut approach to the rectilinear distance location problem. Oper. Res. **29**(3), 616–620 (1981)
16. Kusakari, Y., Nishizeki, T.: An algorithm for finding a region with the minimum total $L1$ from prescribed terminals. In: Leong, H.W., Imai, H., Jain, S. (eds.) ISAAC 1997. LNCS, vol. 1350, pp. 324–333. Springer, Heidelberg (1997). https://doi.org/10.1007/3-540-63890-3_35
17. Laporte, G., Nickel, S., Saldanha-da-Gama, F.: Introduction to location science. In: Laporte, G., Nickel, S., Saldanha da Gama, F. (eds.) Location Science, pp. 1–21. Springer, Cham (2019). https://doi.org/10.1007/978-3-030-32177-2_1
18. Teramoto, S., Demaine, E.D., Uehara, R.: Voronoi game on graphs and its complexity. In: CIG, pp. 265–271 (2006)
19. Wesolowsky, G.O., Love, R.F.: Location of facilities with rectangular distances among point and area destinations. Nav. Res. Log. Q. **18**, 83–90 (1971)

Faster Multi-sided One-Bend Boundary Labelling

Prosenjit Bose[1], Saeed Mehrabi[2(✉)], and Debajyoti Mondal[3]

[1] School of Computer Science,
Carleton University, Ottawa, Canada
jit@scs.carleton.ca
[2] Computer Science Department,
Memorial University, St. John's, Canada
smehrabi@mun.ca
[3] Department of Computer Science,
University of Saskatchewan, Saskatoon, Canada
d.mondal@usask.ca

Abstract. A *1-bend boundary labelling* problem consists of an axis-aligned rectangle B, n points (called *sites*) in the interior, and n points (called *ports*) on the labels along the boundary of B. The goal is to find a set of n axis-aligned curves (called *leaders*), each having at most one bend and connecting one site to one port, such that the leaders are pairwise disjoint. A 1-bend boundary labelling problem is k-sided ($1 \le k \le 4$) if the ports appear on k different sides of B. Kindermann et al. [Algorithmica, 76(1): 225–258, 2016] showed that the 1-bend 4-sided and 3-sided boundary labelling problems can be solved in $O(n^9)$ and $O(n^4)$ time, respectively. Bose et al. [SWAT, 12:1–12:14, 2018] improved the former running time to $O(n^6)$ by reducing the problem to computing maximum independent set in an outerstring graph. In this paper, we improve both previous results by giving new algorithms with running times $O(n^5)$ and $O(n^3 \log n)$ to solve the 1-bend 4-sided and 3-sided boundary labelling problems, respectively.

1 Introduction

Map labelling is a well-known problem in cartography with applications in educational diagrams, system manuals and scientific visualization. Traditional map labelling that places the labels on the map such that each label is incident to its corresponding feature, creates overlap between labels if the features are densely located on the map. This motivated the use of *leaders* [8,11]: line segments that connect features to their labels. As a formal investigation of this approach, Bekos et al. [3] introduced *boundary labelling*: all the labels are required to be placed on the boundary of the map and to be connected to their features using leaders.

The work is partially supported by the Natural Sciences and Engineering Research Council of Canada (NSERC).

R. Uehara et al. (Eds.): WALCOM 2021, LNCS 12635, pp. 116–128, 2021.
https://doi.org/10.1007/978-3-030-68211-8_10

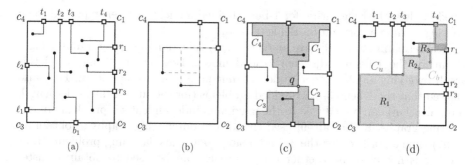

Fig. 1. (a) A 1-bend 4-sided boundary labelling problem with $w = 4, x = 3, y = 1$ and $z = 2$, (b) an instance of the 1-bend 2-sided boundary labelling problem with no planar solution with 1-bend leaders, and (c) a partitioned solution. (d) The curves C_u and C_b as well as the sequence of empty rectangles.

The point where a leader touches the label is called a *port*. Their work initiated a line of research in developing labelling algorithms with different labelling aesthetics, such as minimizing leader crossings, number of bends per leader and sum of leader lengths [1,2,4,6].

In this paper, we study the *1-bend k-sided boundary labelling* problem. The input is an axis-aligned rectangle B and a set of n points (called *sites*) in the interior of B. In addition, the input contains a set of n points on the boundary of B representing the ports on k consecutive sides of B for some $1 \leq k \leq 4$. The objective is to decide whether each site can be connected to a unique port using an axis-aligned leader with at most 1 bend such that the leaders are disjoint and each leader lies entirely in the interior of B, except the endpoint that is attached to a port. Figure 1(a) illustrates a labelling for a 1-bend k-sided boundary labelling instance. If such a solution exists, then we call it a *feasible* solution and say that the problem is *solvable*. Notice that not every instance of the boundary labelling problem is solvable; see Fig. 1(b).

Related work. The boundary labelling problem was first formulated by Bekos et al. (Graph Drawing 2004, see also [3]), who solved the 1-bend 1-sided and 1-bend 2-sided models (when the ports lie on two opposite sides of R) in $O(n \log n)$ time. They also gave an $O(n \log n)$-time algorithm for the 2-bend 4-sided boundary labelling problem (i.e., when each leader can have at most 2 bends).

Kindermann et al. [10] examined k-sided boundary labelling, where the ports appear on adjacent sides. For the 1-bend 2-sided boundary labelling problem, they gave an $O(n^2)$-time algorithm. For 1-bend 3- and 4-sided models of the problem, they gave $O(n^4)$- and $O(n^9)$-time algorithms, respectively.

If a boundary labelling instance admits an affirmative solution, then it is desirable to seek for a labelling that optimizes a labelling aesthetic, such as minimizing the sum of the leader lengths or minimizing the number of bends per leader. For minimizing the sum of leader lengths, Bekos et al. [3] gave an exact $O(n^2)$-time algorithm for the 1-bend 1-sided and 1-bend (opposite) 2-

sided models; their algorithm for 1-bend 1-sided model was later improved to an $O(n \log n)$-time algorithm by Benkert et al. [4].

Kindermann et al. [10] gave an $O(n^8 \log n)$-time dynamic programming algorithm for the 1-bend (adjacent) 2-sided model. Bose et al. [6] improved this result by giving an $O(n^3 \log n)$-time dynamic programming algorithm. They also showed that the 1-bend 3- and 4-sided problems (for the sum of the leader length minimization) can be reduced to the maximum independent set problem on outerstring graphs. Keil et al. [9] used the idea of outerstring graphs to obtain an $O(n^6)$-time algorithm for the 3- and 4-sided boundary labelling problem. However, it is not obvious whether this approach can be used to obtain a faster algorithm for the decision version, where we do not require leader length minimization. Throughout the paper, we consider 1-bend leaders, which are also known as *po-leaders* [3]. Moreover, unless otherwise specified, by 2-sided we mean two adjacent sides.

Our Results. In this paper, we give algorithms with running times $O(n^3 \log n)$ and $O(n^5)$ for the 1-bend 3-sided and 4-sided boundary labelling problems, which improve the previously best-known algorithms.

The fastest known algorithm for the 3-sided model was Kindermann et al.'s [10] $O(n^4)$-time algorithm that reduced the problem into $O(n^2)$ 2-sided boundary labelling problems. While we also use their partitioning technique, our improvement comes from a dynamic programming approach that carefully decomposes 3-sided problems into various simple shapes that are not necessarily rectangular. We prove that such a decomposition can be computed fast using suitable data structures.

For the 4-sided model, the fastest known algorithm was the $O(n^6)$-time algorithm of Bose et al. [6] that reduced the problem into the maximum independent set problem in an outerstring graph. We show that Kindermann et al.'s [10] observation on partitioning 2-sided boundary labelling problems using a xy-monotone curve can be used to find a fast solution for the 4-sided model. Such an approach was previously taken by Bose et al. [6], but it already took $O(n^3 \log n)$ time for the 2-sided model, and they eventually settled with an $O(n^6)$-time algorithm for the 4-sided model. With the four sides involved, designing a decomposition with a few different types of shapes of low complexity becomes challenging. Our improvement results from a systematic decomposition that generates a small number of subproblems at the expense of using a larger size dynamic programming table, resulting in an $O(n^5)$-time algorithm.

2   Preliminaries

In this section, we give some notation and preliminaries that will be used in the rest of the paper. For a point p in the plane, we denote the x- and y-coordinates of p by $x(p)$ and $y(p)$, respectively. Consider the input rectangle B and let $c_1, \ldots, c_4$ denote the corners of B that are named in clockwise order such that c_1 is the top-right corner of B. Let $B_{\text{top}}, B_{\text{bottom}}, B_{\text{left}}$ and B_{right} denote the top, bottom, left and right sides of B, respectively. We refer to a port as a *top port* (resp., *bottom, left* and *right port*), if it lies on B_{top} (resp., $B_{\text{bottom}}, B_{\text{left}}$ and B_{right}). Similarly, we call a leader a *top leader* (resp., *bottom, left* and *right leader*), if it is connected to a top port (resp., bottom, left and right port).

Let w, x, y, z be the number of ports on the top, right, bottom and left side of B; notice that $w + x + y + z = n$. We denote these ports as $t_1, \ldots, t_w$, $r_1, \ldots, r_x, b_1, \ldots, b_y$ and $\ell_1, \ldots, \ell_z$ in clockwise order. See Fig. 1(a) for an example. We assume that the sites are in general position; i.e., the number of sites and ports on every horizontal (similarly, vertical) line that properly intersects B is at most one. For the rest of the paper, whenever we say a rectangle, we mean an axis-aligned rectangle.

For a point q inside B, consider the rectangle $B_i(q)$ that is spanned by q and c_i, where $1 \leq i \leq 4$. Each rectangle $B_i(q)$ contains only two types of ports; e.g., $B_1(q)$ contains only top and right ports. A feasible solution for a solvable instance of a boundary labelling problem is called *partitioned*, if there exists a point q such that for each rectangle $B_i(q)$, there exists an axis-aligned xy-monotone polygonal curve C_i from q to c_i that separates the two types of leaders in $B_i(q)$. That is, every pair of sites in $B_i(q)$ that lie on different sides of C_i are connected to ports that lie on different (but adjacent) sides of B. See Fig. 1(c). We refer to the polygonal curve as the *xy-separating curve*. Kindermann et al. [10] observed that if an instance of a boundary labelling problem admits a feasible solution, then it must admit a partitioned solution.

Lemma 1 (Kindermann et al. [10]). *If there exists a feasible solution for the 1-bend 4-sided boundary labelling problem, then there also exists a partitioned solution for the problem.*

Consider a 2-sided problem, where the ports are on B_{top} and B_{right}, and assume that the problem has a feasible solution. Here, an xy-separating curve C is the one that connects c_3 to c_1. Then (for such an xy-separating curve C), let *above(C)* (resp., *below(C)*) be the polygonal regions above C (resp., below C) that is bounded by B_{top} and B_{left} (resp., by B_{right} and B_{bottom}). Now, let C_u (resp., C_b) be the xy-separating curve for which the area of *above(C_u)* (resp., the area of *below(C_b)*) is minimized. Given C_u and C_b, we construct a sequence of rectangles as follows (see Fig. 1(d)).

- Each rectangle is a maximal rectangle between C_u and C_b.
- The bottom-left corner of R_1 is c_3. Since R_1 is maximal, it is uniquely determined.
- Let $i > 1$. We know that the top and right sides of R_{i-1} are determined by a pair of leaders L^t and L^r, respectively. Let $a \in L^t$ be the rightmost point on the top side of R_{i-1}, and let $b \in L^r$ be the topmost point on the right side of R_{i-1}. Then the rectangle R_i is the maximal empty rectangle whose bottom-left corner is $(x(a), y(b))$ and that is bounded by C_u and C_b.

We say that a subproblem is *balanced* if it contains the same number of sites and ports.

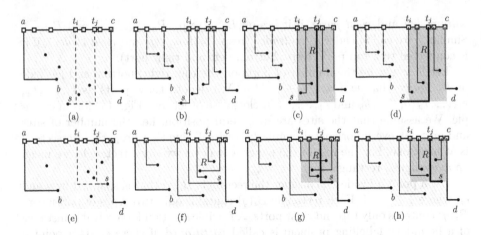

Fig. 2. An illustration for Lemma 2.

Lemma 2. *Let P be a 1-bend 1-sided boundary labelling problem with a constraint that the leftmost and rightmost ports a and c on B_{top} must be connected to a pair of points b and d, respectively. Let s be the rightmost or bottommost site of the problem excluding b and d. If P has a feasible solution satisfying the given constraint, then it also has a solution that connects s to the port t_j such that (i) connecting s to t_j decomposes the problem into two balanced subproblems, and (ii) t_j is the first port (while walking from c to a along the boundary) that provides such a decomposition.*

Proof. First, assume that s is the bottommost point (see Fig. 2(a)–(b)). Assume for a contradiction that connecting s to t_j would not give a feasible solution, whereas there is another port t_i such that connecting s to t_i would yield a feasible solution. Note that t_i lies to the left of t_j. Let L be the leader of t_i. We swap the leaders of t_i and t_j: we connect s to t_j and the site one connected to t_j to t_i. Such a swap may introduce crossings in R: the rectangular region to the right side of L. However, after the swap both sides of the leader of t_j in R are balanced 1-sided problems (see Fig. 2(c)). Such a solution with crossings to a 1-sided boundary labelling problem can always be made planar by local swaps [4]. The proof for the case when s is the rightmost point is the same. Figure 2(e)–(h) illustrate such a scenario. □

3   Three-Sided Boundary Labelling

In this section, we give an $O(n^3 \log n)$-time algorithm for the 3-sided boundary labelling problem. We assume that the ports are located on B_{left}, B_{top} and B_{right}. Kindermann et al. [10] gave an $O(n^4)$-time algorithm for this problem as follows. Consider the grid induced by a horizontal and a vertical line through every port and site. For each node of this grid, they partition the 3-sided problem

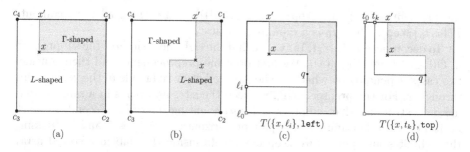

Fig. 3. (a)–(b) An L- and a Γ-shaped problem. (c)–(d) Encoding of the L-shaped problems. The site q that determines the boundary of the L-shape can be recovered from the encoding.

into an L-shaped 2-sided and a Γ-shaped 2-sided problem (Fig. 3(a)–(b)). They showed that the 3-sided problem is solvable if and only if there exists a grid node whose two 2-sided problems both are solvable.

Algorithm Overview. We also start by considering every grid node x, but we employ a dynamic programming approach that expresses the resulting 2-sided subproblems by x and a port. We show that the 2-sided subproblems can have $O(1)$ different types. This gives us $O(1)$ different tables, each of size $O(n^3)$. We show that the running time to fill all the entries is $O(n^3 \log n)$. For a grid node x, let x' be the projection of x onto $B_{\texttt{top}}$. The point x' splits the ports on B into two sets: those to the left of x' and those to the right of it. This means that the line segment xx' can be extended to an axis-aligned curve with two bends that splits the problem into two balanced L- and Γ-*shaped* 2-sided subproblems; see Fig. 3. Kindermann et al. [10] observed that such a balanced partition is unique. Hence if there exists a balanced partition by extending xx' to a 2-bend curve, then there must be a unique position where the number of sites to the left of the curve matches the number of ports to the left of x'.

In the following, we describe the details of solving the L- and Γ-shaped 2-sided problems. W.l.o.g., we assume that the 2-sided problem contains the top-left corner c_4 of B. While decomposing an L-shaped problem, we will reduce it either into a smaller L-shaped problem or a rectangular 2-sided problem. Decomposition of a Γ-shaped problem is more involved.

3.1   Solving an L-Shaped Problem

We denote an L-shaped subproblem with a grid node x and a port b, where b lies either on $B_{\texttt{top}}$ or $B_{\texttt{left}}$. If b is a port on $B_{\texttt{left}}$ (i.e., $b = \ell_i$ for some i), then we denote the subproblem by $T(\{x, \ell_i\}, \texttt{left})$ in which the bottom side is determined by a horizontal line through ℓ_i (e.g., see Fig. 3(c)). Here, the last parameter denotes whether the port belongs to $B_{\texttt{left}}$ or $B_{\texttt{top}}$. Initially, we assume a dummy port ℓ_0 located at c_3 and so our goal is to compute $T(\{x, \ell_0\}, \texttt{left})$. On the other hand, if $b = t_k$ for some k (i.e., it is a port on $B_{\texttt{top}}$), then we define the problem as $T(\{x, t_k\}, \texttt{top})$ (see Fig. 3(d))). Here, the goal is to compute $T(\{x, t_0\}, \texttt{top})$, where t_0 is a dummy port at c_4.

To decompose $T(\{x, \ell_i\}, \texttt{left})$, we find the rightmost site p of the subproblem in $O(\log n)$ time (Fig. 4(a)). We first do some preprocessing, and then consider two cases depending on whether p lies to the left or right side of the vertical line through x. For the preprocessing, we keep the sites and ports in a range count-ing data structure that supports $O(\log n)$ counting query [5]. Second, for each horizontal slab determined by a pair of horizontal grid lines h and h' (passing through sites and ports), we keep the points inside the slab in a sorted array $M(h, h')$, which takes $O(n^3 \log n)$ time. We now use these data structures to find p. We first compute the sites and ports in the rectangle determined by the diagonal $c_4 x$, and then find the number of points needed (for having a balanced subproblem) in $O(\log n)$ time. Finally, we find a site r (that balances the number of sites and points) in $O(\log n)$ time by a binary search in the array M for the slab determined by the horizontal grid lines through ℓ_i and x. If r lies to the right of the vertical line through x, then r is the desired point p. Otherwise, p lies to the left of the vertical line through x, and we can find p by searching in the array M for the slab determined by the horizontal grid lines through ℓ_i and t_0. We will frequently search for such a unique site throughout the paper and so we will use a similar preprocessing.

Case 1 (p Lies to the Right Side of the Vertical Line Through x). We connect p to the "correct" port: after connecting, the resulting subproblems are balanced; i.e., the number of sites in each resulting subproblem is the same as the number of ports of that subproblem. There might be several ports that are right in this sense, but one can apply Lemma 2 (assuming dummy leaders as boundary constraints) to show that the subproblem has a feasible solution if and only if there exists a feasible solution connecting p either to the bottommost or to the rightmost port that satisfies the balanced condition. Once we find p and the appropriate port c for p, there are three possible scenarios as illustrated in Fig. 4. We can decompose the subproblem using the following recursive formula (depending on whether p is connected to $B_{\texttt{left}}$ or $B_{\texttt{top}}$):

$$T(\{x, \ell_i\}, \texttt{left}) = \begin{cases} T(\{x, \ell_j\}, \texttt{left}) \wedge T' & \text{e.g., see Fig. 4(a)--4(b), or} \\ T(\{x, t_j\}, \texttt{top}) \wedge T(\{y, \ell_i\}, \texttt{left}) & \text{e.g., see Fig. 4(c)} \end{cases}$$

Here, $T(\{x, t_j\}, \texttt{top})$ is an L-shaped 1-sided problem and T' is a rectangular 1-sided problem. There always exists a solution for the balanced rectangular 1-sided problem [4], and thus T' can be considered as true. Since $T(\{x, t_j\}, \texttt{top})$ is balanced, we can show that there always exists a solution for $T(\{x, t_j\}, \texttt{top})$, as follows. First assume that the vertical line through x does not pass through a port (see Fig. 5(a)). Let t be the rightmost port in $T(\{x, t_j\}, \texttt{top})$, and let Q be the set of points inside the rectangle R_x determined by diagonal xq. Scale down the rectangle R_x horizontally and translate the rectangle inside the vertical slab determined by the vertical lines through t and x. There always exists a solution for the balanced rectangular 1-sided problem [4], and we can translate the points back to their original position extending the leaders as required. The case when

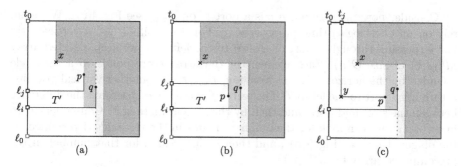

Fig. 4. The decomposition of an L-shaped problem when $b = \ell_i$.

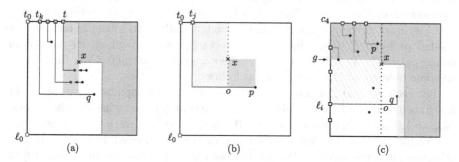

Fig. 5. (a) Reducing $T(\{x, t_j\}, \texttt{top})$ to a rectangular 1-sided problem, and (b) finding the port c for p when $c = t_j$ for some j. (c) Precomputation of the 2-sided problems.

the vertical line through x passes through a port t_x can be processed in the same way, by first connecting t_x to the topmost point of R_x, and then choosing the port immediately to the left of t_x as t. The recurrence formula is as follows.

$$T(\{x, \ell_i\}, \texttt{left}) = \begin{cases} T(\{x, \ell_j\}, \texttt{left}) & \text{if } \ell_j \text{ exists, e.g., see Fig. 4(a)–4(b), or} \\ T(\{y, \ell_i\}, \texttt{left}) & \text{if } t_j \text{ exists, e.g., see Fig. 4(c)} \end{cases}$$

We now show how to find the point p and then decompose $T(\{x, \ell_i\}, \texttt{left})$. The table $T(\{x, \ell_i\}, \texttt{left})$ is of size $O(n^3)$. We will show that each subproblem can be solved by a constant number of table look-ups, and these entries can be found in $O(\log n)$ time. Hence, the overall computation takes $O(n^3 \log n)$ time.

We now show how to find the "right" port c for p. First, assume that c is a port ℓ_j on $B_{\texttt{left}}$ (see Fig. 4(a)–(b)). We need another $O(n^3 \log n)$-time preprocessing as follows. Define a table $M'(s, \ell_i)$, where s is a site, and ℓ_i is a port on $B_{\texttt{left}}$. At the entry $M'(s, \ell_i)$, we store the port ℓ_j such that $j > i$ is the smallest index for which the rectangle defined by $B_{\texttt{left}}$, the horizontal lines through ℓ_i, ℓ_j and the vertical line through s contains exactly $(j - i - 1)$ sites. We set j to 0 when no such port ℓ_j exists. The table M has size $O(n^2)$ and we can fill each entry of the table in $O(n \log n)$ time; hence, we can fill out the entire M' in $O(n^3 \log n)$ time. Observe that the port c for p is stored in $M'(p, \ell_i)$.

Consider now the case when c is a port t_j on B_{top} (see Fig. 4(c)). We again rely on an $O(n^3 \log n)$-time preprocessing. For every site s and a vertical grid line ℓ (passing through a port or a site) to the left of s, we keep a sorted array $M''(s, \ell)$ of size $O(n)$. Each element of the array corresponds to a rectangle bounded by the horizontal line through s, B_{top}, ℓ, and another vertical grid line ℓ' through a port to the left of ℓ. The rectangles are sorted based on the difference between the sites and ports and then by the x-coordinate of ℓ'. Consequently, to find $c (= t_j)$, we can look for the number of sites in the rectangle determined by the diagonal px (see Fig. 5(b)), and then binary search for that number in the precomputed array for $M''(p, x)$.

Case 2 (p Lies to the Left Side of the Vertical Line Through x). Let o be the intersection of the vertical line through x and the leader connecting ℓ_i and q (see Fig. 5(c)). Since p is the rightmost point, it suffices to solve the rectangular 2-sided problem P determined by the rectangle with diagonal oc_4 (shown in falling pattern). We will determine whether a solution exists in $O(1)$ time based on some precomputed information, as follows.

For each vertical grid line, we will precompute the topmost horizontal grid line g such that the 2-sided problem (shown in orange) determined by these lines, B_{top} and B_{left} has a feasible solution. If g lies above ℓ_i, then the 2-sided problem is solvable (as the remaining region determines a balanced 1-sided problem); otherwise, it is not solvable. We will use the known $O(n^2)$-time algorithm [10] to check the feasibility of a 2-sided problem, which looks for a 'partitioned' solution. However, we do not necessarily require our 2-sided problem to be partitioned.

We now show that the precomputation takes $O(n^3 \log n)$ time. For every vertical grid line v, we first compute a sorted array A_v of horizontal grid lines such that the 2-sided problem determined by v and each element of A_v is balanced. This takes $O(n \log n)$ time for v and $O(n^2 \log n)$ time for all vertical grid lines. For each v, we then do a binary search on A_v to find the topmost grid line g such that the corresponding 2-sided problem has a feasible solution. Since computing a solution to the 2-sided problem takes $O(n^2)$ time [10], g can be found in $O(n^2 \log n)$ time for v, and in $O(n^3 \log n)$ time for all vertical grid lines.

Avoiding Γ-Shaped Problem While Decomposing L-Shaped Problems. While decomposing L-shapes, we always find the rightmost point p. If the x-coordinate of p is larger than that of x, then the subproblems $T(\{x, \ell_j\}, \texttt{left})$ and $T(\{y, \ell_i\}, \texttt{left})$ are also L-shaped. Otherwise, the problem reduces to a rectangular 1-sided or 2-sided problem. Hence a Γ-shape problem does not appear during the decomposition of L-shaped problems.

3.2   Solving a Γ-Shaped Problem

Consider the Γ-shaped problem containing the top-left corner c_4; see e.g. Fig. 6(a). Let o be the projection of x onto $B_{\texttt{bottom}}$, and let y be the other bend of the Γ shape. In the following, we refer to the rectangle with diagonal oy as the *forbidden region*.

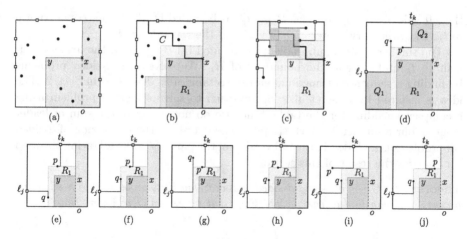

Fig. 6. (a) A Γ-shaped problem. (b)–(c) Three possibilities for the empty rectangle R_1, and a sequence of maximal rectangles corresponding to the separating curve. (d) Rectangle R_1 decomposes the problem into two 1-sided subproblems and a smaller 2-sided subproblem. (e)–(j) All possible cases for the two leaders determining the top and left sides of R_1.

Kindermann et al. [10, Lemma 8] observed that there must exist an axis-aligned xy-monotone curve C that connects c_4 to x such that the sites above C are connected to top ports and the sites below C are connected to left ports. We extend C to o along the vertical line xo (e.g., see Fig. 6(b)). Since no leader can enter the forbidden region, we can now consider C as a separating curve for a 2-sided problem determined by the rectangle with diagonal c_4o. For any partitioned solution, we can compute a sequence of maximal empty rectangles, as we discussed in Sect. 2, such that the lower-right corner of R_1 coincides with o. To decompose the problem, consider the first rectangle R_1 in this sequence. The rectangle R_1 can either cover the forbidden region entirely or partially in three different ways; see Fig. 6(b). Since the separating curve through the grid point x (Fig. 6(a)) determines a partition, the leaders of the Γ-shape must not enter into the forbidden region. Therefore, it suffices to consider an empty rectangle R_1 that entirely covers the forbidden region.

Since R_1 is maximal, the top or left side of R_1 must contain a site (e.g., see Fig. 6(c)); because otherwise, the leaders that determine the top and left side of R_1 will cross. Assume w.l.o.g. that the top side of R_1 contains a site p that is connected to a port $b(= t_k)$. Notice that b cannot be a left port because it contradicts the existence of the curve C. Moreover, the left side of R_1 either contains a site or it is aligned with a leader that connects a port ℓ_j to a site q. The bottom row in Fig. 6 shows all possible scenarios; notice that q cannot be connected to a top port, again because of having C. This decomposes the problem into three subproblems (see Fig. 6(d)). Two 1-sided subproblems Q_1 and Q_2, and a smaller 2-sided subproblem. Q_1 is bounded by the leader connecting ℓ_j to q, the left side of R_1, $B_{\mathtt{bottom}}$ and $B_{\mathtt{left}}$. Q_2 is bounded by the leader connecting p to t_k, $B_{\mathtt{top}}$, the curve determined by x, and the top side of R_1.

The smaller 2-sided subproblem is bounded by $B_{\mathtt{left}}, B_{\mathtt{top}}$, the leaders incident to t_k and ℓ_j, and the boundary of R_1. We solve such 2-sided problems by finding a sequence of maximal empty rectangles as described in Sect. 2. The idea is inspired by Bose et al.'s [6] approach to solve a 2-sided problem using a compact encoding for the table. Since we do not optimize the sum of leader lengths, our approach is much simpler. The details of the processing of 2-sided subproblems is included in the full version of the paper [7]. The following theorem summarizes the result of this section.

Theorem 1. *Given a 1-bend 3-sided boundary labelling problem with n sites and n ports, one can find a feasible labelling (if it exists) in $O(n^3 \log n)$ time.*

4   Four-Sided Boundary Labelling

In this section, we give an $O(n^5)$-time dynamic programming algorithm for the 4-sided boundary labelling. Our dynamic programming solution will search for a set of maximal empty rectangles $R_1, \ldots, R_k$ corresponding to the xy-separating curve, as described in Sect. 2. The intuition is to represent a problem using at most two given leaders. For example, in Fig. 7(a) the first empty rectangle is R_1, with the top and right sides of R_1 determined by the leaders of ℓ_2 and b_2, respectively. The problem will have a feasible solution with these given leaders if and only if the subproblems P_1 and P_2 (shown in falling and rising patterns, respectively) have feasible solutions. Since the subproblems must be balanced, given the two leaders adjacent to P_2 (see Fig. 7(a)), we can determine the boundary of the problem. We will use this idea to encode the subproblems.

It may initially appear that to precisely describe a subproblem, one would need some additional information along with the two given leaders. For example, the dashed boundary in Fig. 7(b). However, such information can be derived from the given leaders. Here, the top-right corner of R_1 (hence, the dashed line) can be recovered using the y-coordinate of the bottommost point of the leader of ℓ_2, and the x-coordinate of the leftmost point of the leader of b_2. We will try all possible choices for the first empty rectangle R_1. In a general step, we will continue searching for the subsequent empty rectangle. For example, consider the subproblem in Fig. 7(b). For a subsequent empty rectangle, we will decompose the problem into at most three new subproblems (Fig. 7(c)). Each of these new subproblems can be represented using at most two leaders.

In the full version of the paper [7], we show how to decompose the subproblems of each type while keeping the total running time bounded by $O(n^5)$. The following theorem summarizes the result of this section.

Theorem 2. *Given a 1-bend 4-sided boundary labelling problem with n sites and n ports, one can find a feasible labelling (if it exists) in $O(n^5)$ time.*

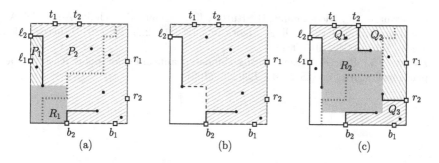

Fig. 7. The idea of decomposing a problem into subproblems.

5   Directions for Future Research

The main direction for future work is to improve the running time of our algorithms. Another direction is to study the fine-grained complexity of the problem; e.g., can we find a non-trivial lower bound on the running time? Even a quadratic lower bound is not known. Note that Bose et al.'s algorithm [6] minimizes the sum of leader lengths, but ours does not. It would also be interesting to seek faster algorithms that minimize the sum of leader lengths.

References

1. Bekos, M.A., et al.: Many-to-one boundary labeling with backbones. J. Graph Algorithms Appl. **19**(3), 779–816 (2015)
2. Bekos, M.A., Kaufmann, M., Nöllenburg, M., Symvonis, A.: Boundary labeling with octilinear leaders. Algorithmica **57**(3), 436–461 (2010). https://doi.org/10.1007/s00453-009-9283-6
3. Bekos, M.A., Kaufmann, M., Symvonis, A., Wolff, A.: Boundary labeling: models and efficient algorithms for rectangular maps. Comput. Geom. **36**(3), 215–236 (2007)
4. Benkert, M., Haverkort, H.J., Kroll, M., Nöllenburg, M.: Algorithms for multi-criteria boundary labeling. J. Graph Algorithms Appl. **13**(3), 289–317 (2009)
5. de Berg, M., Cheong, O., van Kreveld, M., Overmars, M.: Computational Geometry: Algorithms and Applications. Springer, Heidelberg (2008). https://doi.org/10.1007/978-3-540-77974-2
6. Bose, P., Carmi, P., Keil, J.M., Mehrabi, S., Mondal, D.: Boundary labeling for rectangular diagrams. In: 16th Scandinavian Symposium and Workshops on Algorithm Theory, SWAT 2018, 18–20 June 2018, Malmö, Sweden, pp. 12:1–12:14 (2018)
7. Bose, P., Mehrabi, S., Mondal, D.: (Faster) multi-sided boundary labelling. CoRR abs/2002.09740 (2020)
8. Freeman, H., Hitesh Chitalia, S.M.: Automated labeling of soil survey maps. In: ASPRS-ACSM Annual Convention, Baltimore, vol. 1. pp. 51–59 (1996)
9. Keil, J.M., Mitchell, J.S.B., Pradhan, D., Vatshelle, M.: An algorithm for the maximum weight independent set problem on outerstring graphs. Comput. Geom. **60**, 19–25 (2017)

10. Kindermann, P., Niedermann, B., Rutter, I., Schaefer, M., Schulz, A., Wolff, A.: Multi-sided boundary labeling. Algorithmica **76**(1), 225–258 (2016). https://doi.org/10.1007/s00453-015-0028-4
11. Zoraster, S.: Practical results using simulated annealing for point feature label placement. Cartogr. GIS **24**(4), 228–238 (1997)

On the Geometric Red-Blue Set Cover Problem

Raghunath Reddy Madireddy[1], Subhas C. Nandy[2], and Supantha Pandit[3(✉)]

[1] BITS Pilani, Hyderabad Campus,
Hyderabad, Telangana, India
raghunath@hyderabad.bits-pilani.ac.in
[2] Indian Statistical Institute, Kolkata, India
nandysc@isical.ac.in
[3] Dhirubhai Ambani Institute of Information and Communication Technology,
Gandhinagar, Gujarat, India
pantha.pandit@gmail.com

Abstract. We study the variations of the geometric Red-Blue Set Cover (*RBSC*) problem in the plane using various geometric objects. We show that the *RBSC* problem with intervals on the real line is polynomial-time solvable. The problem is NP-hard for rectangles anchored on two parallel lines and rectangles intersecting a horizontal line. The problem admits a polynomial-time algorithm for axis-parallel lines. However, if the objects are horizontal lines and vertical segments, the problem becomes NP-hard. Further, the problem is NP-hard for axis-parallel unit segments.

We introduce a variation of the Red-Blue Set Cover problem with the set system, the *Special-Red-Blue Set Cover* problem, and show that the problem is APX-hard. We then show that several geometric variations of the problem with: (i) axis-parallel rectangles containing the origin in the plane, (ii) axis-parallel strips, (iii) axis-parallel rectangles that are intersecting exactly zero or four times, (iv) axis-parallel line segments, and (v) downward shadows of line segments, are APX-hard by providing encodings of these problems as the Special-Red-Blue Set Cover problem. This is on the same line of the work by Chan and Grant [6], who provided the APX-hardness results of the geometric set cover problem for the above classes of objects.

Keywords: Red-Blue Set Cover · Special Red-Blue Set Cover · Anchored rectangles · Segments · Polynomial time algorithms · APX-hard · NP-hard

1 Introduction

Set Cover is a fundamental and well-studied optimization problem in computer science. In this problem, we are given a set system (U, S) where U is a set of elements and S is a collection of subsets of U. The objective is to find a minimum sub-collection $S' \subseteq S$ that covers all the elements of U. The problem is in the

© Springer Nature Switzerland AG 2021
R. Uehara et al. (Eds.): WALCOM 2021, LNCS 12635, pp. 129–141, 2021.
https://doi.org/10.1007/978-3-030-68211-8_11

list of Karp's 21 NP-hard optimization problems [13]. Several researchers have studied numerous variations of this problem, and one of the well-know variations is the Red-Blue Set Cover Problem, which was introduced by Carr et al. [5]. In this paper, we consider the geometric version of the red-blue set cover problem. The formal definition of the problem is given below:

> **Geometric Red-Blue Set Cover ($RBSC$).** Given three sets R, B, and O where R (resp. B) is the set of red (resp. blue) points and O is the set of geometric objects in the plane, the goal is to find a subset $O' \subseteq O$ of objects such that O' covers all the blue points in B while covering the minimum number of red points in R.

We study the $RBSC$ problem with different classes of geometric objects such as intervals on the real line, axis-parallel rectangles stabbing a horizontal line, axis-parallel rectangles anchored on two parallel horizontal lines, axis-parallel lines, vertical lines and horizontal segments, axis-parallel line segments (both unit and arbitrary lengths), axis-parallel strips, downward shadows of segments.

1.1 Previous Work

The classical Set Cover Problem is NP-hard [13] and it can not approximated better than $O(\log n)$ unless P=NP [15]. The Geometric Set Cover problem has also been studied extensively in the literature. The geometric set cover problem is known to be NP-hard when the objects are unit disks and unit squares [10]. Further, the problem admits PTAS for objects like unit disks [14,18] and axis-parallel unit squares [7,9]. On the other hand, the problem is known to be APX-hard for triangles and circles [12]. Further, Chan and Grant [6] proved that the Geometric Set Cover is APX-hard for several classes of objects such as axis-parallel rectangles contain a common point, axis-parallel strips, axis-parallel rectangles that intersect exactly zero or four times, downward shadows of line segments, and many more.

In the literature, numerous variations of the Set Cover problem (in both classical and geometric settings) are extensively studied (see [2,8,17]). A well-know variation is the Red-Blue Set Cover ($RBSC$) problem that is introduced by Carr et al. [5] in a classical setting. They proved that this problem cannot be approximated in polynomial time within $2^{\log^{1-\delta} n}$ factor, where $\delta = \frac{1}{\log^c \log n}$ and for any constant $c \leq \frac{1}{2}$, unless P=NP, even when each set contains two red and one blue elements (here n is the number of sets). They also provided a $2\sqrt{n}$ factor approximation algorithm for the restricted case where each set contains exactly one blue element.

In a geometric setting, the $RBSC$ problem was first considered by Chan and Hu [7]. They studied this problem with axis-parallel unit squares and proved that the problem is NP-hard and gave a PTAS for the same by using the *mod-one method*. Recently, Shanjani [20] proved that the $RBSC$ problem is APX-hard when the objects are axis-parallel rectangles. She also showed that the

$RBSC$ problem with triangles (or convex objects) cannot be approximated to within $2^{\log^{1-\frac{1}{(\log \log m)^c}} m}$ factor in polynomial time for any constant $c < \frac{1}{2}$, unless P=NP (here m is the number of objects). For the $RBSC$ problem with unit disks, the first constant factor approximation algorithm was proposed in [16].

1.2   Our Contributions

➤ We show that the $RBSC$ problem is APX-hard for (i) axis-parallel rectangles containing the origin in the plane, (ii) axis-parallel strips, axis-parallel rectangles that are intersecting exactly zero or four times, (iii) axis-parallel line segments, and (iv) downward shadows of line segments. To provide these hardness results, we define a special case of the $RBSC$ problem with set systems called the Special Red-Blue Set Cover ($SPECIAL\text{-}RBSC$) problem and prove that this problem is APX-hard (see Sect. 2). The results are inspired from the APX-hardness results given by Chan and Grant [6] for geometric set cover problems with different objects.
➤ For axis-parallel rectangles anchored on two parallel lines, we prove that the $RBSC$ problem is NP-hard. We also prove that the $RBSC$ problem is NP-hard when the objects are axis-parallel rectangles stabbing a horizontal line (see Sect. 3).
➤ For axis-parallel lines, we show that the $RBSC$ problem can be solved in polynomial time, whereas the problem is NP-hard when the objects are horizontal lines and vertical segments. For axis-parallel unit segments in the plane, we also prove that the $RBSC$ problem is NP-hard (see Sect. 4).
➤ We present a sweep-line based polynomial-time dynamic programming algorithm for $RBSC$ problem with intervals on the real line (see Sect. 5).

1.3   Preliminaries

The 3-SAT problem [11] is a CNF formula ϕ containing n variables $x_1, x_2, \ldots, x_n$ and m clauses $C_1, C_2, \ldots, C_m$ where each clause C_i contains exactly 3 literals. The objective is to decide whether a truth assignment to the variables exist that satisfies ϕ. A clause is said to be a positive clause if it contains all positive literals, otherwise if a clause contains all negative

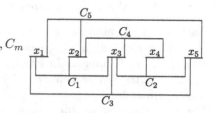

Fig. 1. An instance of the Planar Monotone Rectilinear 3 SAT problem.

then the clause is said to be a negative clause. A Monotone 3-SAT ($M3SAT$) problem [11] is a 3-SAT problem where the formula contains either positive or negative clauses. Both 3-SAT and $M3SAT$ problems are NP-complete [11]. In a Planar Monotone Rectilinear 3-SAT ($PMR3SAT$) problem [3,4], a 3-CNF formula ϕ is given that contains n variables $x_1, x_2, \ldots, x_n$ and m clauses $C_1, C_2, \ldots, C_m$ where each clause is either positive or negative and contains exactly 3 literals. For each variable or clause, take a horizontal segment. The variable segments are arranged on a horizontal line (may be on the x-axis) in a order from left to right and all the positive (resp. negative) clauses are connected by three legs to the 3 literals it contains from above (resp. below). The clauses are placed in different levels in the y-direction. Further, the drawing is planar, i.e., no segments and legs can intersect with each other. See Fig. 1 for an instance ϕ of the *PMR3SAT* problem. The objective is to decide whether ϕ is satisfiable or not. de Berg and Khosravi [3,4] proved that this problem is NP-complete.

The variables $x_1, x_2, \ldots, x_n$ are ordered from left to right on the x-axis. For a negative clause C_l containing x_i, x_j, and x_k in this order on the x-axis, we say that x_i is the *left*, x_j is the *middle*, and x_k is the *right* variables. Let us consider a variable x_i. Let $1, 2, \ldots$ be the left to right order of the legs of negative clauses that connects to x_i. We interpret the clause corresponding to the κ-th leg as the (κ, i)-th clause. For example, for the variable x_3, C_1 is $(1, 3)$-th, C_3 is $(2, 3)$-th, and C_2 is $(3, 3)$-th clause respectively. A similar interpretation can be given for positive clauses by taking the *PMR3SAT* embedding (Fig. 1) upside down.

2   APX-hardness Results

In this section, we show that the *RBSC* problem is APX-hard for several classes of objects (see Theorem 2 for the exact list). For this purpose, we first define a variation of red-blue set cover with the set system, *SPECIAL-RBSC*, and show that the problem is APX-hard. Then, we give encodings of the *RBSC* problem with different classes of objects to the *SPECIAL-RBSC* problem.

Special-Red-Blue Set Cover ($SPECIAL$-$RBSC$) Problem Let $(R \cup B, S)$ be a range space where $R = \{r_1, r_2, \ldots, r_n, r_1^1, r_2^1, \ldots, r_m^1, r_1^2, r_2^2, \ldots, r_m^2\}$ and $B = \{b_1^1, b_2^1, \ldots, b_m^1, b_1^2, b_2^2, \ldots, b_m^2, b_1^3, b_2^3, \ldots, b_m^3\}$ be sets of red and blue elements respectively, and let S be a collection of $6m$ subsets of $R \cup B$ satisfying the following properties:

1. Each set in S contains exactly one blue and one red elements.
2. Each blue element in B appears in exactly two sets in S, every red element $r_i \in R$ appears in exactly three sets in S, and for each $i = 1, 2, \ldots, m$, the red elements $r_i^1, r_i^2 \in R$ appear in exactly two sets in S. Hence, $2m = 3n$.
3. For each $1 \leq t \leq m$, there exist two positive integers i and j, where $1 \leq i < j \leq n$, such that S contains the sets $\{r_i, b_t^1\}$, $\{b_t^1, r_t^1\}$, $\{r_t^1, b_t^2\}$, $\{b_t^2, r_t^2\}$, $\{r_t^2, b_t^3\}$, and $\{b_t^3, r_j\}$.

The goal is to find a sub-collection $S' \subseteq S$, which covers all blue elements in B while covering the minimum number of red elements in R.

In the following, we use the *L-reduction* [19] to show that the *SPECIAL-RBSC* problem is APX-hard. Let A and B be two optimization problems and let g be a polynomial-time computable function from A to B. For an instance

τ of A, let $OPT(\tau)$ and $OPT(g(\tau))$ be the sizes of the optimal solutions of τ and $g(\tau)$ respectively. Function g is said to be an L-reduction if there exists two positive constants α and β (usually 1) such that the following two properties hold for each instance τ of A:

P1: $OPT(g(\tau)) \le \alpha \cdot OPT(\tau)$.

P2: For any given feasible solution of $g(\tau)$ with cost $C(g(\tau))$, there exists a polynomial-time algorithm that finds a feasible solution of τ with cost $C(\tau)$ such that $|C(\tau) - OPT(\tau)| \le \beta \cdot |C(g(\tau)) - OPT(g(\tau))|$.

Theorem 1. *The SPECIAL-RBSC problem is* **APX**-*hard.*

Proof. We give a L-reduction [19] from a known **APX**-hard problem, the *vertex cover on cubic graphs* [1]. In this problem, a graph $G = (V, E)$ is given, $V = \{v_1, \ldots, v_n\}$ and $E = \{e_1, \ldots, e_m\}$, and the degree of each vertex is exactly three, the goal is to find a minimum size subset $V^* \subseteq V$ of vertices that covers all the edges of G. We construct an instance of the *SPECIAL-RBSC* problem as follows:

1. For each vertex $v_i \in V$, place a red element r_i in a set R_1. Thus, $R_1 = \{r_1, \ldots, r_n\}$.

2. For each edge $e_t = (v_i, v_j)$ where $i < j$ ($1 \le t \le m$), consider three blue elements b_t^1, b_t^2, and b_t^3 and two red elements r_t^1 and r_t^2. Further, consider six sets $\{r_i, b_t^1\}$, $\{b_t^1, r_t^1\}$, $\{r_t^1, b_t^2\}$, $\{b_t^2, r_t^2\}$, $\{r_t^2, b_t^3\}$, and $\{b_t^3, r_j\}$ in S.

3. Let $B = \{b_t^1, b_t^2, b_t^3 \mid$ for all $t = 1, 2, \ldots, m\}$. Further, let $R_2 = \{r_t^1, r_t^2 \mid$ for all $t = 1, 2, \ldots, m\}$. Finally, let $R = R_1 \cup R_2$.

We now show that this is a L-reduction with $\alpha = 4$ and $\beta = 1$. For any set $S' \subseteq S$, let $cost(S')$ denote the number of red elements in R that are covered by the sets in S'. Let $V^* \subseteq V$ be an optimal vertex cover of G. In the following, we give a procedure to compute an optimal solution $S^* \subseteq S$ for the above instance of the *SPECIAL-RBSC* problem. Let (v_i, v_j) be the t-th edge in E, for $1 \le i < j \le n$. Since V^* is an optimal vertex cover for G, we have three cases:

i) **Both $v_i, v_j \in V^*$:** Pick both sets $\{r_i, b_t^1\}$ and $\{b_t^3, r_j\}$ in S^*. Further, to cover the blue element b_t^2, pick exactly one set among $\{r_t^1, b_t^2\}$ and $\{b_t^2, r_t^2\}$ in S^*.

ii) **$v_i \in V^*$ and $v_j \notin V^*$:** Pick set $\{r_i, b_t^1\}$ in S^*. Further, to cover the blue elements b_t^2 and b_t^3, pick sets $\{b_t^2, r_t^2\}$ and $\{r_t^2, b_t^3\}$ in S^*.

iii) **$v_j \in V^*$ but $v_i \notin V^*$:** Pick three sets $\{b_t^3, r_j\}$, $\{b_t^1, r_t^1\}$, and $\{r_t^1, b_t^2\}$ in S^*.

Do the same process for all edges in E. One can observe that the set S^* is an optimal solution for the instance of *SPECIAL-RBSC* problem. Further, $cost(S^*) = |V^*| + |E|$.

Proof of $\alpha = 4$: Since the given graph G is a cubic graph, $|V^*| \ge |E|/3$, which implies that $cost(S^*) \le 4|V^*|$. Hence, $\alpha = 4$.

Proof of $\beta = 1$**:** Let $S' \subseteq S$ be a solution of the *SPECIAL-RBSC* problem. We now obtain a vertex cover $V' \subseteq V$ of G, in polynomial time such that $||V'| - |V^*|| \leq |cost(S') - cost(S^*)|$. Let $R' \subseteq R$ be the set of red elements covered by the sets in S'. For each $t = 1, 2, \ldots, m$, we do the following: let (v_i, v_j) be the t-th edge in G for $1 \leq i < j \leq n$. There are three possible cases:

a) *Both r_i and r_j are in R'.* Pick both v_i and v_j in V'. Note that in this case, at least one red element among r_t^1 and r_t^2 is in R'.

b) *Exactly one of r_i and r_j is in R'.* Assume that $r_i \in R'$ (other case is similar). Note that r_t^2 is also in R'. In this case, place v_i in V'.

c) *None of r_i and r_j is in R'.* In this case, both r_t^1 and r_t^2 must be in R'. Pick any one of the vertices v_i and v_j in V'.

Observe that V' is a vertex cover for G. Further, for each $t = 1, 2, \ldots, m$, there exists at least one red element in R' for which we have not picked a vertex in V'. Hence, $|V'| \leq |R'| - |E| = cost(S') - |E|$. Thus, $||V'| - |V^*|| \leq |cost(S') - |E| - |V^*|| = |cost(S') - cost(S^*)|$. Therefore, $\beta = 1$.   □

Theorem 2. *The RBSC problem is* **APX**-*hard for following classes of objects:*

    C1: *Rectangles containing the origin of the plane.*
    C2: *Downward shadows of segments.*
    C3: *Axis-parallel strips.*
    C4: *Rectangles intersecting exactly zero or four times.*
    C5: *Axis-parallel line segments.*

Proof. We prove the theorem by giving an encoding of each class of problems from the *SPECIAL-RBSC* problem. For every blue (resp. red) element in an instance of *SPECIAL-RBSC* problem, we consider a blue (resp. red) point in the plane, and for every set in the instance of *SPECIAL-RBSC* problem, we consider a geometric object. The detailed embedding is given below for each class of objects.

C1: See the encoding in Fig. 2(a). Place the red points $r_1, r_2, \ldots, r_n$, in the order from bottom to top, on a line parallel to $y = x-1$ in the fourth quadrant. Further,

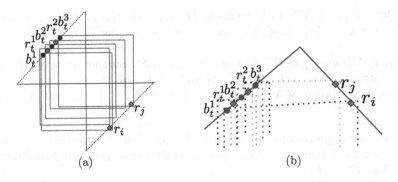

                     (a)   (b)

Fig. 2. Encoding of *SPECIAL-RBSC* problem (a) Class C1. (b) Class C2. (Color figure online)

place all the blue points $b_1^1, b_2^1, \ldots, b_m^1, b_1^2, b_2^2, \ldots, b_m^2, b_1^3, b_2^3, \ldots, b_m^3$ and red points $r_1^1, r_2^1, \ldots, r_m^1, r_1^2, r_2^2, \ldots, r_m^2$ on a line parallel to $y = x+1$ in the second quadrant such that for each $t = 1, 2, \ldots, m$, the five points $b_t^1, r_t^1, b_t^2, r_t^2,$ and b_t^3 are always together in the same order from bottom to top. For each $t = 1, 2, \ldots, m$, place a rectangle for each set $\{r_i, b_t^1\}, \{b_t^1, r_t^1\}, \{r_t^1, b_t^2\}, \{b_t^2, r_t^2\}, \{r_t^2, b_t^3\}$ and $\{b_t^3, r_j\}$, where $1 \le i < j \le n$, that cover the respective points.

C2: See the encoding in Fig. 2(b). Place the red points $r_1, r_2, \ldots, r_n$ in the same order, from bottom to top, on a line parallel to $y = -x$. Further, the placement of blue points and other red points are similar to C1. Place the objects (segments and its shadows) as show in Fig. 2(b).

C3: Place the red points $r_1, r_2, \ldots, r_n$, on a diagonal line with sufficiently large gap between every two consecutive points (see Fig. 3(a)). We know that each red point r_i ($i = 1, 2, \ldots, n$) is in exactly three sets. We place blue points b_t^1, b_t^2, b_t^3, other red points $r_t^1, r_t^2, 1 \le t \le m$, and strips as follows: If $\{r_i, b_t^1\}$ is the first set in which r_i appears, then place b_t^1 to the left of r_i below the diagonal and cover both of them with a vertical strip (see Fig. 3(b)). If $\{r_i, b_t^1\}$ is the second set in which r_i appears, then place b_t^1 to the right of r_i below the diagonal and cover both of them with a vertical strip. If $\{r_i, b_t^1\}$ is the third set in which r_i appears, then place b_t^1 to the left of r_i above the diagonal and cover both points with a horizontal strip (see Fig. 3(c)). The same is true for b_t^3. Now place the red point r_t^1 and blue point b_t^2 as follows: If b_t^1 is placed below the diagonal line and is to the left (resp. right) of r_i, place r_t^1 to the left (resp. right) of b_t^1 such that a horizontal strip can cover only b_t^1 and r_t^1. Further, place b_t^2 vertically below r_t^1 and cover the points with a horizontal strip. If b_t^1 is placed above the diagonal, then place r_t^1 vertically above b_t^1 and cover the points with a horizontal strip. Further, place b_t^2 vertically above r_t^1 and cover the points with a horizontal strip. Finally, place the point r_t^2 appropriately and cover with an axis-parallel strip (see Figs. 3(b) and 3(c)).

C4: This case is similar to the class C3.

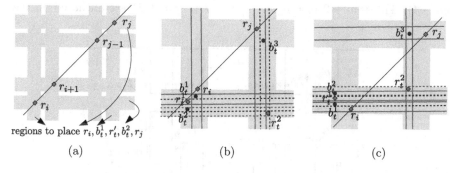

regions to place $r_i, b_t^1, r_t^1, b_t^2, r_j$

(a)   (b)   (c)

Fig. 3.   Class C3 (a) Position of red points $r_1, r_2, \ldots, r_n$ (b) $\{r_i, b_t^1\}$ is the first set containing r_i for some t. (c) $\{r_i, b_t^1\}$ is the third set containing r_i for some t. (Color figure online)

C5: This case is also similar to classes C3 and C4. We place the red points $r_1, r_2, \ldots, r_n$ on a diagonal line from bottom to top with sufficiently large gap between every two consecutive points. See Fig. 3(a). We place the segments for the six sets $\{r_i, b_t^1\}$, $\{b_t^1, r_t^1\}$, $\{r_t^1, b_t^2\}$, $\{b_t^2, r_t^2\}$, $\{r_t^2, b_t^3\}$, and $\{b_t^3, r_j\}$ as follows.

1. *Suppose t is the smallest integer such that r_i appear in a set $\{r_i, b_t^1\}$. Then,* place a blue point b_t^1 just vertically above r_i. Further, place the vertical segment that covers only points r_i and b_t^1 for set $\{r_i, b_t^1\}$. Place r_t^1 vertically above b_t^1 and place b_t^2 vertically above r_t^1. Further, place the smallest vertical segments to cover the sets $\{b_t^1, r_t^1\}$ and $\{r_t^1, b_t^2\}$ covering the respective points.
2. *Suppose r_i appears in exactly one set $\{r_i, b_{t'}^1\}$ for $t' < t$.* Then place the point b_t^1 just horizontally right to r_i. Further, place the smallest horizontal segment covering r_i and b_t^1 for set $\{r_i, b_t^1\}$. Place r_t^1 to the right b_t^1 and place b_t^2 to the right of r_t^1. Further, place the smallest horizontal segments to cover the sets $\{b_t^1, r_t^1\}$ and $\{r_t^1, b_t^2\}$ covering the respective points.
3. *Suppose r_i appears in exactly two sets $\{r_i, b_{t'}^1\}$ and $\{r_i, b_{t''}^1\}$, for $t' < t'' < t$.* This case is symmetric to the first case.

Similarly, the blue point b_t^3 is placed based on the red point r_j and place an appropriate axis-parallel segment for the set $\{b_t^3, r_j\}$. Finally, one can place the red point r_t^2 at an appropriate position and connect it with b_t^2 and b_t^3 by using axis-parallel segments which do not cover any other points.   □

3   Rectangles Anchored on Two Parallel Lines

We prove that the *RBSC* problem with axis-parallel rectangles anchored on two parallel lines (*RBSC-RATPL*) is NP-hard. We give a reduction from the *PMR3SAT* problem. From an instance ϕ of the *PMR3SAT* problem, we create an instance I of the *RBSC-RATPL* problem as follows.

Variable Gadget: Consider two infinite horizontal lines L_1 and L_2. For each variable x_i, the gadget (see Fig. 4) consists of $36m + 4$ rectangles $\{t_j^i | 1 \leq j \leq 36m + 4\}$. The $18m + 2$ rectangles $\{t_j^i | 1 \leq j \leq 18m + 2\}$ are anchored on the

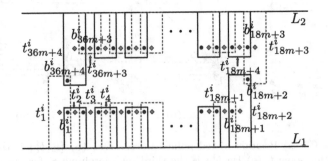

Fig. 4. Variable gadget (Color figure online)

line L_1 and the remaining $18m + 2$ rectangles $\{t^i_j | 18m + 3 \le j \le 36m + 4\}$ are anchored on the line L_2. The rectangle t^i_1 contains two blue points b^i_1 and b^i_{36m+4} and the rectangle t^i_j contains the two blue points b^i_{j-1} and b^i_j for $2 \le j \le 36m+4$. Further, the rectangle t^i_j contains the red point r^i_j, for $1 \le j \le 36m + 4$. See Fig. 4 for the structure of the gadget for x_i.

It is easy to observe that, there are exactly two disjoint sets of rectangles $T^i_0 = \{t^i_2, t^i_4, \ldots, t^i_{36m+4}\}$ and $T^i_1 = \{t^i_1, t^i_3, \ldots, t^i_{36m+3}\}$ such that each set covers all the blue points and also covers the minimum possible number, i.e., $18m + 2$, of red points. This two sets correspond to the truth value of the variable x_i; T^i_1 corresponds x_i is true and T^i_0 corresponds x_i is false.

Clause Gadget and Variable-Clause Interaction: We first give a schematic overview of the structure and position of the variable and clause gadgets (rectangles, red, and blue points). See Fig. 5, for a schematic diagram of the variable and clause gadgets. Each variable has a designated area, the variable area. The gadget (Fig. 4) is fully contained inside its corresponding area. Each variable area has a special area, the variable region where the points corresponding to the variable gadget are placed. Similarly, each clause has a designated area, the clause area. The clause gadget is fully contained inside that area. Each clause has a special area called the clause region. We place some red and blue points (other than the points of the variables) in that region.

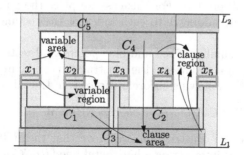

Fig. 5. Schematic diagram of clause gadget (Color figure online)

Let C_l be a negative clause that contains variables x_i, x_j and x_k in the order from left to right (Fig. 1). For this clause we take 11 blue points $\{a^l_i, a^l_j, a^l_k, c^l_1, c^l_2, c^l_3, c^l_4, c^l_5, c^l_6, c^l_7, c^l_8\}$, 8 red points $\{q^l_1, q^l_2, \ldots, q^l_8\}$, and 4 rectangles $\{s^l_1, s^l_2, s^l_3, s^l_4\}$ anchored on L_1. Now we describe the placement of these points and rectangles.

Note that for clause C_l, $x_i, x_j,$ and x_k are left, middle, and right variables respectively. Let us assume that C_l is a (κ_1, i)-th, (κ_2, j)-th, and (κ_3, k)-th clause for the variables x_i, x_j, and x_k, respectively.

Since x_i is a left variable that is negatively present in C_l, place the point a^l_i inside the rectangle $18\kappa_1 + 4$. Similarly, since x_k is a right variable that is negatively present in C_l, place the point a^l_k inside the rectangle $18\kappa_3 + 4$.

The variable x_j is a middle variable that is negatively present in C_l, place the 17 points $q_1^l, q_2^l, c_1^l, c_2^l, c_3^l, c_4^l, q_3^l, q_4^l, a_j^l, q_5^l, q_6^l, c_5^l, c_6^l, c_7^l, c_8^l, q_7^l, q_8^l$ in the left to right order inside the rectangles $18\kappa_2 + 2, 18\kappa_2 + 3, \ldots, 18\kappa_2 + 18$, respectively.

The rectangle s_1^l covers the points $a_i^l, q_1^l, q_2^l, c_1^l, c_2^l, c_3^l, c_4^l$, the rectangle s_2^l covers the points $c_1^l, c_2^l, c_3^l, c_4^l, q_3^l, q_4^l, a_j^l$, the rectangle s_3^l covers the points $a_j^l, q_5^l, q_6^l, c_5^l, c_6^l, c_7^l, c_8^l$, and the rectangle s_4^l covers $c_5^l, c_6^l, c_7^l, c_8^l, q_7^l, q_8^l, a_k^l$ (see Fig. 6).

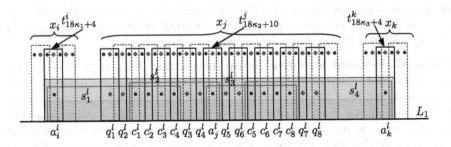

Fig. 6. Clause gadget and variable-clause interaction

For a positive clause the construction is similar by looking at the *PMR3SAT* instance upside down. The only difference is now the literals that a clause contains are all positive. For this case we place the points of the clauses in an identical way by adding 1 to each index of the rectangles of the clause gadget.

Observe that the number of red and blue points, and rectangles in I is polynomial with respect to n and m and the construction can be made in polynomial time. We conclude the following theorem (proof is in the full version).

Theorem 3. *The RBSC-RATPL problem is* NP-*hard.*

Modifying the above construction, we prove that the *RBSC* problem with axis-parallel rectangles intersecting a horizontal line (*RBSC-RSHL*) is NP-hard.

4 Axis-Parallel Lines and Segments

The *RBSC* problem can be solved in polynomial time when the objects are axis-parallel lines (see full version of the paper). Further, it can be proved that the problem is NP-hard for axis-parallel unit segments (see full version of the paper).

We prove that the *RBSC* problem is NP-hard for horizontal lines and vertical segments (*RBSC-HLVS*) by a polynomial-time reduction from the *M3SAT* problem. The reduction is similar to the NP-hard reduction of the maximum independent coverage problem with horizontal lines and vertical segments [8]. We generate an instance I of the *RBSC-HLVS* from an instance ϕ of *M3SAT*.

Overall Structure: The overall structure is in Fig. 7(a). The variable gadgets are placed horizontally, one after another, from top to bottom. Each variable gadget has an area within which the gadget is placed. Inside each area, there is a portion called the variable region where the red and blue points of the variable gadget are placed. The clause gadgets are placed vertically, one after another, from left to right after the variable regions. Clause gadgets are fully contained inside its area. There is a horizontal space called the inter-variable area between two variable areas where some blue points of clause gadgets are placed.

Variable Gadget: The gadget (Fig. 7(b)) of variable x_i contains two horizontal lines h_1^i and h_4^i, four vertical segments h_2^i, h_3^i, h_5^i, and h_6^i, six blue points $b_1^i, b_2^i, \ldots, b_6^i$, and 6 red points $r_1^i, r_2^i, \ldots, r_6^i$. The blue points b_1^i and b_6^i are covered by h_6^i and the blue points b_j^i, b_{j+1}^i are covered by h_j^i, for $1 \le j \le 5$. Further, h_j^i covers the red point r_j^i, $1 \le j \le 6$. Observe that, there are two ways to cover the blue points while minimizing the red points: either $H_1^i = \{h_1^i, h_3^i, h_5^i\}$ or $H_0^i = \{h_2^i, h_4^i, h_6^i\}$ each covering exactly 3 red points. The set H_0^i represents x_i is true and H_1^i represents x_i is false.

Clause Gadget and Variable-Clause Interaction: The gadget for the clause C_l consists of four vertical segments g_1^l, g_2^l, g_3^l, and g_4^l, five blue points $c_1^l, c_2^l, c_3^l, c_4^l$, and c_5^l, and four red points q_1^l, q_2^l, q_3^l, and q_4^l. The vertical segments are on a vertical line and inside the clause area. The segment g_j^l covers two blue points c_j^l and c_{j+1}^l, and one red point q_j^l, $1 \le j \le 4$. Let C_l be a positive clause that contains variables x_i, x_j, and x_k. Then we place the blue points c_1^l, c_3^l, and c_5^l on segments h_4^i, h_4^j, and h_4^k respectively. If C_l be a negative clause that contains variables x_i, x_j, and x_k. Then we place the blue points c_1^l, c_3^l, and c_5^l on segments h_1^i, h_1^j, and h_1^k, respectively. See Fig. 7(b) for this construction.

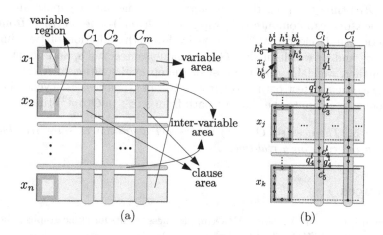

Fig. 7. (a) Overall structure of the construction. (b) Structure of variable and clause gadgets and their interaction. (Color figure online)

This completes the construction. Clearly, the number of horizontal lines, vertical segments, and blue and red points in I is polynomial in n and m and the construction can be performed in polynomial time. We now have the following theorem (proof is in the full version).

Theorem 4. *The RBSC-HLVS problem is NP-hard.*

5   Intervals on the Real Line

In this section, we consider that all blue and red points are on the real line and the objects are arbitrary size intervals. For this case, we give an exact algorithm with $O(nm^2)$-time where $n = |R \cup B|$ and $m = |S|$. Each i-th interval is denoted by $[l_i, r_i]$ where l_i (left endpoint) and r_i (right endpoint) are some real numbers.

Let $O^* \subseteq O$ be an optimal set of intervals which cover all blue points while covering the minimum number of red points. Without loss of generality, we assume that no interval in O^* is completely contained inside an another interval in O^*. In the following, we reduce the problem to a shortest path problem in a directed acyclic graph G. Let $l = \min\{l_1, l_2, \ldots, l_m\}$ and $r = \max\{r_1, r_2, \ldots, r_m\}$. We add two dummy intervals d_1 and d_2 where d_1 completely lies to the left of $x = l$ and d_2 completely lies to the right of $x = r$. Note that the dummy intervals d_1 and d_2 will not cover any blue and red points.

Consider a node for each interval in S, i.e., $V(G) = \{v_i \mid i$ is an interval in $S\}$. We place directed edges (v_i, v_j) if the following conditions are satisfied: (i) $l_i < l_j$ and $r_i < r_j$ and (ii) all the blue points in $[l_i, l_j]$ are covered by interval $[l_i, r_i]$. The cost of the edge (v_i, v_j) is the number of red points covered by $[l_j, r_j]$ which are not covered by $[l_i, r_i]$.

We note that graph G has $O(m)$ vertices and $O(m^2)$ edges. Further, we require $O(n)$-time to check the feasibility of an edge and compute an edge's cost. Hence, one can compute the graph G in $O(nm^2)$-time. Further, from the construction, it is clear that G is acyclic. From the addition of two dummy vertices, d_1 and d_2 we claim that G has unique source, say s, and unique sink, say t. Further, we claim that the cost of the optimal solution of $RBSC$ is also the same as the shortest path's cost from s to t in G. Since G is acyclic directed graph, one can compute the shortest path from s to t in $O(nm^2)$-time.

Theorem 5. *There exists $O(nm^2)$-time exact algorithm to solve red-blue set cover with intervals on a real-line.*

References

1. Alimonti, P., Kann, V.: Some APX-completeness results for cubic graphs. Theoret. Comput. Sci. **237**(1), 123–134 (2000)
2. Bereg, S., Cabello, S., Díaz-Báñez, J.M., Pérez-Lantero, P., Seara, C., Ventura, I.: The class cover problem with boxes. Comput. Geom. **45**(7), 294–304 (2012)
3. de Berg, M., Khosravi, A.: Optimal binary space partitions in the plane. In: Thai, M.T., Sahni, S. (eds.) COCOON 2010. LNCS, vol. 6196, pp. 216–225. Springer, Heidelberg (2010). https://doi.org/10.1007/978-3-642-14031-0_25

4. de Berg, M., Khosravi, A.: Optimal binary space partitions for segments in the plane. Int. J. Comput. Geom. Appl. **22**(3), 187–206 (2012)
5. Carr, R.D., Doddi, S., Konjevod, G., Marathe, M.: On the red-blue set cover problem. In: SODA, pp. 345–353 (2000)
6. Chan, T.M., Grant, E.: Exact algorithms and APX-hardness results for geometric packing and covering problems. Comput. Geom. **47**(2), 112–124 (2014)
7. Chan, T.M., Hu, N.: Geometric red blue set cover for unit squares and related problems. Comput. Geom. **48**(5), 380–385 (2015)
8. Dhar, A.K., Madireddy, R.R., Pandit, S., Singh, J.: Maximum independent and disjoint coverage. J. Comb. Optim. **39**(4), 1017–1037 (2020)
9. Erlebach, T., van Leeuwen, E.J.: PTAS for weighted set cover on unit squares. In: Serna, M., Shaltiel, R., Jansen, K., Rolim, J. (eds.) APPROX/RANDOM -2010. LNCS, vol. 6302, pp. 166–177. Springer, Heidelberg (2010). https://doi.org/10. 1007/978-3-642-15369-3_13
10. Fowler, R.J., Paterson, M.S., Tanimoto, S.L.: Optimal packing and covering in the plane are NP-complete. Inf. Process. Lett. **12**(3), 133–137 (1981)
11. Garey, M.R., Johnson, D.S.: Computers and Intractability: A Guide to the Theory of NP-Completeness. W. H. Freeman & Co., New York (1979)
12. Har-Peled, S.: Being Fat and Friendly is Not Enough. CoRR abs/0908.2369 (2009)
13. Karp, R.M.: Reducibility among combinatorial problems. In: Miller, R.E., Thatcher, J.W., Bohlinger, J.D. (eds.) Complexity of Computer Computations. IRSS, pp. 85–103. Springer, Boston (1972). https://doi.org/10.1007/978-1-4684-2001-2_9
14. Li, J., Jin, Y.: A PTAS for the weighted unit disk cover problem. In: Halldórsson, M.M., Iwama, K., Kobayashi, N., Speckmann, B. (eds.) ICALP 2015. LNCS, vol. 9134, pp. 898–909. Springer, Heidelberg (2015). https://doi.org/10.1007/978-3-662-47672-7_73
15. Lund, C., Yannakakis, M.: On the hardness of approximating minimization problems. J. ACM **41**(5), 960–981 (1994)
16. Madireddy, R.R., Mudgal, A.: A constant-factor approximation algorithm for red-blue set cover with unit disks. In: WAOA (2020, to be appeared)
17. Mehrabi, S.: Geometric unique set cover on unit disks and unit squares. In: CCCG, pp. 195–200 (2016)
18. Mustafa, N.H., Ray, S.: Improved results on geometric hitting set problems. Discrete Comput. Geom. **44**(4), 883–895 (2010). https://doi.org/10.1007/s00454-010-9285-9
19. Papadimitriou, C.H., Yannakakis, M.: Optimization, approximation, and complexity classes. J. Comput. Syst. Sci. **43**(3), 425–440 (1991)
20. Shanjani, S.H.: Hardness of approximation for red-blue covering. In: CCCG (2020)

Fixed-Treewidth-Efficient Algorithms for Edge-Deletion to Interval Graph Classes

Toshiki Saitoh[1]([✉])([iD]), Ryo Yoshinaka[2]([iD]), and Hans L. Bodlaender[3]([iD])

[1] Kyushu Institute of Technology, Kitakyushu, Japan
toshikis@ces.kyutech.ac.jp
[2] Tohoku University, Sendai, Japan
ryoshinaka@tohoku.ac.jp
[3] Utrecht University, Utrecht, The Netherlands
H.L.Bodlaender@uu.nl

Abstract. For a graph class $\mathcal{C}$, the $\mathcal{C}$-EDGE-DELETION problem asks for a given graph G to delete the minimum number of edges from G in order to obtain a graph in $\mathcal{C}$. We study the $\mathcal{C}$-EDGE-DELETION problem for $\mathcal{C}$ the class of interval graphs and other related graph classes. It follows from Courcelle's Theorem that these problems are fixed parameter tractable when parameterized by treewidth. In this paper, we present concrete FPT algorithms for these problems. By giving explicit algorithms and analyzing these in detail, we obtain algorithms that are significantly faster than the algorithms obtained by using Courcelle's theorem.

Keywords: Parameterized algorithms · Treewidth · Edge-Deletion · Interval graphs

1 Introduction

Intersection graphs are represented by geometric objects aligned in certain ways so that each object corresponds to a vertex and two objects intersect if and only if the corresponding vertices are adjacent. Intersection graphs are well-studied in the area of graph algorithms since there are many important applications and we can solve many NP-hard problems in general graphs in polynomial time on such graph classes. Interval graphs are intersection graphs which are represented by intervals on a line. CLIQUE, INDEPENDENT SET, and COLORING on interval graphs can be solved in linear time and interval graphs have many applications in bioinformatics, scheduling, and so on. See [3,14,26] for more details of interval graphs and other intersection graphs.

Graph modification problems on a graph class $\mathcal{C}$ are to find a graph in $\mathcal{C}$ by modifying a given graph in certain ways. $\mathcal{C}$-VERTEX-DELETION, $\mathcal{C}$-EDGE-DELETION, and $\mathcal{C}$-COMPLETION are to find a graph in $\mathcal{C}$ by deleting vertices, deleting edges, and adding edges, respectively, with the minimum cost.

© Springer Nature Switzerland AG 2021
R. Uehara et al. (Eds.): WALCOM 2021, LNCS 12635, pp. 142–153, 2021.
https://doi.org/10.1007/978-3-030-68211-8_12

These problems can be seen as generalizations of many NP-hard problems. CLIQUE is equivalent to COMPLETE-VERTEX-DELETION: we find a complete graph by deleting the smallest number of vertices. Modification problems on intersection graph classes also have many applications. For example, INTERVAL-VERTEX/EDGE-DELETION problems have applications to DNA (physical) mapping [12,13,27]. Lewis and Yannakakis showed that C-VERTEX-DELETION is NP-complete for any nontrivial hereditary graph class [18]. A graph class C is hereditary if for any graph in C, every induced subgraph of the graph is also in C. Since the class of intersection graphs are hereditary, C-VERTEX DELETION is NP-complete for any nontrivial intersection graph class C. The problems C-EDGE-DELETION are also NP-hard when C is the class of perfect, chordal, split, circular arc, chain [24], interval, proper interval [13], trivially perfect [25], threshold [21], permutation, weakly chordal, or circle graphs [4]. See the lists in [4,20].

Parameterized complexity is well-studied in the area of computer science. A problem with a parameter k is *fixed parameter tractable*, *FPT* for short, if there is an algorithm running in $f(k)n^c$ time where n is the size of input, f is a computable function and c is a constant. Such an algorithm is called an *FPT algorithm*. The *treewidth* $tw(G)$ of a graph G represents treelikeness and is one of the most important parameters in parameterized complexity concerning graph algorithms. For many NP-hard problems in general, there are tons of FPT algorithms with parameter $tw(G)$ by dynamic programming on tree decompositions. Finding the treewidth of an input graph is NP-hard and it is known that CHORDAL-COMPLETION with minimizing the size of the smallest maximum clique is equivalent to the problem. There is an FPT algorithm for computing the treewidth of a graph by Bodlaender [2] which runs in $O(f(tw(G))(n + m))$ time where n and m are the numbers of vertices and edges of a given graph: i.e., the running time is linear in the size of input. Courcelle showed that every problem that can be expressed in monadic second order logic (MSO_2) has a linear time algorithm on graphs of bounded treewidth [9]. Some intersection graph classes, for example interval graphs, proper interval graphs, chordal graphs, and permutation graphs, can be represented by MSO_2 [8] and thus there are FPT algorithms for EDGE-DELETION problems on such graph classes. However, the algorithms obtained by Courcelle's theorem have a very large hidden constant factor even when the treewidth is very small, since the running time is the exponential tower of the coding size of the MSO_2 expression.

Our Results: We propose concrete FPT algorithms for EDGE-DELETION to interval graphs and other related graph classes, when parameterized by the treewidth of the input graph. Our algorithms virtually compute a set of edges S with the minimum size such that $G - S$ is in a graph class C by using dynamic programming on a tree-decomposition. We maintain possible alignments of geometric objects corresponding to vertices in the bag of each node of the tree-decomposition. Alignments of the objects of forgotten vertices are remembered only relatively to the objects of the current bag. If two forgotten objects have the same relative position to the objects of the current bag, we remember only the fact that there is at least one forgotten object at that position. In this way, we achieve the fixed-parameter-tractability, while guaranteeing that no object pairs of non-adjacent vertices of the input graph will intersect in our dynamic programming algorithm. Our algorithms run in $O(f(tw(G)) \cdot (n+m))$ time where n and m are the numbers of vertices and edges of the input graph. Our explicit algorithms are significantly faster than those obtained by using Courcelle's theorem. We also analyze the time complexity of our algorithms parameterized by pathwidth which is analogous to treewidth. The relation among the graph classes for which this paper provides C-EDGE-DELETION algorithms is shown in Fig. 1.

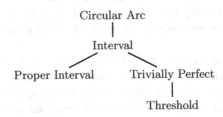

Fig. 1. The graph classes of which this paper presents algorithms for the edge-deletion problems.

Related Works: Another kind of common parameters considered in parameterized complexity of graph modification problems is the number of vertices or edges to be removed or to be added. Here we review preceding studies on those problems for intersection graphs with those parameters.

Concerning parameterized complexity of C-VERTEX-DELETION, Hof et al. proposed an FPT algorithm for PROPER-INTERVAL-VERTEX-DELETION [16], and Marx proposed an FPT algorithm for CHORDAL-VERTEX-DELETION [22]. Heggernes et al. showed PERFECT-VERTEX-DELETION and WEAKLY-CHORDAL-VERTEX-DELETION are W[2]-hard [15]. Cai showed that C-VERTEX/EDGE-DELETION are FPT when C is characterized by a finite set of forbidden induced subgraphs [5].

For modification problems on interval graphs, Villanger et al. presented an FPT algorithm for INTERVAL-COMPLETION [28], and Cao and Marx presented an FPT algorithm for INTERVAL-VERTEX-DELETION [7]. Cao improved these algorithms and developed an FPT algorithm for EDGE-DELETION [6].

It is known that THRESHOLD-EDGE-DELETION, CHAIN-EDGE-DELETION and TRIVIALLY-PERFECT-EDGE-DELETION are FPT, since threshold graphs, chain graphs and trivially perfect graphs are characterized by a finite set of forbidden induced subgraphs [5]. Nastos and Gao presented faster algorithms for the problems [23], and Liu et al. improved their algorithms to $O(2.57^k(n+m))$ and $O(2.42^k(n+m))$ using modular decomposition trees [19], where k is the number of deleted edges. There are algorithms to find a polynomial kernel for CHAIN-EDGE-DELETION and TRIVIALLY PERFECT-EDGE-DELETION [1,11].

Organization of this Article: Section 2 prepares the notation and definitions used in this paper. We propose an FPT algorithm for INTERVAL-EDGE-DELETION in Sect. 3. We then extend the algorithm related to the interval graphs in Sect. 4. We conclude this paper and provide some open questions in Sect. 5.

2 Preliminaries

For a set X, its cardinality is denoted by $|X|$. A *partition* of X is a tuple $(X_1, \ldots, X_k)$ of subsets of X such that $X = X_1 \cup \cdots \cup X_k$ and $X_i \cap X_j = \emptyset$ if $1 \le i < j \le k$, where we allow some of the subsets to be empty. For entities $x, y, z \in X$, we let $x[y/z] = y$ if $x = z$, and $x[y/z] = x$ otherwise. For a subset $Y \subseteq X$, define $Y[y/z] = \{ x[y/z] \mid x \in Y \}$.

For a linear order π over a finite set X, by $\mathsf{suc}_\pi(x)$ we denote the successor of $x \in X$ w.r.t. π, i.e., $x <_\pi \mathsf{suc}_\pi(x)$ and $\mathsf{suc}_\pi(x) \le_\pi y$ for all y with $x <_\pi y$. The maximum element of X w.r.t. π is denoted by $\max_\pi X$. Note that $\mathsf{suc}_\pi(\max_\pi X)$ is undefined. Similarly $\mathsf{pred}_\pi(x)$ is the predecessor of x and $\min_\pi X$ is the least element of X.

A simple graph $G = (V, E)$ is a pair of vertex and edge sets, where each element of E is a subset of V consisting of exactly two elements.

A *tree-decomposition* of $G = (V, E)$ is a tree T such that[1]

- to each node of T a subset of V is assigned,
- if the assigned sets of two nodes of T contain a vertex $u \in V$, then so does every node on the path between the two nodes,
- for each $\{u, v\} \in E$, there is a node of T whose assigned set includes both u and v.

The *width* of a tree-decomposition is the maximum cardinality of the assigned sets minus one and the *treewidth* of a graph is the smallest width of its tree-decompositions. A tree-decomposition is said to be *nice* if it is rooted, the root is assigned the empty set, and its nodes are grouped into the following four:

- *leaf nodes*, which have no children and are assigned the empty set,
- *introduce nodes*, each of which has just one child, where the set assigned to the parent has one more vertex than the child's set,
- *forget nodes*, each of which has just one child, where the set assigned to the parent has one less vertex than the child's set,
- *join nodes*, each of which has just two children, where the same vertex set is assigned to the parent and its two children.

It is known that every tree-decomposition has a nice tree-decomposition of the same width whose size is $O(k|V|)$, where k is the treewidth of the tree-decomposition [10,17]. Hereafter, under a fixed graph G and a fixed nice tree-decomposition T, we let X_s denote the subset of V assigned to a node s of a tree-decomposition and $X_{\le s}$ denote the union of all the subsets assigned to the

---

[1] We use the terms "vertices" for an input graph and "nodes" for a tree-decomposition.

node s and its descendant nodes. We call vertices in X_s and in $X_{\leq s} - X_s$ *active* and *forgotten*, respectively. Moreover, we define $E_s = \{\{u, v\} \in E \mid u, v \in X_s\}$ and $E_{\leq s} = \{\{u, v\} \in E \mid u, v \in X_{\leq s}\}$. Given a tree decomposition of treewidth k, we can compute a nice tree-decomposition with treewidth k and $O(kn)$ nodes in $O(k^2(n + m))$ time [10].

A tree-decomposition is called a *path-decomposition* if the tree is a path. The *pathwidth* of a graph is the smallest width of its path-decompositions. Every path-decomposition has a nice path-decomposition of the same pathwidth, which consists of leaf, introduce, forget, but not join nodes.

The problem we tackle in this paper is given as follows.

Definition 1. *For a graph class C, the C-EDGE-DELETION is a problem to find the minimum natural number c such that there is a subgraph $G' = (V, E')$ of G with $G' \in C$ and $|E| - |E'| = c$ for an input simple graph $G = (V, E)$.*

In the succeeding sections, for different classes C of intersection graphs, we present algorithms for C-EDGE-DELETION that run in linear time in the input graph size when the treewidth is bounded. We assume that the algorithm takes a nice tree-decomposition T of G in addition as input [2,17]. Our algorithms are dynamic programming algorithms that recursively compute solutions (and some auxiliary information) for the subproblems on $(X_{\leq s}, E_{\leq s})$ for each node s in the given tree-decomposition from leaves to the root.

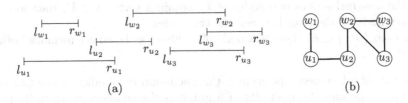

(a)   (b)

Fig. 2. (a) Interval representation ρ. (b) Interval graph G_ρ. We have $\mathscr{A}(\rho, s) = (\pi, I, J, K, c)$ for $X_s = \{u_1, u_2, u_3\}$ and $X_{\leq s} - X_s = \{w_1, w_2, w_3\}$ where $I = \{l_{u_1}, r_{u_1}, r_{u_2}, l_{u_3}, r_{u_3}\}$, $J = \{r_{u_2}, l_{u_3}, r_{u_3}\}$, and $K = \{r_{u_2}, l_{u_3}\}$.

3   Finding a Largest Interval Subgraph

An *interval representation* π over a set X is a linear order over the set $LR_X = L_X \cup R_X \cup \{\bot, \top\}$ with $L_X = \{l_x \mid x \in X\}$ and $R_X = \{r_x \mid x \in X\}$ such that $\bot <_\pi l_x <_\pi r_x <_\pi \top$ for all $x \in X$. Let $\langle\!\langle p_1, p_2 \rangle\!\rangle_\pi = \{q \in LR_X \mid p_1 \leq_\pi q <_\pi p_2\}$ and for $Y \subseteq X$, let

$$[\![Y]\!]_\pi = \bigcup_{u \in Y} \langle\!\langle l_u, r_u \rangle\!\rangle_\pi = \{q \in LR_X \mid l_u \leq_\pi q <_\pi r_u \text{ for some } u \in Y\},$$

which may contain some elements of $LR_X - LR_Y$. The *interval graph* G_π of an *interval representation* π on V is (V, E_π) where

$$E_\pi = \{\, \{u, v\} \subseteq V \mid \langle\!\langle l_u, r_u \rangle\!\rangle_\pi \cap \langle\!\langle l_v, r_v \rangle\!\rangle_\pi \neq \emptyset \text{ and } u \neq v \,\}.$$

Figure 2 shows an example of an interval graph.

This section presents an FPT algorithm for the interval edge deletion problem w.r.t. the treewidth. Let $G = (V, E)$ be an input graph and s a node of a nice tree-decomposition T of G. On each node s of T, for each interval representation ρ over $X_{\leq s}$ that gives an interval subgraph of $(X_{\leq s}, E_{\leq s})$, we would like to remember some pieces of information about ρ, which we call the "abstraction" of ρ. The abstraction includes the linear order π over LR_{X_s} that restricts ρ. In addition, we remember how the intervals of the forgotten vertices intersect the points of the active vertices using three sets $I, J, K \subseteq LR_X$. Figure 3 explains the meaning of those sets. When $p \in I$, $(p, \mathsf{suc}_\pi(p))$ intersects with a forgotten interval. If $p \in J$, p is within a forgotten interval. Moreover if $p \in K$, then the interval $(p, \mathsf{suc}_\pi(p))$ is properly covered by some forgotten intervals. Using those sets, we can introduce new interval without making it intersect with forgotten intervals.

More formally, for an interval representation ρ over $X_{\leq s}$ such that G_ρ is a subgraph of $(X_{\leq s}, E_{\leq s})$, we define the *abstraction* $\mathscr{A}(\rho, s)$ of ρ *for* s to be the quintuple (π, I, J, K, c) such that

- π is the restriction of ρ to LR_{X_s},
- $I = \{\, p \in LR_{X_s} \mid \langle\!\langle p, \mathsf{suc}_\pi(p) \rangle\!\rangle_\rho \cap [\![X_{\leq s} - X_s]\!]_\rho \neq \emptyset \,\}$,
- $J = \{\, p \in LR_{X_s} \mid p \in [\![X_{\leq s} - X_s]\!]_\rho \,\}$,
- $K = \{\, p \in LR_{X_s} \mid \langle\!\langle p, \mathsf{suc}_\pi(p) \rangle\!\rangle_\rho \subseteq [\![X_{\leq s} - X_s]\!]_\rho \,\}$,
- $c = |E_{\leq s} - E_\rho - E_s|$
  $\quad = |\{\, \{u, v\} \in E_{\leq s} \mid u \notin X_s \text{ and } \langle\!\langle l_u, r_u \rangle\!\rangle \cap \langle\!\langle l_v, r_v \rangle\!\rangle_\rho = \emptyset \,\}|.$

Note that $p \in K$ implies $p, \mathsf{suc}_\pi(p) \in J$. Moreover, if $p \in J$ or $\mathsf{suc}_\pi(p) \in J$, then $p \in I$. We say that $\mathscr{A}(\rho', s) = (\pi', I', J', K', c')$ *dominates* $\mathscr{A}(\rho, s) = (\pi, I, J, K, c)$ iff $\pi' = \pi$, $I' \subseteq I$, $J' \subseteq J$, $K' \subseteq K$, and $c' \leq c$. In this case, every possible way of introducing new intervals to ρ is also possible for ρ' by cheaper or equivalent cost. Therefore, it is enough to remember $\mathscr{A}(\rho', s)$ discarding $\mathscr{A}(\rho, s)$. We call a set of abstractions *reduced* if it has no pair of distinct elements such that one dominates the other.

Our algorithm calculates a reduced set $\mathscr{I}_s$ of abstractions of interval representations of interval subgraphs of $(X_{\leq s}, E_{\leq s})$ for each node s of T which satisfies the following invariant.

Condition 1

- *Every element* $(\pi, I, J, K, c) \in \mathscr{I}_s$ *is the abstraction of some interval representation of an interval subgraph of* $(X_{\leq s}, E_{\leq s})$ *for* X_s,
- *Any interval representation* ρ *of any interval subgraph of* $(X_{\leq s}, E_{\leq s})$ *has an element of* $\mathscr{I}_s$ *that dominates its abstraction* $\mathscr{A}(\rho, X_s)$.

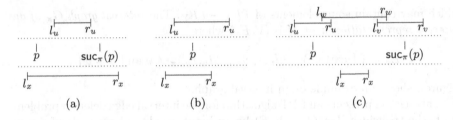

Fig. 3. Typical situations with (a) $p \in I$, (b) $p \in J$, (c) $p \in K$, where $\mathscr{A}(\rho, s) = (\pi, I, J, K, c)$ and $u, v, w \in X_{\leq s} - X_s$ are forgotten vertices. In the respective cases, we cannot introduce a new interval (l_x, r_x) that (a) covers $(p, \mathrm{suc}_\pi(p))$, (b) includes p, (c) overlaps $(p, \mathrm{suc}_\pi(p))$.

Since $\mathscr{I}_s$ is reduced, if $X_s = \emptyset$, we have $\mathscr{I}_s = \{(o, I, \emptyset, \emptyset, c)\}$ for some $I \subseteq \{\bot\}$ and $c \in \mathbb{N}$, where o is the trivial order such that $\bot <_o \top$. Particularly for the root node s, the number c is the least number such that one can obtain an interval subgraph by removing c edges from G. That is, c is the solution to our problem. If s is a leaf, $\mathscr{I}_s = \{(o, \emptyset, \emptyset, \emptyset, 0)\}$ by definition. It remains to show how to calculate $\mathscr{I}_s$ from the child(ren) of s, while preserving the invariant (Condition 1).

Introduce Node: Suppose that s has just one child t such that $X_s = X_t \cup \{x\}$. For (π, I, J, K, c) in $\mathscr{I}_t$, we say an extension π' of π to X_s *respects* E, I, J, K if

- $\{x, u\} \notin E$ for $u \in X_t$ implies $\langle\!\langle l_x, r_x \rangle\!\rangle_{\pi'} \cap \langle\!\langle l_u, r_u \rangle\!\rangle_{\pi'} = \emptyset$,
- $p \in I$ implies $\langle\!\langle p, \mathrm{suc}_\pi(p) \rangle\!\rangle_{\pi'} \not\subseteq \langle\!\langle l_x, r_x \rangle\!\rangle_{\pi'}$,
- $p \in J$ implies $p \notin \langle\!\langle l_x, r_x \rangle\!\rangle_{\pi'}$,
- $p \in K$ implies $\langle\!\langle p, \mathrm{suc}_\pi(p) \rangle\!\rangle_{\pi'} \cap \langle\!\langle l_x, r_x \rangle\!\rangle_{\pi'} = \emptyset$,

respectively. If π' does not respect some of E, I, J, K, then it means that we are creating an edge between two vertices which are not connected in the input graph G. For each interval representation π' extending π to LR_{X_s} that respects E, I, J, K, we put one or two elements into $\mathscr{I}'_s$ by the following manner. If $r_x \neq \mathrm{suc}_{\pi'}(l_x)$, we add $(\pi', I[r_x/\mathrm{pred}_{\pi'}(r_x)], J, K, c)$ to $\mathscr{I}'_s$ (see Fig. 4 (a)). Otherwise, let $p = \mathrm{pred}_{\pi'}(l_x)$, for which it holds that $p <_{\pi'} l_x <_{\pi'} r_x <_{\pi'} \mathrm{suc}_\pi(p)$. We have four exhaustive cases shown in Fig. 4 (b). If either $p \notin I$ or $J \cap \{p, \mathrm{suc}_\pi(p)\} = \{p\}$, we add (π', I, J, K, c) to $\mathscr{I}'_s$. If $\mathrm{suc}_\pi(p) \in J$, we add $(\pi', I \cup \{r_x\}, J, K, c)$ to $\mathscr{I}'_s$. If $p \in I$ and $J \cap \{p, \mathrm{suc}_\pi(p)\} = \emptyset$, we add both (π', I, J, K, c) and $(\pi', I[r_x/p], J, K, c)$ to $\mathscr{I}'_s$. Those exhaust all the possibilities. We then obtain $\mathscr{I}_s$ by reducing $\mathscr{I}'_s$.

Forget Node: Suppose that s has just one child t such that $X_t = X_s \cup \{x\}$. For each (π, I, J, K, c) in $\mathscr{I}_t$, in accordance with the definition of abstractions, we add to $\mathscr{I}'_s$ the quintuple (π', I', J', K', c) where

- π' is the restriction of π,
- $I' = \{p \in LR_{X_s} \mid \langle\!\langle p, \mathrm{suc}_{\pi'}(p) \rangle\!\rangle_\pi \cap (I \cup \langle\!\langle l_x, r_x \rangle\!\rangle_\pi) \neq \emptyset\}$,
- $J' = \{p \in LR_{X_s} \mid p \in J \cup \langle\!\langle l_x, r_x \rangle\!\rangle_\pi\}$,
- $K' = \{p \in LR_{X_s} \mid \langle\!\langle p, \mathrm{suc}_{\pi'}(p) \rangle\!\rangle_\pi \subseteq K \cup \langle\!\langle l_x, r_x \rangle\!\rangle_\pi\}$,

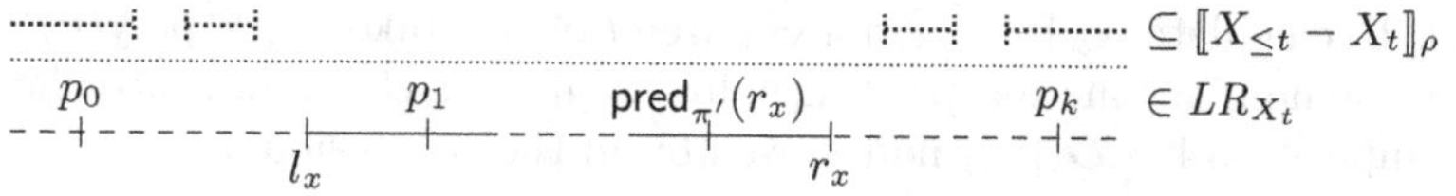

$$\subseteq [\![X_{\leq t} - X_t]\!]_\rho$$
$$\in LR_{X_t}$$

(a) When $r_x \neq \mathsf{suc}_{\pi'}(l_x)$. We assume possible forgotten intervals should appear left to l_x and right to r_x to prevent (l_x, r_x) from intersecting forgotten intervals.

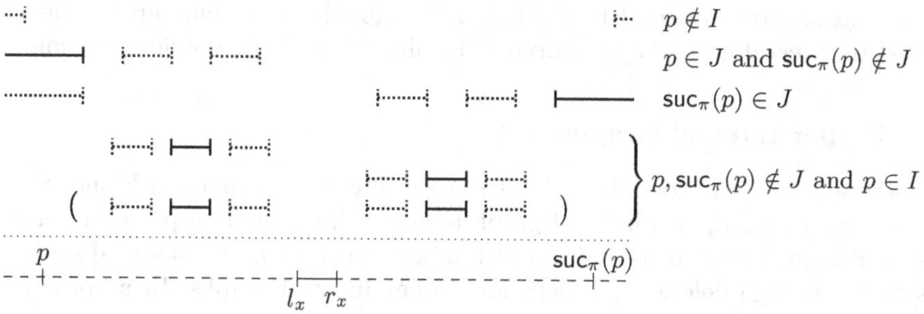

$$p \notin I$$
$$p \in J \text{ and } \mathsf{suc}_\pi(p) \notin J$$
$$\mathsf{suc}_\pi(p) \in J$$
$$\left.\vphantom{\begin{array}{c}1\\1\\1\end{array}}\right\} p, \mathsf{suc}_\pi(p) \notin J \text{ and } p \in I$$

(b) When $r_x = \mathsf{suc}_{\pi'}(l_x)$. Our algorithm does not consider all the admissible extensions ρ' of π'. For example, we do not put $(\pi, I \cup \{p, r_x\}, J, K, c)$ into $\mathscr{I}'_s$ as illustrated in the parentheses above, since it is dominated by other possibilities and will be absent in $\mathscr{I}_s$ anyway.

Fig. 4. Illustrating how we insert l_x and r_x into previously determined interval representation. Thick lines illustrate $[\![X_{\leq t} - X_t]\!]_\rho$ for a possible extension ρ such that $\mathscr{A}(\rho, X_t) = (\pi, I, J, K, c)$.

- $c' = c + |\{\, \{x, u\} \in E \mid \langle\!\langle l_u, r_u \rangle\!\rangle_\pi \cap \langle\!\langle l_x, r_x \rangle\!\rangle_\pi = \emptyset \text{ and } u \in X_s \,\}|.$

Then we obtain $\mathscr{I}_s$ by reducing $\mathscr{I}'_s$.

Join Node: Suppose that s has two children t_1 and t_2, where $X_s = X_{t_1} = X_{t_2}$. We say that $A_1 = (\pi_1, I_1, J_1, K_1, c_1) \in \mathscr{I}_{t_1}$ and $A_2 = (\pi_2, I_2, J_2, K_2, c_2) \in \mathscr{I}_{t_2}$ are *compatible* if $\pi_1 = \pi_2$ and $J_1 \cap J_2 = I_1 \cap K_2 = K_1 \cap I_2 = \emptyset$. If A_1 and A_2 are not compatible, any interval representation ρ on $X_{\leq s}$ which extends ρ_1 and ρ_2 will connect two vertices which are not adjacent in the input graph G for any interval representations ρ_i on $X_{\leq t_i}$ of which A_i is the abstraction for $i = 1, 2$. For each compatible pair (A_1, A_2), one can find an interval representation ρ on $X_{\leq s}$ that forms a subgraph of $(X_{\leq s}, E_{\leq s})$ which extends some ρ_1 and ρ_2 whose abstractions are A_1 and A_2, respectively. Then we add the quintuple $(\pi_1, I_1 \cup I_2, J_1 \cup J_2, K_1 \cup K_2, c_1 + c_2)$ to $\mathscr{I}'_s$. We obtain $\mathscr{I}_s$ by reducing $\mathscr{I}'_s$.

Theorem 1. *The edge deletion problem for interval graphs can be solved in* $O(|V|N^2\mathrm{poly}(k))$ *time where* $N = (2k)! \cdot 2^{2k}$ *for the treewidth k of G. If k is the pathwidth, it can be solved in* $O(|V|N\mathrm{poly}(k))$ *time.*

Proof. Let k be the maximum size of the assigned set X_s to a node of a nice tree-decomposition. Each $\mathscr{I}_s$ may contain at most $N = (2k)!/2^k \cdot (2^k)^3 = (2k)! \cdot 2^{2k}$

elements. To calculate $\mathscr{I}_s$ from $\mathscr{I}_t$ for children t of s, it takes $O(N^\ell \text{poly}(k))$ time for some polynomial function poly, if it has at most ℓ children. Since the nice tree-decomposition has $O(|V|)$ nodes, we obtain the conclusion.

4 Algorithms for Other Graph Classes

We in this section present EDGE-DELETION algorithms on some graph classes related to interval graphs by modifying the algorithm in the previous section.

4.1 Proper Interval Graphs

An interval representation π is said to be *proper* if there are no $u, v \in V$ such that $l_u <_\pi l_v <_\pi r_v <_\pi r_u$. An interval graph is *proper* if it admits a proper interval representation. The algorithm presented in Sect. 3 can easily be modified so that it solves the edge deletion problem for proper interval graphs. In accordance with the definition of a proper interval representation, we simply require π' in *Introduce Node* to be a proper interval representation. Under the restriction, we see that $I = \{ p \mid p \in J \text{ or } \text{suc}_\pi(p) \in J \}$ if $(\pi, I, J, K, c) \in \mathscr{I}_s$. Then, I can be discarded from each abstraction.

Corollary 1. *The edge deletion problem for proper interval graphs can be solved in $O(|V|N^2\text{poly}(k))$ time where $N = (2k)! \cdot 2^k$ for the treewidth k of G. If k is the pathwidth, it can be solved in $O(|V|N\text{poly}(k))$ time.*

4.2 Trivially Perfect Graphs

An interval representation π is said to be *nested* if there are no $u, v \in V$, such that $l_u <_\pi l_v <_\pi r_u <_\pi r_v$. A *trivially perfect graph* is an interval graph that admits a nested interval representation. The algorithm presented in Sect. 3 can easily be modified so that it solves the edge deletion problem for trivially perfect graphs. In accordance with the definition of the graph class, we simply require π' in *Introduce Node* to be a nested interval representation. Under the restriction, we see that $l_v \in J$ if and only if $r_v \in J$ if and only if $l_v \in K$. Therefore, we do not need to have the set J any more.

Corollary 2. *The edge deletion problem for trivially perfect graphs can be solved in $O(|V|N^2\text{poly}(k))$ time where $N = (2k)! \cdot 2^k$ for the treewidth k of G. If k is the pathwidth, it can be solved in $O(|V|N\text{poly}(k))$ time.*

4.3 Circular-Arc Graphs

Circular-arc graphs are a generalization of interval graphs which have a "circular" interval representation. For a linear order π over a set S, we let

$$\langle\!\langle p_1, p_2 \rangle\!\rangle_\pi = \begin{cases} \{ q \in S \mid p_1 \leq_\pi q <_\pi p_2 \} & \text{if } p_1 \leq_\pi p_2, \\ \{ q \in S \mid p_1 \leq_\pi q \vee q <_\pi p_2 \} & \text{if } p_2 <_\pi p_1. \end{cases}$$

A *circular-arc graph* is a graph $G_\pi = (V, E)$ such that

$$E = \{\, \{u, v\} \subseteq V \mid \langle\!\langle l_u, r_u \rangle\!\rangle_\pi \cap \langle\!\langle l_v, r_v \rangle\!\rangle_\pi \neq \emptyset \,\}$$

for some linear order π over $L_V \cup R_V$. Note that this set contains neither $\top$ nor $\bot$. The algorithm presented in Sect. 3 can easily be modified so that it solves the edge deletion problem for circular-arc graphs by replacing the definition of $\langle\!\langle p_1, p_2 \rangle\!\rangle_\pi$ as above, and defining $\mathsf{suc}_\pi(\max_\pi LR_V) = \min_\pi LR_V$ and $\mathsf{pred}_\pi(\min_\pi LR_V) = \max_\pi LR_V$. Since we allow $r_u <_\pi l_u$, the number of admissible circular interval representations is bigger than that of (ordinary) interval representations. This affects the computational complexity.

Corollary 3. *The edge deletion problem for circular-arc graphs can be solved in $O(|V|N^2\mathrm{poly}(k))$ time where $N = (2k)! \cdot 2^{3k}$ for the treewidth k of G. If k is the pathwidth, it can be solved in $O(|V|N\mathrm{poly}(k))$ time.*

4.4 Threshold Graphs

Threshold graphs are special cases of trivially perfect graphs, which can be defined in several different ways. Here we use a pair of a vertex subset $W \subseteq V$ and a linear order π over R_V as a *threshold interval representation*. We say that vertices u and v *intersect on (W, π)* if and only if $u \in W$ and $r_v <_\pi r_u$ or the other way around. A *threshold graph* is a graph $G_{W,\pi} = (V, E_{W,\pi})$ where (W, π) is a threshold interval representation on V and $E_{W,\pi} = \{\, \{u, v\} \subseteq V \mid u$ and v intersect on $(W, \pi)\,\}$. By extending π to π' over LR_V so that $l_w <_{\pi'} r_v$ for all $w \in W$ and $v \in V$ and $\mathsf{suc}_{\pi'}(l_u) = r_u$ for all $u \in V - W$, then the induced interval graph coincides with the threshold graph. To attain drastic improvement on the complexity, we design an algorithm for the edge deletion problem for threshold graphs from scratch, rather than modifying the one for interval graphs.

For a threshold representation (Y, ρ) of a subgraph $G_{Y,\rho} = (X_{\leq s}, E_{Y,\rho})$, we define its abstraction $\mathscr{A}((Y, \rho), s) = (Y', \pi, b, p, c)$ as follows: (1) π is the restriction of ρ to R_{X_s}, (2) $Y' = Y \cap X_s$, (3) if $Y' = Y$, then $b = 0$ and $p = \max_\pi\{p \in R_{X_s} \mid p <_\rho r_y$ for all $y \in X_{\leq s} - X_s\}$, (4) if $Y' \neq Y$, then $b = 1$ and $p = \max_\pi\{p \in R_{X_s} \mid p <_\rho r_y$ for some $y \in Y - Y'\}$, and (5) $c = |E_{\leq s} - E_{Y,\rho} - E_s| = |\{\, \{u, v\} \in E \mid \{u, v\} \not\subseteq X_s$ and u and v do not intersect on $(Y, \rho)\,\}|$.

We say that (Y', π', b', p', c') *dominates* (Y, π, b, p, c) if $Y' = Y$, $\pi' = \pi$, $c' \leq c$, and either (a) $b' = b = 0$ and $p' \geq_\pi p$, (b) $b' = b = 1$ and $p' \leq_\pi p$, or (c) $b' = 0$ and $b = 1$. Using the above invariant, we can provide an algorithm for THRESHOLD-EDGE-DELETION.

Theorem 2. *The edge deletion problem for threshold graphs can be solved in $O(|V|N^2\mathrm{poly}(k))$ time where $N = k! \cdot 2^k$ for the treewidth k of G. If k is the pathwidth, it can be solved in $O(|V|N\mathrm{poly}(k))$ time.*

5   Conclusion

We propose FPT algorithms for EDGE-DELETION to some intersection graphs parameterized by treewidth in this paper. Our algorithms maintain partial intersection models on a node of a tree decomposition with some restrictions and extend the models consistently for the restrictions in the next step. We expect that the ideas in our algorithms can be applied to other intersection graphs whose intersection models can be represented as linear-orders, for example circle graphs, chain graphs and so on, and to VERTEX-DELETION of intersection graphs.

We have the following questions as future work:

– Do there exist single exponential time algorithms for the considered problems, that is, $O^*(2^{tw(G)})$ time, or can we show matching lower bounds assuming the Exponential Time Hypothesis?
– Are there FPT algorithms parameterized by treewidth for $\mathcal{C}$-COMPLETION which is to find the minimum number of adding edges to obtain a graph in an intersection graph class $\mathcal{C}$? We can naturally apply the idea of our algorithms to $\mathcal{C}$-COMPLETION problems. While $\mathcal{C}$-EDGE-DELETION algorithms do not allow introduced objects to intersect with forgotten objects, $\mathcal{C}$-COMPLETION algorithms do allow it with the cost of addition of new edges. Thus $\mathcal{C}$-COMPLETION algorithms based on this naive approach will be XP algorithms since we have to remember the number of forgotten objects in the representation to count the number of intersections between the introduced objects and forgotten objects.
– Are there FPT algorithms for EDGE-DELETION to intersection graphs defined using objects on a plane, like unit disk graphs? The intersection graph classes discussed in this paper are all defined using objects aligned on a line. Going up to a geometric space of higher dimension is a challenging topic.

Acknowledgement. This work was supported in part by JSPS KAKENHI Grant Numbers JP18H04091 and JP19K12098.

References

1. Bessy, S., Perez, A.: Polynomial kernels for proper interval completion and related problems. Inf. Comput. **231**, 89–108 (2013)
2. Bodlaender, H.L.: A linear-time algorithm for finding tree-decompositions of small treewidth. SIAM J. Comput. **25**(6), 1305–1317 (1996)
3. Brandstädt, A., Le, V.B., Spinrad, J.P.: Graph Classes: A Survey. Society for Industrial and Applied Mathematics, Philadelphia, PA, USA (1999)
4. Burzyn, P., Bonomo, F., Durán, G.: NP-completeness results for edge modification problems. Discrete Appl. Math. **154**(13), 1824–1844 (2006)
5. Cai, L.: Fixed-parameter tractability of graph modification problems for hereditary properties. Inf. Process. Lett. **58**(4), 171–176 (1996)
6. Cao, Y.: Linear recognition of almost interval graphs. In: Proceedings of the Twenty-Seventh Annual ACM-SIAM Symposium on Discrete Algorithms, SODA 2016, Arlington, VA, USA, 10–12 January 2016, pp. 1096–1115 (2016)

7. Cao, Y., Marx, D.: Interval deletion is fixed-parameter tractable. ACM Trans. Algorithms **11**(3), 21:1–21:35 (2015)
8. Courcelle, B.: The monadic second-order logic of graphs XV: on a conjecture by D. Seese. J. Appl. Logic **4**(1), 79–114 (2006)
9. Courcelle, B., Engelfriet, J.: Graph Structure and Monadic Second-Order Logic - A Language-Theoretic Approach. Encyclopedia of Mathematics and Its Applications, vol. 138. Cambridge University Press (2012)
10. Cygan, M., et al.: Lower bounds for kernelization. In: Cygan, M., et al. (eds.) Parameterized Algorithms, pp. 523–555. Springer, Cham (2015). https://doi.org/10.1007/978-3-319-21275-3_15
11. Drange, P.G., Pilipczuk, M.: A polynomial kernel for trivially perfect editing. Algorithmica **80**(12), 3481–3524 (2018). https://doi.org/10.1007/s00453-017-0401-6
12. Frenkel, Z., Paux, E., Mester, D.I., Feuillet, C., Korol, A.B.: LTC: a novel algorithm to improve the efficiency of contig assembly for physical mapping in complex genomes. BMC Bioinform. **11**, 584 (2010). https://doi.org/10.1186/1471-2105-11-584
13. Goldberg, P.W., Golumbic, M.C., Kaplan, H., Shamir, R.: Four strikes against physical mapping of DNA. J. Comput. Biol. **2**(1), 139–152 (1995)
14. Golumbic, M.C.: Algorithmic Graph Theory and Perfect Graphs (Annals of Discrete Mathematics, vol. 57). North-Holland Publishing Co., Amsterdam (2004)
15. Heggernes, P., van't Hof, P., Jansen, B.M.P., Kratsch, S., Villanger, Y.: Parameterized complexity of vertex deletion into perfect graph classes. Theor. Comput. Sci. **511**, 172–180 (2013)
16. van't Hof, P., Villanger, Y.: Proper interval vertex deletion. Algorithmica **65**(4), 845–867 (2013). https://doi.org/10.1007/s00453-012-9661-31
17. Kloks, T. (ed.): Treewidth, Computations and Approximations. LNCS, vol. 842. Springer, Heidelberg (1994). https://doi.org/10.1007/BFb0045375
18. Lewis, J.M., Yannakakis, M.: The node-deletion problem for hereditary properties is NP-complete. J. Comput. Syst. Sci. **20**(2), 219–230 (1980)
19. Liu, Y., Wang, J., You, J., Chen, J., Cao, Y.: Edge deletion problems: Branching facilitated by modular decomposition. Theor. Comput. Sci. **573**, 63–70 (2015)
20. Mancini, F.: Graph modification problems related to graph classes. Ph.D. thesis, University of Bergen, May 2008
21. Margot, F.: Some complexity results about threshold graphs. Discrete Appl. Math. **49**(1–3), 299–308 (1994)
22. Marx, D.: Chordal deletion is fixed-parameter tractable. Algorithmica **57**(4), 747–768 (2010). https://doi.org/10.1007/s00453-008-9233-8
23. Nastos, J., Gao, Y.: Bounded search tree algorithms for parametrized cograph deletion: Efficient branching rules by exploiting structures of special graph classes. Discrete Math., Alg. and Appl. **4**(1) (2012)
24. Natanzon, A., Shamir, R., Sharan, R.: Complexity classification of some edge modification problems. Discrete Appl. Math. **113**(1), 109–128 (2001)
25. Sharan, R.: Graph Modification Problems and their Applications to Genomic Research. Ph.D. thesis, University of Bergen, May 2008
26. Spinrad, J.P.: Efficient Graph Representations. Fields Institute Monographs. American Mathematical Society, Providence (2003)
27. Waterman, S., Griggs, M.R.: Interval graphs and maps of DNA. J. Bull. Math. Biol. **48**, 189–195 (1986)
28. Villanger, Y., Heggernes, P., Paul, C., Telle, J.A.: Interval completion is fixed parameter tractable. SIAM J. Comput. **38**(5), 2007–2020 (2009)

r-Gathering Problems on Spiders: Hardness, FPT Algorithms, and PTASes

Soh Kumabe[1,2(✉)] and Takanori Maehara[2]

[1] The University of Tokyo, Tokyo, Japan
soh_kumabe@mist.i.u-tokyo.ac.jp
[2] RIKEN AIP, Tokyo, Japan
takanori.maehara@riken.jp

Abstract. We consider the *min-max r-gathering problem* described as follows: We are given a set of users and facilities in a metric space. We open some of the facilities and assign each user to an opened facility such that each facility has at least r users. The goal is to minimize the maximum distance between the users and the assigned facility. We also consider the *min-max r-gather clustering problem*, which is a special case of the *r*-gathering problem in which the facilities are located everywhere. In this paper, we study the tractability and the hardness when the underlying metric space is a *spider*, which answers the open question posed by Ahmed et al. [WALCOM'19]. First, we show that the problems are NP-hard even if the underlying space is a spider. Then, we propose FPT algorithms parameterized by the degree d of the center. This improves the previous algorithms because they are parameterized by both r and d. Finally, we propose PTASes to the problems. These are best possible because there are no FPTASes unless P = NP.

1 Introduction

Background and Motivation. We consider the following problem, called the *min-max r-gathering problem* (*r*-gathering problem, for short) [5].

Problem 1 (*r*-gathering problem). *We are given a set $\mathcal{U}$ of n users, a set $\mathcal{F}$ of m facilities on a metric space $(\mathcal{M}, \mathrm{dist})$, and a positive integer r. We open a subset of the facilities and assign each user to an opened facility such that all opened facilities have at least r users. The objective is to minimize the maximum distance from the users to the assigned facilities. Formally, the problem is written as follows.*

$$\begin{aligned}
&\text{minimize } \max_{u \in \mathcal{U}}(\mathrm{dist}(u, \pi(u))) \\
&\text{such that } \pi(u) \in \mathcal{F} \text{ for all } u \in \mathcal{U}, \\
&\qquad |\pi^{-1}(f)| = 0 \text{ or } |\pi^{-1}(f)| \ge r \text{ for all } f \in \mathcal{F}.
\end{aligned} \tag{1}$$

We also consider the *r-gather clustering problem* [2,3,8], which is a variant of the *r*-gathering problem in which the facilities are located everywhere.

© Springer Nature Switzerland AG 2021
R. Uehara et al. (Eds.): WALCOM 2021, LNCS 12635, pp. 154–165, 2021.
https://doi.org/10.1007/978-3-030-68211-8_13

Problem 2 (r-gather clustering problem). *We are given a set $\mathcal{U}$ of n users on a metric space $(\mathcal{M}, \text{dist})$ and a positive integer r. We partition $\mathcal{U}$ into arbitrarily many clusters $C_1, \ldots, C_k$ such that each user is contained in exactly one cluster, and all clusters contain at least r users. The objective is to minimize the maximum diameter (distance between the farthest pair) of the clusters. Formally the problem is written as follows.*

$$\begin{aligned}
&minimize \ \max\{\text{diam}(C_1), \ldots, \text{diam}(C_k)\} \\
&such\ that\ \{C_1, \ldots, C_k\}\ is\ a\ partition\ of\ \mathcal{U}, \\
&\hspace{2.5cm} |C_1| \geq r, \ldots, |C_k| \geq r.
\end{aligned} \tag{2}$$

These problems have several practical applications, with privacy protection [10] being a typical one. Imagine a company that publishes clustered data about their customers. If there is a tiny cluster, each individual of the cluster can be easily identified. Thus, to guarantee anonymity, the company requires the clusters to have at least r individuals; this criterion is called the r-*anonymity*. Such clusters are obtained by solving the r-gather clustering problem. Another typical problem is the sport-team formation problem [2]. Imagine that a town has n football players and m football courts. We want to divide the players into several teams, each of which contains at least eleven people and assign a court to each team such that the distance from their homes to the court is minimized. Such an assignment is obtained by solving the 11-gathering problem.

In theory, because the r-gathering problem and r-gather clustering problem are some of the simplest versions of the constrained facility location problems [6], several studies have been conducted, and many tractability and intractability results have been obtained so far. If $\mathcal{M}$ is a general metric space, there is a 3-approximation algorithm for the r-gathering problem, and no algorithm can achieve a better approximation ratio unless $P = NP$ [5]. If the set of locations of the users is a subset of that of the facilities[1], there is a 2-approximation algorithm [1], and no algorithm can achieve a better approximation ratio unless $P = NP$ [5]. If $\mathcal{M}$ is a line, there are polynomial-time exact algorithms by *dynamic programming* (DP) for the r-gathering problem [3,4,7,8], where the fastest algorithm runs in linear time [4]. The same technique can be implemented to the r-gather clustering problem.

If $\mathcal{M}$ is a *spider*, which is a metric space constructed by joining d half-line-shaped metrics together at endpoints[2] there are *fixed-parameter tractable* (FPT) algorithms parameterized by both r and d [2]—More precisely, the running time of their algorithm is $O(n + m + r^d 2^d (r + d)d)$ time[3]. Note that this is not an FPT algorithm parameterized *only* by d because it has a factor of r^d. They also posed an open problem that demands a reduction in the complexity of the r-gathering problem on a spider.

---

[1] This version of the problem is originally called the r-gather clustering problem [1].

[2] Ahmed et al. [2] called this metric space "star." In this paper, we followed https://www.graphclasses.org/classes/gc_536.html, a part of Information System of Graph Classes and their Inclusions (ISGCI).

[3] Ahmed et al. [2]'s original algorithm runs in $O(n + r^2 m + r^d 2^d (r + d)d)$ time, but by combining Sarker [4]'s linear-time algorithm on a line, we obtain this running time.

Our Contribution. In this study, we answer the open question that asks the complexity of r-gathering problem, posed by Ahmed et al. [2] by closing the gap between the tractability and intractability of the problems on a spider.

First, we prove that the problems are NP-hard, even on a spider as follows.

Theorem 1. *The min-max r-gather clustering problem and min-max r-gathering problem are NP-hard even if the input is a spider.*

The proof outline appears in Sect. 3. See the full version for the full proof. This implies that some parameterization, such as by the degree d of the center of the spider, is necessary to obtain FPT algorithms for the problems.

Second, we propose FPT algorithms parameterized by d as follows.

Theorem 2. *There is an algorithm to solve the r-gather clustering problem on a spider in $O(2^d r^4 d^5 + n)$ time, where d is the degree of the center of the spider. Similarly, there is an algorithm to solve the r-gathering problem on a spider in $O(2^d r^4 d^5 + n + m)$ time.*

The proof appears in Sect. 4. This result is the best possible in the sense of the number of parameters with superpolynomial dependence because at least one parameter (e.g., degree) is necessary according to Theorem 1. Our algorithms have lower parameter dependencies than previous algorithms [2] because they are parameterized by both r and d. More concretely, our algorithms have no $O(r^d)$ factors.

Finally, we propose *polynomial-time approximation schemes* (PTASes) to the problems.

Theorem 3. *There are PTASes to the r-gather clustering problem and r-gathering problem on a spider.*

The proof appears in Sect. 5. This result is also the best possible because Theorem 1 implies that there are no *fully polynomial-time approximation schemes* (FPTASes) unless P = NP (Corollary 1). These PTASes can be generalized to the r-gather clustering problem and r-gathering problem on a tree (see the full version).

2 Preliminaries

A *spider* $\mathcal{L} = \{l_1, \ldots, l_d\}$ is a set of half-lines that share the endpoint o (see Figure (a)). Each half-line is called a *leg* and o is the *center*. The point on leg l, whose distance from the center is x, is denoted by $(l, x) \in \mathcal{L} \times \mathbb{R}_+$. It should be noted that $(l, 0)$ is the center for all l. $\mathcal{L}$ induces a metric space whose distance is defined by $\text{dist}((l, x), (l', x')) = |x - x'|$ if $l = l'$ and $x + x'$ if $l \neq l'$.

Let $\mathcal{U} = \{u_1, \ldots, u_n\}$ be a set of n users on $\mathcal{L}$. A *cluster* C is a subset of users. The *diameter* of C is the distance between two farthest users in the cluster, i.e., $\text{diam}(C) = \max_{u_i, u_j \in C} \text{dist}(u_i, u_j)$.

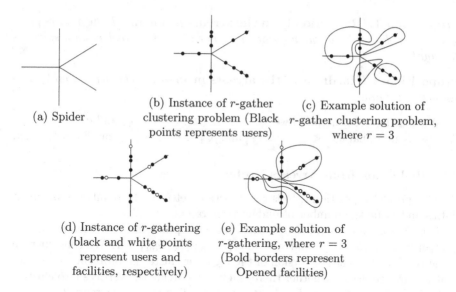

(a) Spider

(b) Instance of r-gather
clustering problem (Black
points represents users)

(c) Example solution of
r-gather clustering problem,
where $r = 3$

(d) Instance of r-gathering
(black and white points
represent users and
facilities, respectively)

(e) Example solution of
r-gathering, where $r = 3$
(Bold borders represent
Opened facilities)

3   NP-Hardness of r-Gather Clustering on Spider

We prove Theorem 1 by showing the r-gather clustering problem is NP-hard even
on a spider. The NP-hardness of the r-gathering problem immediately follows
from this result because the r-gathering problem is reduced to the r-gather
clustering problem by putting facilities on the midpoints of the pairs of users.

The strategy for the proof is as follows. We first introduce the *arrears problem*
(Problem 3) as an intermediate problem. Then, we reduce the arrears problem
to the r-gather clustering problem on a spider. Finally, we prove the strong
NP-hardness of the arrears problem.

The arrears problem is the following decision problem.

Problem 3 (Arrears Problem). *We are given n sets $S_1, \ldots, S_n$ of pairs of
integers, i.e., $S_i = \{(a_{i,1}, p_{i,1}), \ldots, (a_{i,|S_i|}, p_{i,|S_i|})\}$ for all $i = 1, \ldots, n$, and m
pairs of integers $(b_1, q_1), \ldots, (b_m, q_m)$. The task is to decide whether there are n
integers $z_1, \ldots, z_n$ such that the following inequality holds for all $j = 1, \ldots, m$:*

$$\sum_{a_{i,z_i} \leq b_j} p_{i,z_i} \leq q_j. \tag{3}$$

The name of the "arrears problem" comes from the following interpretation.
Imagine a person who has pending arrears in his n *payment duties* $S_1, \ldots, S_n$.
Each payment duty S_i has multiple options $(a_{i,1}, p_{i,1}), \ldots, (a_{i,|S_i|}, p_{i,|S_i|})$ such
that he can choose a *payment amount* of $\$p_{i,k}$ with the *payment date* $a_{i,k}$ for
some k. Each pair (b_j, q_j) corresponds to his *budget constraint* such that he can
pay at most $\$q_j$ until the b_j-th day.

The arrears problem itself may be an interesting problem, but here we use
this problem as a milestone to prove the hardness of the r-gather clustering
problem on a spider. The proof follows the following two propositions.

Proposition 1 (Reduction from the arrears problem). *If the arrears problem is strongly NP-hard, the min-max r-gather clustering problem on a spider is NP-hard.*

Proposition 2 (Hardness of the arrears problem). *The arrears problem is strongly NP-hard.*

Without loss of generality, we assume that $b_1 < \cdots < b_m$ and $q_1 < \cdots < q_m$. We also assume that $a_{i,1} < \cdots < a_{i,|S_i|}$ and $p_{i,1} < \cdots < p_{i,|S_i|}$ for all $i = 1, \ldots, n$.

3.1   Reduction from Arrears Problem

We first prove Proposition 1. In this subsection, let n be the number of payment duties and m be the number of budget constraints.

Let $\mathcal{I}$ be an instance of the arrears problem. We define $L = \max\{\max_i a_{i,|S_i|}, b_m\} + 1$ and $r = \max\{\max_i p_{i,|S_i|}, q_m\} + 1$. We construct an instance $\mathcal{I}'$ of the decision version of the r-gather clustering problem on a spider that requires to decide whether there is a way to divide the vertices into clusters each of which has the size of at least r and the diameters of at most $2L$.

In the construction, we distinguish the legs into two types — *long* and *short*. Each long leg corresponds to a payment duty and each short leg corresponds to a budget constraint. For each payment duty S_i, we define a long leg i. We first put r users on $(i, 4L - a_{i,|S_i|} + 1)$. Then, we put $p_{i,k+1} - p_{i,k}$ users on $(i, 2L - a_{i,k})$ for all $k = 1, \ldots, |S_i| - 1$. Finally, we put $r - p_{i,|S_i|}$ users on $(i, 2L - a_{i,|S_i|})$. Each short leg has only one user. The distance from the center to the user is referred to as the *length* of the short leg. For each $j = 1, \ldots, m$, we define $q_j - q_{j-1}$ short legs of length $b_{j-1} + 1$, where we set $q_0 = b_0 = 0$. We also define r short legs of length L. This construction is done in *pseudo-polynomial* time.

Now, we prove that $\mathcal{I}'$ has a feasible solution if and only if $\mathcal{I}$ is a YES-instance of the arrears problem. We first observe a basic structure of clusters in a feasible solution of $\mathcal{I}'$. The following lemma ensures that the choices of the payment dates on different payment duties are independent of each other.

Lemma 1. *In a feasible solution to $\mathcal{I}'$, there is no cluster that contains users from two different long legs.*

Proof. By definition, the distance between the center and a user on a long leg is larger than L. Therefore, the distance between users from two different long legs exceeds $2L$, indicating that they cannot be in the same cluster.

An *end cluster* of long leg i is a cluster that contains the farthest user of i. The above lemma implies that in a feasible solution, any end cluster of a different long leg is different. Intuitively, the "border" of the end cluster of long leg i corresponds to the choice from the options of payment duty S_i.

Lemma 2. *For each long leg i, the following three statements hold. (a) An end cluster of i only contains the users from i. (b) There is exactly one end cluster of i, and no other cluster consists of only users from leg i. (c) Some users on i are not present in the end cluster.*

Proof. (a) The endpoint of i is distant by more than $2L$ from the center. (b) There are less than $2r$ users on i; therefore, they cannot form more then one clusters alone. (c) Users on the point $(i, 2L - a_{i,|S_i|})$ are distant from the endpoint of i by more than $2L$; thus they cannot be in the same cluster.

Lemma 1 and Statement (c) of Lemma 2 imply that the users on a long leg who are not contained in end clusters should form a cluster together with users from short legs. Now, we prove Proposition 1.

Proof. (Proof of Proposition 1). Suppose that we have a feasible solution to the instance of the r-gathering problem on a spider that is constructed as mentioned above. For each long leg i, let u_i be the last user that is not contained in end clusters, and C_i be the cluster that contains u_i. Then, the location of u_i is represented as $(i, 2L - a_{i,z_i})$ using an integer z_i. We choose the payment date a_{i,z_i} for payment duty i. We prove that these choices of payment dates are a feasible solution to the arrears problem.

As described above, C_i consists of users from leg i and short legs. Because there are only $(r - p_{i,|S_i|}) + (p_{i,|S_i|} - p_{i,|S_i|-1}) + \cdots + (p_{i,z_i+1} - p_{i,z_i}) = r - p_{i,z_i}$ users in leg i on the path from the center to the location of u_i, C_i should contain at least p_{i,z_i} users on short legs with an at most length of a_{i,z_i}. For the j-th budget constraint, by the rule of construction, there are $(q_1 - q_0) + \cdots + (q_j - q_{j-1}) = q_j$ users on short legs whose length is at most b_j. Suppose $a_{i,z_i} \le b_j$. We use at least p_{i,z_i} users on short legs whose lengths are at most $a_{i,z_i} \le b_j$ in the cluster C_i. Thus, the sum of p_{i,z_i} among all i with $a_{i,z_i} \le b_j$ is at most the number of users on short legs whose length is at most b_j, that is, q_j. This implies that the budget constraint is valid.

Conversely, suppose that we are given a feasible solution to the instance $\mathcal{I}$ of the arrears problem. First, for each payment duty i we make a cluster with all users located between $(i, 4L - a_{i,|S_i|} + 1)$ and $(i, 2L - a_{i,z_i} + 1)$, inclusively. This cluster contains at least r users because there are r users on point $(i, 4L - a_{i,|S_i|} + 1)$ with a diameter of at most $2L$. We renumber the payment duties in the non-decreasing order of a_{i,z_i} and proceed them through the order of indices: for a payment duty $i = 1, 2, \ldots, n$, we make a cluster C_i using all remaining users on leg i and all users from the remaining p_{i,z_i} shortest short legs. By the construction, these clusters have exactly r users. We show that the diameter of C_i is at most $2L$. The diameter is spanned by a long leg and the longest short leg. The distance to the long leg in C_i is $2L - a_{i,z_i}$. The longest short leg in C_i is the $p_{1,z_1} + \cdots + p_{i,z_i}$-th shortest short leg. We take the smallest j such that $a_{i,z_i} \le b_j$. Then, because the given solution is a feasible solution to $\mathcal{I}$, $p_{1,z_1} + \cdots + p_{i,z_i} \le q_j$ holds. Because there are q_j users on short legs with a length of less than $b_{j-1} + 1 \le a_{i,z_i}$, the length of the longest short leg in C_i is at most a_{i,z_i}. This gives the diameter of C_i to be at most $2L$. Finally, we make a cluster with all remaining users. Because there are r short legs of length L and all these users are located within the distance L from the center, we can put them into a cluster. Then, we obtain a feasible solution to $\mathcal{I}'$.

3.2   Strong NP-Hardness of Arrears Problem

Now we give a proof outline of Proposition 2; the full proof is given in full version. We reduce the 1-IN-3SAT problem, which is known to be NP-complete [9].

Problem 4 (1-IN-3 SAT problem [9]). *We are given a set of clauses, each of which contains exactly three literals. Decide whether there is a truth assignment such that all clauses have exactly one true literal.*

Proof (Proof Outline of Proposition 2). Let n and m be the number of boolean variables and clauses, respectively. For each variable x_i, we prepare $N = 3m(m+2) + 1$ items T_i for a positive literal x_i and N items $\bar{T}_i$ for a negative literal $\bar{x}_i$. Let $T = \bigcup_i (T_i \cup \bar{T}_i)$ be the set of all items. Each item $y \in T$ corresponds to a payment duty $\{(a_{y,1}, p_{y,1}), (a_{y,2}, p_{y,2})\}$ of two options. Then, a solution to the arrears problem is specified by a set $X \subseteq T$ of items y such that $a_{y,2}$ is chosen. The complement of X is denoted by $\bar{X} = T \setminus X$. We want to construct a solution to the 1-IN-3SAT problem from a solution X to the arrears problem by $x_i = \texttt{true}$ if $y \in X$ for some $y \in T_i$; otherwise $x_i = \texttt{false}$. We define the payment dates and the amounts suitably to make this construction valid as follows. The payment days consist of two periods: the first period is $\{1, \ldots, n\}$ and the second period is $\{n+1, \ldots, n+m+2\}$. For each item y, $a_{y,1}$ belongs to the first period and $a_{y,2}$ belongs to the second period. Let $i = a_{y,1}$ and $j = a_{y,2} - (n+1)$. Then, the payment amount $p_{y,1}$ is given in the form of $B^4 + \alpha_y B^3 + iB^2 + i\alpha_y B + j$ where B is a sufficiently large integer, and α_y is a non-negative integer, where $\sum_{y \in T_i} \alpha_y = \sum_{y \in \bar{T}_i} \alpha_y = N$ holds for all i. We define $p_{y,2} = 2p_{y,1}$ for all $y \in T$.

Let $R = (1/2) \sum_{y \in T} p_{y,1} = nNB^4 + nNB^3 + n(n+1)/2NB^2 + n(n+1)/2NB + \cdots$. We make two budget constraints (n, R) and $(n+m+2, 3R)$. Then, these constraints hold in equality: Let $x \leq R$ be the total payment until n. Then, the total payment until $n+m+2$ is $x + 2(2R - x) = 4R - x \leq 3R$. These inequalities imply that $x = R$. (see the full version).

We use the first period to ensure that the truth assignment produced by X is well-defined, i.e., if $y \in X$ for some $y \in T_i$, then $y' \in X$ for all $y' \in T_i$. First, for each $i = 1, \ldots, n$, we add a budget constraint $(i, iNB^4 + iNB^3 + (B^3 - 1))$. By comparing the coefficients of B^4 and B^3, we have

$$\sum_{y \in \bar{X} \cap \bigcup_{j=1}^{i}(T_j \cup \bar{T}_j)} (B^4 + a_y B^3) \leq iNB^4 + iNB^3. \tag{4}$$

We can prove that for all i, these inequalities hold in equality, i.e.,

$$\sum_{y \in y \in \bar{X} \cap (T_i \cup \bar{T}_i)} (B^4 + a_y B^3) = NB^4 + NB \tag{5}$$

for all i as follows. Using the relation between the coefficients of $p_{y,1}$, we have $\sum_{y \in \bar{X}} (iB^2 + i a_y B) \geq \frac{n(n+1)}{2} NB^2 + \frac{n(n+1)}{2} NB$ (see the full version). Because the budget constraint (n, R) is fulfilled in equality, and the coefficients of B^2 and B in R are both $\frac{n(n+1)}{2} N$, this inequality holds in equality, which implies Eq. (5).

Then, we define the values of α_y appropriately so that only $X \cap (T_i \cap \bar{T}_i) = T_i$ or $X \cap (T_i \cap \bar{T}_i) = \bar{T}_i$ satisfies Eq. (5) (see the full version). This ensures the well-definedness of the truth assignment.

The second period represents the clauses. Let Z_i be the set of items with $a_{y,2} = i$. We put a budget constraint $(i, (nN + 2\sum_{j=n+1}^{i} K_j)B^4 + (B^4 - 1))$ for each $i = n + 1, \ldots, n + m + 2$, where $K_{n+1}, \ldots, K_{n+m+2}$ are non-negative integers determined later. Then, as similar to the first period, we can prove that

$$|\bar{X}| + 2|X \cap (Z_{n+1} \cup \cdots \cup Z_i)| = nN + 2 \sum_{j=n+1}^{i} K_j \qquad (6)$$

for each $i = n+1, \ldots, n+m+2$ (see the full version). This implies that $|X \cap Z_i| = K_i$ for each $i = n + 1, \ldots, n + m + 2$. The budget constraint on day $i \geq n + 3$ corresponds to the $i - (n + 2)$-th clause. For $i = n + 3, \ldots, n + m + 2$, we set $K_i = 1$. Then, we have $|X \cap Z_i| = 1$, i.e., exactly one literal in the $i - (n + 2)$-th clause is true. The budget constraints on day $n + 1$ and $n + 2$ are used for the adjustment. Because $\{Z_{n+1}, \ldots, Z_{n+m+2}\}$ forms a partition of items, we have $|X \cap Z_{n+1}| + |X \cap Z_{n+2}| = |X| - (|X \cap Z_{n+3}| + \cdots + |X \cap Z_{n+m+2}|) = N - m$. Moreover, because the constant term $e_{y,0}$ of $p_{y,1}$ is $e_{y,0} = i - (n+1)$ for all $y \in Z_i$ and $i = n+1, \ldots, n+m+2$, we have $\sum_{y \in X} e_{y,0} = \sum_{i=n+1}^{n+m+2}(i-(n+1))|X \cap Z_i|$. By solving these equations, we obtain $K_{n+1} = |X \cap Z_{n+1}|$ and $K_{n+1} = |X \cap Z_{n+2}|$. Because all values appearing in $\mathcal{I}'$ are at most $2B^4$, we can take B in a polynomial of n, m. Thus, the hardness proof is completed.

The following is a consequence of the construction.

Corollary 1 *The r-gather clustering problem on a spider does not admit an FPTAS unless $P = NP$.*

Proof. The diameter of the constructed spider is bounded by $O(n + m)$. Let us take such an instance. If there is an FPTAS for the r-gathering problem on a spider, by taking $\epsilon = 1/(c(n + m))$ for a sufficiently large constant c, we get an optimal solution because the optimal value is an integer at most $O(n+m)$. This contradicts the hardness.

4   FPT Algorithm for R-Gather Clustering and r-Gathering on Spider

We prove Theorem 2 by obtaining FPT algorithms to solve the r-gather clustering problem and r-gathering problem on a spider parameterized by the number d of legs. Due to the space limitation, we put all the pseudocodes in the full version.

First, we exploit the structure of optimal solutions. After that, we give a brute-force algorithm. Finally, we accelerate it by DP.

We denote the coordinate of user u by $(l(u), x(u))$. Without loss of generality, we assume that $x(u_1) \leq \cdots \leq x(u_n)$. We use this order to explain a set of users;

for example, "the first (resp. last) k users on leg l" indicates the users with k smallest (resp. largest) index among all users on leg l. We choose an arbitrary leg and consider all the users on the center as being located on this leg.

We introduce a basic lemma about the structure of a solution. A cluster is *single-leg* if it contains users from a single leg; otherwise, it is *multi-leg*. Ahmed et al. [2] showed that there is an optimal solution that has a specific single-leg/multi-leg structure as follows.

Lemma 3 ([2, Lemma 2]). *For both r-gather clustering problem and r-gathering problem, there is an optimal solution such that for all leg l, some users from the beginning (with respect to the order described above) are contained in multi-leg clusters, and the rest of them are contained in single-leg clusters.*

Now, we concentrate on the structure of multi-leg clusters. Let C be a multi-leg cluster. Let u_i be the last user in C and u_j be the last user with $l(u_i) \neq l(u_j)$ in C. A *ball part* of C is the set of users in C whose indices are at most j and a *segment part* of C is the set of the remaining users in C. C is *special* if C contains all users on $l(u_i)$ and the ball part is $\{u_1, \ldots, u_k\}$ for some integer k. The list of multi-leg clusters $\{C_1, \ldots, C_t\}$ are *suffix-special* if for all $1 \leq i \leq t$, C_i is a special when we only consider the users in $C_i, \ldots, C_t$.

The following lemma is the key to our algorithm. We omit the proof because it is a reformulation of Lemma 3 and Lemma 8 in [2] using Lemma 2 in [8].

Lemma 4 (Reformulation of [2, Lemmas 3 and 8] by [8, Lemma 2]). *Suppose $|\mathcal{U}| \geq 1$ and there exists an optimal solution without any single-leg cluster. Then, there is an optimal solution such that all clusters contain at most $2r - 1$ users, and there exists a special cluster.*

By definition, the segment part of a cluster is non-empty and contains users from a single leg. By removing a special cluster and applying the lemma repeatedly, we can state that there is an optimal solution consisting of a suffix-special family of multi-leg clusters.

We give a brute-force algorithm that enumerates all suffix-special families of multi-leg clusters in the full version, whose correctness is clear from the definition. For each enumerated clusters, we fix them and consider the remaining problem, which consists of single-leg clusters. Thus, the optimal solution is obtained by solving the line case problems independently for each leg.

Now, we accelerate this algorithm by DP. We observe that instead of remembering all data of C, it is sufficient to remember (1) the size of C (to avoid creating too-small clusters) and (2) the index of the last user in the ball part of C (to calculate the diameter/cost of the cluster). Here, (2) implies that if we know the last user u in the ball part of C and last user v in the segment part of C, the diameter/cost of C is computed because C is spanned by u and v. Below, we denote the diameter/cost of the multi-leg cluster by $\mathrm{Cost}(v, u)$ for both problems.

We also accelerate the process for single-leg clusters. As pre-processing for all leg l and all integers k from 0 to the number of users on leg l, we first compute the optimal value of the problem that only considers the last k users on leg l.

For each user u_i, we denote the optimal objective value for the set of users on leg $l(u_i)$ whose indices are greater than i and no less than i by $R^+(u_i)$ and $R^-(u_i)$, respectively. All these values can be computed in linear time for both r-gather clustering problem and r-gathering problem using the same technique as in [4] (see the full version for details).

The complete algorithm is presented in the full version. The correctness is clear from the construction. Thus, we analyze the time complexity. A naive implementation of the algorithm requires $O(2^d n^2 r^2 d)$ evaluations of Cost and preprocessing for R^+ and R^-. Each evaluation of Cost requires $O(1)$ time for the r-gather clustering problem and $O(m)$ time for the r-gathering problem. The preprocessing requires $O(n)$ time for the r-gather clustering problem and $O(n + m)$ time for the r-gathering problem [4]. Thus, the time complexities are $O(2^d n^2 r^2 d)$ for the r-gather clustering problem and $O(2^d n^2 r^2 dm)$ for the r-gathering problem.

We can further improve the complexities of the algorithms. The loop for i is bounded to look only the first $(2r-1)d$ users from each leg because other users cannot be contained in the ball part of multi-leg clusters. Thus, we can reduce n to rd^2 in the complexity so as we obtain the complexities $O(2^d r^4 d^5 + n)$ for the r-gather clustering problem and $O(2^d r^4 d^5 m + n)$ for the r-gathering problem. This proves Theorem 2 for the r-gather clustering problem. In the r-gathering problem, we can further improve the complexity by improving the algorithm to calculate Cost (see the full version). This reduces the complexity to $O(2^d r^4 d^5 + n + m)$, which proves Theorem 2 for the r-gathering problem.

5   PTAS for r-Gathering Problem

We prove Theorem 3 by demonstrating a PTAS for the r-gathering problem. The technique in this section can be extended to the case that the input is a tree; see the full version.

As mentioned at the beginning of Sect. 3, the r-gather clustering problem is reduced to the r-gathering problem. This establishes the existence of a PTAS for the r-gather clustering problem.

Given an instance $\mathcal{I}$ and a positive number $\epsilon > 0$, our algorithm outputs a solution whose cost is at most $(1+\epsilon)\mathrm{OPT}(\mathcal{I})$. Without loss of generality, we can assume that $\epsilon \leq 1$. First, we guess the optimal value. We can try all candidates of the optimal values because the optimal value is the distance between a user and a facility. We then solve the corresponding (relaxed) feasibility problem whose objective value is at most the guessed optimal value.

Now we consider implementing the following oracle $\mathtt{Solve}(\mathcal{I}, b, \delta)$: Given an instance $\mathcal{I}$, a threshold b, and a positive number δ, it reports YES if $\mathrm{OPT}(\mathcal{I}) \leq (1+\delta)b$ and NO if $\mathrm{OPT}(\mathcal{I}) > b$. If $b < \mathrm{OPT}(\mathcal{I}) \leq (1+\delta)b$ then both answers are acceptable. Our oracle also outputs the corresponding solution as a certificate if it returns YES. It should be noted that we cannot set $\delta = 0$ because it gets reduced to the decision version of the r-gathering problem, which is NP-hard on a spider (Theorem 1). To obtain a PTAS, we call the oracle $\mathtt{Solve}(\mathcal{I}, b, \epsilon)$ for each candidate of the optimal value b. We return the smallest b that the oracle returns YES.

5.1 Algorithm Part 1: Rounding Distance

This and the next subsections present an implementation of $\mathtt{Solve}(\mathcal{I}, b, \delta)$. Our algorithm is a DP that maintains distance information in the indices of the DP table. For this purpose, we round the distances so that the distances from the center to the vertices (thus, the users and facilities) are multiples of a positive number t as follows: For each user or facility on point (l, x), we move it to the coordinate $(l, \lceil x/t \rceil)$ to make a rounded instance $\mathcal{I}'$. Intuitively, this moves all users and facilities "toward the center" and regularizes the edge lengths into integers. Then, we define the rounded distance d' on $\mathcal{I}'$. This rounding process changes the optimal value only slightly as follows.

Lemma 5. *For any pair of points v and w, we have $d(v, w) \le d'(v, w)t \le d(v, w) + 2t$. Especially, $|OPT(\mathcal{I}) - OPT(\mathcal{I}')t| \le 2t$.*

Proof. Let o be the center of the spider. Then, $d(v, w) = d(v, o) + d(o, w)$ holds. By definition, $d(v, o) \le d'(v, o)t \le d(v, o) + t$ and $d(o, w) \le d'(o, w)t \le d(o, w) + t$. Adding them yields the desired inequality.

This lemma implies that an algorithm that determines whether $\mathcal{I}'$ has a solution whose cost is at most b/t works as an oracle $\mathtt{Solve}(\mathcal{I}, b, \epsilon)$ by taking $t = b\delta/2$.

5.2 Algorithm Part 2: Dynamic Programming

Now we propose an algorithm to determine whether $\mathcal{I}'$ has a solution whose cost is at most b/t. Because all distances between the users and the facilities of $\mathcal{I}'$ are integral, we can replace the threshold by $K = \lfloor b/t \rfloor$. An important observation is that K is bounded by a constant because of $K \le b/t = 2/\delta$.

Now, we establish a DP. We define a multi-dimensional table $\mathtt{S}$ of boolean values such that for each integer i and integer arrays $P = (p_0, \ldots, p_K)$ and $Q = (q_0, \ldots, q_K)$, $\mathtt{S}[i][P][Q]$ is $\mathtt{true}$ if and only if there is a way to

- open some facilities on $l_{\le i} := l_1 \cup \cdots \cup l_i$, and
- assign some users on $l_{\le i}$ to the opened facilities so that
  - for all $j = 0, \ldots, K$, there are p_j unassigned users in $l_{\le i}$ who are distant from the center by distance j and no other users are unassigned, and
  - for all $j = 0, \ldots, K$, we will assign q_j users out of $l_{\le i}$ who are distant from the center by distance j to the opened facilities in $l_{\le i}$.

Then, $\mathtt{S}[d][(0, \ldots, 0)][(0, \ldots, 0)]$ is the output of the $\mathtt{Solve}$ oracle. The elements of P and Q are non-negative integers at most n; thus, the size of the DP table is $O(d \times n^{2(K+1)})$, which is polynomial in the size of input.

To fill the table $\mathtt{S}$, we use an auxiliary boolean table $\mathtt{R}$ that only considers the i-th leg, i.e., for each integer i and integer arrays $P = (p_0, \ldots, p_K)$ and $Q = (q_0, \ldots, q_K)$, $\mathtt{R}[i][P][Q]$ is $\mathtt{true}$ if and only if there is a way to

- open some facilities on l_i, and
- assign some users on l_i to the opened facilities so that
  - for all $j = 0, \ldots, K$, there are p_j unassigned users in l_i who are distant from the center by distance j and no other users are unassigned, and
  - for all $j = 0, \ldots, K$, we will assign q_j users out of l_i who are distant from the center by distance j to the opened facilities in l_i.

We can fill the table R in polynomial time, and if we have the table R, we can compute the table S. Therefore we have Theorem 3. Due to the space limitation, the proof is given in the full version.

References

1. Aggarwal, G., et al.: Achieving anonymity via clustering. ACM Trans. Algorithms **6**(3), 49:1–49:19 (2010)
2. Ahmed, S., Nakano, S., Rahman, M.S.: r-gatherings on a star. In: Das, G.K., Mandal, P.S., Mukhopadhyaya, K., Nakano, S. (eds.) WALCOM 2019. LNCS, vol. 11355, pp. 31–42. Springer, Cham (2019). https://doi.org/10.1007/978-3-030-10564-8_3
3. Akagi, T., Nakano, S.: On r-gatherings on the line. In: Wang, J., Yap, C. (eds.) FAW 2015. LNCS, vol. 9130, pp. 25–32. Springer, Cham (2015). https://doi.org/10.1007/978-3-319-19647-3_3
4. Sarker, A., Sung, W., Rahman, M.S.: A linear time algorithm for the r-gathering problem on the line (extended abstract). In: Das, G.K., Mandal, P.S., Mukhopadhyaya, K., Nakano, S. (eds.) WALCOM 2019. LNCS, vol. 11355, pp. 56–66. Springer, Cham (2019). https://doi.org/10.1007/978-3-030-10564-8_5
5. Armon, A.: On min-max r-gatherings. Theoret. Comput. Sci. **412**(7), 573–582 (2011)
6. Drezner, Z., Hamacher, H.W.: Facility Location: Applications and Theory. Springer (2001)
7. Han, Y., Nakano, S.I.: On r-gatherings on the line. In: Proceedings of International Conference on Foundations of Computer Science, pp. 99–104 (2016)
8. Nakano, S.: A simple algorithm for r-gatherings on the line. In: Rahman, M.S., Sung, W.-K., Uehara, R. (eds.) WALCOM 2018. LNCS, vol. 10755, pp. 1–7. Springer, Cham (2018). https://doi.org/10.1007/978-3-319-75172-6_1
9. Schaefer, T.J.: The complexity of satisfiability problems. In: Proceedings of the tenth annual ACM symposium on Theory of computing, pp. 216–226. ACM (1978)
10. Sweeney, L.: k-anonymity: a model for protecting privacy. Int. J. Uncertainty, Fuzziness and Knowl.-Based Syst. **10**(05), 557–570 (2002)

An Improvement of Reed's Treewidth Approximation

Mahdi Belbasi[(✉)] and Martin Fürer

Department of Computer Science and Engineering,
Pennsylvania State University, University Park, PA 16802, USA
{belbasi,furer}@cse.psu.edu

Abstract. We present a new approximation algorithm for the treewidth problem which constructs a corresponding tree decomposition as well. Our algorithm is a faster variation of Reed's classical algorithm. For the benefit of the reader, and to be able to compare these two algorithms, we start with a detailed time analysis for Reed's algorithm. We fill in many details that have been omitted in Reed's paper. Computing tree decompositions parameterized by the treewidth k is fixed parameter tractable (FPT), meaning that there are algorithms running in time $O(f(k)g(n))$ where f is a computable function, $g(n)$ is polynomial in n, and n is the number of vertices. An analysis of Reed's algorithm shows $f(k) = 2^{O(k \log k)}$ and $g(n) = n \log n$ for a 5-approximation. Reed simply claims time $O(n \log n)$ for bounded k for his constant factor approximation algorithm, but the bound of $2^{\Omega(k \log k)} n \log n$ is well known. From a practical point of view, we notice that the time of Reed's algorithm also contains a term of $O(k^2 2^{24k} n \log n)$, which for small k is much worse than the asymptotically leading term of $2^{O(k \log k)} n \log n$. We analyze $f(k)$ more precisely, because the purpose of this paper is to improve the running times for all reasonably small values of k.

Our algorithm runs in $\mathcal{O}(f(k)n \log n)$ too, but with a much smaller dependence on k. In our case, $f(k) = 2^{O(k)}$. This algorithm is simple and fast, especially for small values of k. We should mention that Bodlaender et al. [2016] have an asymptotically faster algorithm running in time $2^{O(k)}n$. It relies on a very sophisticated data structure and does not claim to be useful for small values of k.

1 Introduction

Since the 1970s and early 1980s, when the notions of treewidth and tree decomposition were introduced [3, 10, 13], they have played important roles in computer science [5]. In a nutshell, treewidth is a parameter of a graph that measures how similar it is to a tree. One of the main reasons that the tree decomposition is widely studied is that many NP-complete problems have efficient algorithms for graphs with small treewidth. A graph problem is fixed parameter tractable (FPT) if it can be solved in time $\mathcal{O}(f(k)poly(n))$, where f is a computable function, k is a parameter of the graph, and n is the number of vertices. In fact,

© Springer Nature Switzerland AG 2021
R. Uehara et al. (Eds.): WALCOM 2021, LNCS 12635, pp. 166–181, 2021.
https://doi.org/10.1007/978-3-030-68211-8_14

Courcelle's metatheorem states that every graph property definable in monadic second-order logic of graphs can be solved in linear time on graphs of bounded treewidth [7]. The first step of solving such problems is to find a good or optimal tree decomposition. However, finding an optimal tree decomposition itself is NP-hard [2]. In this work, we propose an algorithm which is based on Reed's algorithm [12] to approximate the treewidth and find an approximately optimal tree decomposition.

1.1   Previously Known Results

In this work, we are interested in algorithms which run fast (polynomial in terms of the number of vertices) for graphs with bounded treewidth. One of the first algorithms given for this problem goes back to the same paper where treewidth has been shown to be NP-complete. Arnborg et al. [2] gave an algorithm which runs in time $\mathcal{O}(n^{k+2})$. In 1986, Robertson and Seymour [14] gave a quadratic time FPT approximation algorithm. Later on, Lagergren introduced an 8-approximation algorithm with the time complexity of $2^{\mathcal{O}(k \log k)} n \log^2 n$ [11]. In 1992, Reed [12] improved these algorithms to have an algorithm running in time $\mathcal{O}(n \log n)$, for fixed k. In this paper, we show that the approximation ratio is 7 or 5, depending on the frequency of the split by volume. We show that this algorithm runs in time $2^{\Theta(k \log k)} n \log n$, in order to be able to compare it to our algorithm. Like the algorithms of Robertson and Seymour, Reed's algorithm is recursive. In [14], they find a separator that partitions G into two parts but they do not force the separator to partition the entire graph in a balanced fashion. Reed finds a separator which partitions the graph in a balanced way to obtain time $\mathcal{O}(n \log n)$ for bounded k. This paper focuses on this algorithm. Later, Bodlaender gave an exact algorithm which runs in $2^{\mathcal{O}(k^3)} n$ [4]. Even though we focus only on constant-factor approximation algorithms but it is worth mentioning the two $f(k)$-approximation algorithms [8], and [1]. Finally, Bodlaender et al. [6] gave two constant factor approximation algorithms which run in $2^{\mathcal{O}(k)} \mathcal{O}(n \log n)$ and $2^{\mathcal{O}(k)} \mathcal{O}(n)$. The former one is a 3-approximation and latter one is a 5-approximation. Although it is a great result from a theoretical point of view. It uses a sophisticated data structure and the constant factor hidden in $\mathcal{O}(k)$ is not claimed to be practical. That is why we focus on Reed's simple and elegant $\mathcal{O}(n \log n)$ algorithm [12] here.

1.2   Our Contribution

First, we analyze Reed's algorithm [12] in detail. Reed has focused on the dependence on n because he wanted to come up with an algorithm which runs faster than $\mathcal{O}(n^2)$ (or even $\mathcal{O}(n \log^2 n)$ in [11]), for fixed k. We show that the dependence on k is of the form $2^{\Omega(k \log k)}$. Furthermore, we give a proof for the approximation ratio of Reed's algorithm by filling in the details.

Then, we propose two improvements and prove that the approximation ratio stays 5. One of our improvements focuses on the notion of "balanced split".

We call a split balanced, if we get two parts of volume $1 - \epsilon$ and ϵ (or better). Then, the running time of our algorithm has another factor of $1/\epsilon$. For instance, if we set $\epsilon = \frac{1}{100}$, a generous estimation shows that the dependence on k in our $\mathcal{O}(f(k)n\log n)$-time algorithm is $k^2\,2^{8.87k}$, instead of $2^{24k}(k+1)!$ in Reed's algorithm (here the asymptotic notation is a bit misleading from a practical point of view, as $2^{24k} = o(k!)$, even though $k!$ is reasonable for small k, while 2^{24k} is not). In the end, the main aim of this paper is to produce an algorithm that runs in time $2^{ck}n\log n$ with c as small as possible.

2   Preliminaries

2.1   Tree Decomposition

Definition 1. *A* tree decomposition *of a graph* $G = (V, E)$, *is a tree* $T = (V_T, E_T)$ *such that each node* x *in* V_T *is associated with a set* B_x *(called the bag of* x*) of vertices in* G, *and such that* T *has the following properties:*

- *The union of all bags is equal to* V. *In other words, for each* $v \in V$, *there exists at least one node* $x \in V_T$ *with* B_x *containing* v.
- *For every edge* $\{u, v\} \in E$, *there exists a node* x *such that* $u, v \in B_x$.
- *For any nodes* $x, y \in V_T$, *and any node* $z \in V_T$ *belonging to the path connecting* x *and* y *in* T, $B_x \cap B_y \subseteq B_z$.

In this paper we use a variation of tree decomposition where the adjacent bags differ in at most one vertex (converting can happen in linear time).

The *width of a tree decomposition* is the size of its largest bag minus one. The *treewidth* of a graph G is the minimum width over all tree decompositions of G called $tw(G)$. Observe that the treewidth of a tree is 1. In the following, we reserve the letter k for the treewidth+1. As we mentioned earlier, in 1992, Reed gave an algorithm which solves this problem in a nice balanced recursive way. That is why the running time of his algorithm is $\mathcal{O}(f(k)\,n\log n)$, for some computable function f. Reed does not specify f but an analysis of his algorithm shows it to be $k!$. This algorithm was a huge improvement in this field. Before, the fastest algorithm used quadratic time.

We have to mention that Bodlaender et al. [6] filled in some details on Reed's algorithm. We need to be more detailed because we do not use Reed's algorithm as a black box. That is why first we analyze Reed's algorithm precisely (Sect. 3) and then introduce improvements of his algorithm (Sect. 4).

3   Analysis of Reed's Algorithm

In 1992, Reed gave an elegant algorithm [12] to either construct a tree decomposition of width at most $7k$ or $5k$ of a given graph G, or declare that the treewidth is greater than k and outputs a subgraph which is a bottleneck (no separator of size $\leq k$).

3.1   Summary of Reed's Algorithm

In Reed's algorithm, one of the main tasks is to find a "balanced" separator S that splits the graph $G - S$ into two subgraphs with sets of vertices $X, Y \subseteq V(G)$. Once a balanced separator is found, the algorithm recursively finds a tree decomposition for $G[X \cup S]$ (the subgraph induced by $X \cup S$) and $G[Y \cup S]$.

The main task is to find a balanced separator. Instead of branching on every vertex (going to X, Y, or S, which will be exponential in n), Reed groups vertices together and works with the representatives of the groups. Then, he branches on the representatives.

Reed does a DFS and finds the deepest vertex v whose subtree has at least $\frac{n}{24k}$ vertices (later, we talk about this threshold). We call such a vertex a "representative", and he defines weight of the representative ($w(v)$) as the size of its subtree. We call these types of subtrees "small subtrees". The idea here, is that if a representative goes to either X or Y, most of its descendants will go to the same set. The reason is that if a descendant goes to another set, the path connecting the representative to the descendant should have at least one vertex in the separator. However, we know that the separator cannot have more than k vertices. So, not many vertices will go to the wrong set (not more than $\frac{n}{24}$ vertices, in total. This is because every subtree that partially goes to the other side should go through the separator and have one vertex there. Hence, not more than k small subtrees can go through the separator, which results in at most $\frac{n}{24}$ vertices on the wrong side). This nice property allows Reed to work with the set of representatives (which is much smaller) rather than all the vertices.

Now, one might think that why not just check all the possibilities of the representatives going to X, Y, or S. The reason that this simple idea does not work is that if a representative goes to the separator, its entire subtree of arbitrary size can go anywhere and we do not have any control over them. Reed handles this problem by deciding if any representative is going to the separator, at the very beginning of the algorithm. If so, he just places such a representative (namely v) into S (and not its subtree) and starts forming a new group of representatives by running a new DFS on $G - \{v\}$. So, the other representatives might change. Also, since one vertex has been placed into the separator, now $k \leftarrow k - 1$. However, if none of the representatives goes to the separator, he branches on placing them left (X) or right (Y). This is the high-level idea of Reed's algorithm.

Our main modification improves the running time (the dependence on k) significantly. We do not decide in the beginning whether any representative is going to the separator. Instead, we follow a sequential process and handle vertices one at a time, in a serial fashion. Whenever we find a small subtree and its representative v, we decide whether v goes into the separator or not. If it does not go to the separator, we consider both possibilities of that vertex going to X or Y. But once a representative (namely v) is to go to the separator, we do not start from scratch, and we do not do DFS for the entire $G - \{v\}$. We place v into S and undo the DFS for the small subtree rooted at v (unmark the vertices in its corresponding small subtree) and continue the DFS (for the remaining tree). This was the high-level idea of one of our improvements which we discuss and analyze later in detail. Let's start with presenting and reviewing some definitions.

3.2   Centroids and Separators

For an undirected graph $G = (V, E)$ and a subset W of the vertices, $G[W]$ is the subgraph induced by W. For the sake of simplicity throughout this paper, let $G - W$ be $G[V \setminus W]$ and $G - v$ be $G - \{v\}$ for any $W \subseteq V(G)$ and any $v \in V(G)$.

Also, in a weighted graph, a non-negative integer weight $w(v)$ is defined for each vertex v. For a subset W of the vertices, the weight $w(W)$ is simply the sum of the weights of all vertices in W. Furthermore, the total weight or the weight of G is the weight of V.

Definition 2. *A centroid of a weighted tree T is a node x such that none of the trees in the forest $T - x$ has more than half the total weight.*

For a tree decompositions with the adjacent bags differing in at most one node (call it "good tree decomposition"), we choose a stronger version of centroid.

Definition 3. *A strong centroid of a good tree decomposition τ of a graph $G = (V, E)$ with respect to $W \subseteq V$ is a node x of τ such that none of the connected components of $G - B_x$ contains more than $\frac{1}{2}|W \setminus B_x|$ vertices of W.*

The following lemma shows the existence of a strong centroid for any given $W \subseteq V$.

Lemma 1. *For every good tree decomposition $(T, \{B_x : x \in V_T\})$ of a graph $G = (V, E)$ and every subset $W \subseteq V$, there exist a strong centroid with respect to W (proved in the Arxiv version).*

We use the definitions of balanced W-separator and weakly balanced W separation from the book of Flum and Grohe [9].

Definition 4. *Let $G = (V, E)$ be a graph and $W \subseteq V$. A balanced W-separator is a set $S \subseteq V$ such that every connected component of $G - S$ has at most $\frac{1}{2}|W|$ vertices.*

Lemma 2 [9, Lemma 11.16]. *Let $G = (V, E)$ be a graph of treewidth at most $k - 1$ and $W \subseteq V$. Then there exists a balanced W-separator of G of size at most k.*

We say that a separator S separates $X \subseteq V$ from $Y \subseteq V$ if $C \cap X = \emptyset$ or $C \cap Y = \emptyset$ for every connected component C of $G - S$.

Definition 5. *Let $G = (V, E)$ be a graph and $W \subseteq V$. A weakly balanced separation of W is a triple (X, S, Y), where $X, Y \subseteq W$, $S \subseteq V$ are pairwise disjoint sets such that:*

- $W = X \cup (S \cap W) \cup Y$.
- S separates X from Y.
- $0 < |X|, |Y| \leq \frac{2}{3}|W|$.

Lemma 3 [9, Lemma 11.19]. *For $k \geq 3$, let $G = (V, E)$ be a graph of treewidth at most $k - 1$ and $W \subseteq V$ with $|W| \geq 2k + 1$. Then there exists a weakly balanced separation of W of size at most k.*

Theorem 1 [9, Corollary 11.22]. *For a graph of treewidth at most $k - 1$ with a given set $W \subseteq V$ of size $|W| = 3k - 2$, a weakly balanced separation of W can be found in time $O(2^{3k}k^2 n)$.*

3.3   Algorithm to Find a Weakly Balanced Separation

Separation(G, k) is the main part of Reed's algorithm. It finds a separator of size at most k in G using the procedures Split(G, X, Y, k) and DFS-Trees(G, k). We explain each of these procedures (check the Arxiv version for the pseudocodes).

Split(G, X, Y, k). For X, Y disjoint subsets of V, Split(G, X, Y, k) finds a separator S of size at most k in G which is strictly between X and Y. Split reports failure if no such separator exits (described in Lemma 11.20 of [9]).

DFS-Trees(G, k). DFS-Trees(G, k) (Algorithm 1 in the Arxiv version) selects a set $W' \subseteq V$ in a DFS (Depth-First Search) tree of G such that:

- the size (number of vertices) of the selected subtree T rooted at any vertex $v \in W'$, with all the subtrees rooted in any vertex $v' \in V_T \cap W'$ removed, is at least $n/24k$, and
- there is no vertex $v' \neq v$ of $V_T \cap W'$ with this property.

W' is a set of roots (representatives) of (intended to be) small DFS trees. The children of the vertices in W' are roots of the subtrees of size less than $\frac{n}{24k}$.

The weight $w[v]$ for $v \in W'$ is the number of vertices in the small tree with root v.

Separation(G, k). Separation(G, k) is the recursive procedure that splits according to the number of vertices (Algorithm 3). Note that when any vertex v is placed into the separator S, then the procedure Separation removes that vertex v from the graph and starts from scratch. The idea is that when we place a root of a small tree (a representative) left or right, then we want to put the whole small tree there. But when a representative is placed into the separator, then its tree does not go there. At this point a new collection of trees is formed.

3.4   Running Time of Reed's Algorithm

Let $T(n, k)$ be the running time of the procedure $SEPARATION(G, k)$ for $G = (V, E)$ and $n = |V|$. Let n' and k' be the current bound on the graph size and current separator capacity. Initially $n' = n$ and $k' = k$. We have the following recurrence for Reed's algorithm.

$$T(n', k') \leq 24k'T(n' - 1, k' - 1) + 2^{24k'} \underbrace{c(k' + 1)kn'}_{\text{flow algorithm}}, \tag{1}$$

for some $c > 0$. It is difficult to obtain a good solution, but by induction on k' we get the following loose upper bound.

$$T(n', k') \leq c2^{24k'} k'! \, kn. \tag{2}$$

For $k' = 0$ and $k' = 1$, this bound is valid. For $k' \geq 2$, we have:

$$
\begin{aligned}
T(n', k') &\leq 24k'T(n' - 1, k' - 1) + c2^{24k'} k'kn' \\
&\leq 24k'c2^{24(k'-1)}(k' - 1)! \, kn + c2^{24k'} k'kn' \qquad \text{by induction hypothesis} \\
&= c2^{24k'} kn\left(\frac{24k'!}{2^{24}} + k'\right) \leq c2^{24k'} knk'! \left(\frac{24}{2^{24}} + \frac{1}{(k' - 1)!}\right) \\
&\leq c2^{24k'} k'! \, kn.
\end{aligned}
$$

Even though, this is not a tight bound, we have $T(n, k) \geq c'24^k k!(n - k)$, which is $2^{\Omega(k \log k)} n$.

$$
\begin{aligned}
T(n, 0) &\geq c'n \\
T(n, k) &\geq 24kT(n - 1, k - 1) \\
&\geq c'24k24^{k-1}(k - 1)! \, (n - k) \text{ by inductive hypothesis} \\
&\geq c'24^k k! \, (n - k)
\end{aligned}
$$

Check the Arxiv version for the proof of correctness of the Reed's algorithm.

4   Our Improved Algorithm

In this section, we discuss how to improve Reed's algorithm. The dependence on k in the running time of Reed's algorithm is huge. We decrease this factor significantly to make the algorithm more applicable. We introduce two main modifications. First, we work with a larger cut-off threshold than Reed's $\frac{|V|}{24k}$. Such an improvement can be achieved by replacing the arbitrary 3/4 bound by $1 - \epsilon$. But even more is possible by arguing about the weights of connected components instead of the weights of the parts of a bipartition.

The second improvement is to avoid branching on whether there is a representative going into the separator or not. Reed branches on these two cases at the beginning, while we branch 3-fold for every representative.

4.1   Relax the Balancing Requirement

Reed's argument starts with a weakly balanced separation by volume that is known to exist. The larger side has at most 2/3 of the volume, but it might have up to $2/3 + 1/24$ of the weight. The algorithmic split by this weight partition might pick a set with another 1/24 fraction more volume. Thus the worst kind of volume split found is now 3/4 to 1/4. Recall that these differences are bounded this way for the following reason. When the weight carrying root of a tree is on one side, some of its small subtrees rooted at the children can be partially on the other side. But the separator prevents more than k small subtrees to have any part on a different side than the root, and each small subtree contains less than $\frac{|V|}{24k}$ vertices. Instead of 3/4, one can choose any number strictly between 2/3 and 1. If $1 - \epsilon$ is chosen, then the constant 24 is replaced by $\frac{1}{((1-\epsilon)-2/3)/2} = \frac{6}{1-3\epsilon} \leq 6 + 24\epsilon$ for $\epsilon < 1/12$.

More improvement is possible by a modification of the analysis. We start with a restricted balanced separator by volume. No connected component of $G - S$ has more than half the vertices. We can afford it to contain up to a fraction of $3/4 - \epsilon/2$ of the weight. There is still a partition of $G - S$ into left and right with a ratio of at most $3/4 - \epsilon/2$ to $1/4 - \epsilon/2$ by weight. The separator found for such a partition by weight creates a partition by volume with a worst case ratio of $1 - \epsilon$ to ϵ. This time, the constant 24 is replaced by $\frac{1}{((1-\epsilon)-1/2)/2} = \frac{4}{1-2\epsilon} \leq 4 + 12\epsilon$ for $\epsilon < 1/6$.

4.2   Main Improvement

The other improvement is to allow the representatives (the roots of the subtrees) to go either, left, right, or into the separator. Once a representative v goes into the separator, we change its weight to 0. We also delete v from G and unmark all of the vertices in its subtree, so that they can be searched again. Then, we continue the DFS from the parent of v. There is one complication here that we have to take care of; what happens if $G - v$ gets disconnected?

Three cases might happen. We cover all the cases and show that they can be handled. Let x, y, and z be three types of children of v with the small subtrees (with size $\leq \frac{n}{Ck}$) τ_x, τ_y, and τ_z, rooted respectively (Fig. 1(a)). The cases are:

- There is a back edge from a vertex in τ_x to an ancestor of v. If we delete v from the tree, the vertices of τ_x can still be searched because they are connected to an ancestor of v. So, we do not need to worry about this case.
- There is no back edge from any vertex of τ_y to an ancestor of v and further-more, there is a leaf p in τ_y with a small subtree τ_q attached to it with q as its root such that q is the last representative below τ_y which has not gone into the separator (this is a bottom-up approach). Even though this case seems to be troublesome, we can fix it. Let $\tau_{p,y}$ be τ_y rooted at p (dangling from p). Make p a child of q. Now, the problem has been fixed (Fig. 1(b)). The same reasoning applies when v is a root and its subtree is too small.
- There is no back edge from τ_z to an ancestor of v, and there is no small subtree below. So deleting v makes τ_z disconnected from the entire tree, and this is only for our advantage.

The main difference here is that Reed branches in the beginning and considers two cases. In the first case, none of the representatives goes to the separator, and in the second case at least one goes to the separator. In the second case, Reed's algorithm branches into at most $24k$ (upper bound for the number of subtrees). This affects the running time a lot. We want to avoid these branches and each

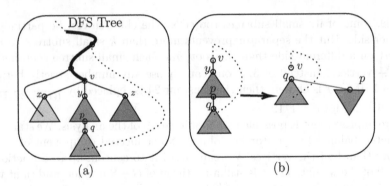

Fig. 1. How to fix the cases after a representative goes to the separator

time only branch into three cases. Assume we want to decide where to put v (a representative with weight $w(v)$). Let L, S, and R be the left, the separator, and the right sets, respectively. If we put v into L (or R), usually most of the vertices in its subtree will be in L (or R) as well. In case v goes to S, we release the other vertices of its tree to be searched again. We have to mention that unlike Reed, we do not decide at the beginning if at least one vertex is going to the separator. Instead, we consider this case for every representative only when we handle that representative (also, we handle the representative in a serial way).

5 Running Time of Our Algorithm

Let G, k, and t be the initial graph, the bound on the size of the separator, and the number of representatives, respectively. t is at most $C_\epsilon k$ since the cut-off threshold for the volume of the subtrees was $\frac{n}{C_\epsilon k}$. While proceeding with the algorithm at each step, let G', k', and t' be the current graph, the current bound on the size (capacity) of the separator, and the current number of representatives, respectively. Each time we send some vertex to the separator, we decrease the capacity by one. The recurrence for the running time to find a separator of size at most k' in G' is:

$$T(t', k') \le T(t'\frac{k'-1}{k'}, k'-1) + 2T(t'-1, k') + Qkn + \mathcal{O}(1), \text{ for } t', k' > 0, \quad (3)$$

where Q is the constant factor of the DFS. On the R.H.S., the first term handles the case where the representative goes to the separator. Therefore, k' decreases by 1, and the number of subtrees becomes at most $t'\frac{k'-1}{k'}$ (delete that vertex and continue the DFS). The second term handles the case that the current representative does not go into the separator but left or right. Here, the capacity of the separator is unchanged but the number of subtrees decreases by 1. The third term is the upper bound of the exact running time of the DFS. And the last term $\mathcal{O}(1)$ is the overhead to make the calls. The base cases of the recurrence:

$T(0, k') \leq Q(k'+1)kn$ to continue $k'+1$ DFSs. We find k' paths from X to Y, and search for augmenting paths

$T(t', 0) \leq Qkn$ We have to test whether S is indeed a separator.

The recurrence in 3 might seem hopeless, so we simplify it (generous bound).

$$T(t', k') \leq T(t', k'-1) + 2T(t'-1, k') + Qk'n + \mathcal{O}(1), \ t', k' > 0 \qquad (4)$$

Now, we have to solve this recurrence. Our recursion tree starts from the root $T(t', k')$ and has two children $T(t', k'-1)$ and $T(t'-1, k')$, left and right, respectively. This is an unbalanced binary tree. Each strand terminates when one of the arguments of $T(\cdot, \cdot)$ becomes zero. Each time we choose the left branch (putting one representative into the separator), we decrease k' by 1. Otherwise (putting the representative and its subtree to the right or left set of the separator), we decrease t' by 1 and multiply the value by 2. Let $\#(t'-i, 0)$ be the number of leaves with the first argument $t'-i$ and the second argument $k' = 0$, for $0 \leq i < t'$ (analogous notation for $\#(0, k'-j)$ for $0 \leq j < k'$). Observe that $\#(t'-i, 0) = \binom{t'+k'-i}{k'}$, and $\#(0, k'-i) = \binom{t'+k'-i}{t'}$. The first two terms of the Eq. 4 can be computed at the leaves and the other two terms are spent in every vertex of the recursion tree. Let us compute the first part.

$$\sum_{i=0}^{t'-1} \left(\#(t'-i, 0) \underbrace{2^i}_{i \text{ right branches}} T(t'-i, 0) \right) + \sum_{i=0}^{k'-1} \left(\#(0, k'-i) \underbrace{2^{t'}}_{t' \text{ right branches}} T(0, k'-i) \right)$$

$$\leq \sum_{i=0}^{t'} \left(\binom{k'+i}{i} 2^i \underbrace{Qk'kn}_{\text{Eq. 4}} \right) + \sum_{i=0}^{k'} \left(\binom{t'+i}{i} 2^{t'} \underbrace{Qkn}_{\text{Eq. 4}} \right)$$

$$\leq 2^{t'} Qk'kn \left(\sum_{i=0}^{t'} \binom{k'+i}{i} + \sum_{i=0}^{k'} \binom{t'+i}{i} \right)$$

$$= 2^{t'} Qk'kn \left(\binom{k'+t'+1}{t'+1} + \binom{k'+t'+1}{k'+1} \right) = 2^{t'} Qk'kn \binom{k'+t'+2}{k'+1}$$

Now, we have to compute the second part of the Eq. 4 where we should look at every internal vertex of the tree. We have $\binom{t'+k'}{k'}$ internal vertices and in each vertex we spend at most $Qk'kn + \mathcal{O}(1) \leq Qk'kn(1 + \mathcal{O}(\frac{1}{n}))$. Hence,

$$T(t', k') \leq 2^{t'} Qk'kn \binom{k'+t'+2}{k'+1} + Qk'kn(1 + \mathcal{O}(\frac{1}{n})) \binom{t'+k'}{k'}$$
$$= Qk'kn \left(2^{t'} \frac{k'+t'+2}{k'+1} \binom{k'+t'+1}{k'} + (1 + \mathcal{O}(\frac{1}{n})) \binom{t'+k'}{k'} \right) \qquad (5)$$

Notice that $T(\cdot, k')$ is monotonic (due to the definition of T) and use the fact that $t' \leq C_\epsilon k'$. Now, we simplify the Eq. 5 by bounding $T(t', k')$ with $T(C_\epsilon k', k')$

due to the monotonicity of $T(\cdot, k')$ and the fact that $t' \leq C_\epsilon k'$ (below, let $U_n = (1 + \mathcal{O}(\frac{1}{n}))$).

$$T(t', k') \leq T(C_\epsilon k', k')$$

$$\leq Qk'kn \left(2^{C_\epsilon k'} \frac{(C_\epsilon + 1) k' + 2}{k' + 1} \left(\frac{(C_\epsilon + 1) k' + 1}{k'} \right) + U_n \left(\frac{(C_\epsilon + 1) k'}{k'} \right) \right)$$

$$\leq Qk'kne^{k'} \left(2^{C_\epsilon k'} \frac{(C_\epsilon + 1) k' + 2}{k' + 1} \left(\frac{(C_\epsilon + 1) k' + 1}{k'} \right)^{k'} + U_n \left(\frac{(C_\epsilon + 1) k'}{k'} \right)^{k'} \right)$$

Here, in order to simplify the closed form, we bound the above running time very generously and give a very loose bound for now.

$$T(t', k') \leq Qk'kne^{k'} (C_\epsilon + 1)^{k'} \left(2^{(C_\epsilon+1)k'} (C_\epsilon + 1) + 2 \right) \tag{6}$$

Now, we compute the running time (T_V), when we split based on V.

$$T_V(n, k) = T_V(\epsilon n + k, k) + T_V((1 - \epsilon)n + k, k) + T(t', k)$$

$$\leq T_V(\epsilon n + k, k) + T_V((1 - \epsilon)n + k, k) + Qk^2 ne^k (C_\epsilon + 1)^k \left(2^{(C_\epsilon+1)k} (C_\epsilon + 1) + 2 \right) \tag{7}$$

$$\leq Qk^2 e^k (C_\epsilon + 1)^k \left(2^{(C_\epsilon+1)k} (C_\epsilon + 1) + 2 \right) \frac{1}{\epsilon} n \ln n$$

In the above equation, we use the following lemma (Lemma 4). Furthermore, the reason that we add k to both recursive calls is that we add the separator to the both subproblems.

Lemma 4. *Assume* $0 < \epsilon \leq \frac{1}{2}$, $0 < c' \leq c$, $2 \leq k$, *and* $n_1 + n_2 = n$. *Then the recurrence*

$$f(n + k) \leq \begin{cases} c'(n + k) & \text{if } n \leq 4k \\ f(n_1 + k) + f(n_2 + k) + c(n + k) & \text{otherwise,} \end{cases}$$

where $\frac{1}{2}n \leq n_1 \leq (1-\epsilon)n$ *has a solution with* $f(n+k) \leq \frac{c}{\epsilon} n \ln n - ck$, *for* $n \geq 2k$.

The proof can be found in Arxiv version.

Corollary. Under the conditions of the Lemma 4,

$$f(n) \leq \frac{c}{\epsilon} n \ln n.$$

Now, the total running time of the algorithm (T_t) is:

$$T_t(n, k) \leq Qk^2 e^k (C_\epsilon + 1)^k \left(2^{(C_\epsilon+1)k} (C_\epsilon + 1) + 2 \right) \frac{\left(1 + \log_{\frac{3}{2}} k \right)}{\epsilon} n \ln n,$$

where it takes $\log_{\frac{3}{2}} k$ steps so that the k_{excess} (check the Arxiv version for the details) drops to zero (that is when we need to split by V once more).

As we mentioned in Sect. 4.1, $C_\epsilon = \frac{4}{1-2\epsilon}$. We plug that into Eq. 7.

$$T_t(n,k) \le Qk^2 e^k (\frac{4}{1-2\epsilon}+1)^k \left(2^{(\frac{4}{1-2\epsilon}+1)k}(\frac{4}{1-2\epsilon}+1)+2\right)\frac{\left(1+\log_{\frac{3}{2}}k\right)}{\epsilon} n \ln n,$$
$$(8)$$

for any positive constant $\epsilon \le \frac{1}{6}$. For instance, if we set $\epsilon = \frac{1}{100}$, the running time would be at most $353\,Q \log_{\frac{3}{2}}(k)\,k^2\,2^{8.87k}\,n \ln n$. Looking at the limit for $\epsilon \to 0$, we have shown the following theoretical result.

Theorem 2. *Let* $C_0 = \log_2 e + \log_2 5 + 5 < 8.765$. *For every* $C > C_0$, *a 5-approximation of the treewidth can be computed in time* $O(2^{Ck} n \log n)$.

Our algorithm with ϵ sufficiently small such that the exponent in Ineq. (8) is less than C has this running time. This shows that the dependence on k in our algorithm is much smaller than in Reed's algorithm.

A   Appendix

The following is the proof of Lemma 1.

Proof. If a node x is not a strong centroid with respect to W, then let C_x be the set of vertices in the unique connected component of $G - B_x$ containing more than $\frac{1}{2}|W \setminus B_x|$ vertices of X. In the forest obtained from the tree $\mathcal{T}$ by removing x, there is a tree $\mathcal{T}_x$ with the property that the union of all bags in $\mathcal{T}_x$ contains all the vertices of C_x.

Now, we define a set F of directed tree edges by $(x,y) \in F$, if all the following conditions hold:

- x is not a centroid
- y is a neighbor of x in $\mathcal{T}$,
- y is a node in $\mathcal{T}_x$.

Now we show that there is a node x with out-degree 0 in $(V_\mathcal{T}, F)$. Such an x is a centroid, and we are done. Otherwise, F contains (x,y) and (y,x) for some $x,y \in V_\mathcal{T}$. W.l.o.g., $B_y = B_x \cup \{v\}$ for some $v \in V \setminus B_x$. Note that $\mathcal{T}_x$ and $\mathcal{T}_y$ are disjoint. Furthermore, any vertex that is in a bag of $\mathcal{T}_x$ and in a bag of $\mathcal{T}_y$ is also in B_x and B_y. Thus also C_x and C_y are disjoint.

Furthermore, $|W \cap C_x| > \frac{1}{2}|W \setminus B_x|$ implies $|W \cap C_x| > |W \cap C_y|$, because $[W \cap C_y] \subseteq [W \setminus B_x]$. Likewise, $|W \cap C_y| > \frac{1}{2}|W \setminus B_y|$ implies $|W \cap C_y| > |W \cap (C_x \setminus \{v\})|$, because $[W \cap (C_x \setminus \{v\})] \subseteq [W \setminus B_y]$. Thus we have $|W \cap C_x| > |W \cap C_y| > |W \cap (C_x \setminus \{v\})|$. Since these are all integers, and the difference between the first and the last number is at most 1, it is a contradiction. Hence, there exists a node x which is a strong centroid.

Now, we present the pseudocodes of Sect. 3.3.

Algorithm 1: Construct Small DFS-Trees

Result: Roots of DFS-Trees with sizes of their strict subtrees $\leq |V|/(24k)$

Procedure DFS-Trees(G, k) `// G is a connected graph.`

* $s = \frac{|V|}{24k}$ `// s: the size bound for splitting off a small tree.`
* $W' = \emptyset$
* **for** *all* $v \in V$ **do**
  | * color$[u] =$ WHITE
  end
* count $= 0$
* Pick any vertex u of G.
* DFS-visit(G, u)
* Add count to $w[v]$, where v is the vertex last included in W'.
* **return** W' and $w[v]$ *for all* $v \in W'$

End Procedure

Algorithm 2: Main recursive procedure of DFS-Trees

Procedure DFS-Visit(G, u)

* color$[u] =$ GRAY
* **for** *all* v *adjacent to* u **do**
  | * **if** *color*$[v] ==$ *WHITE*   `// The white vertex v has been discovered.`
  | **then**
  | | * DFS-Visit(G, v)
  | **end**
  end
* count $=$ count$+1$
* **if** *count* $\geq s$ `// u is a root of a small tree.`
then
  | * $W' = W' \cup \{u\}$
  | * $w[u] = count$
  | * $count = 0$
end

End Procedure

A.1   The Correctness of Reed's Algorithm

If the treewidth is at most $k-1$, then there is a good tree decomposition of G of width $k-1$. Let x be a centroid in it. The connected components of $G[V \setminus B_x]$ can be partitioned into 2 parts L and R, such that no part has more than $\frac{2}{3}|V|$ vertices.

Note that for the correctness proof, we do not have to find this tree decomposition. It is sufficient to know that it exists. We can assume, that we have fixed such a tree decomposition, a centroid x and the sets L and R.

We know that the set W' is partitioned into parts in L, the separator $S = B_x$, and R.

One of the many branches of the procedure Separation(G, k) will try this partition of W' and will succeed. First, it decides which part of W' goes into

---

Algorithm 3: Main recursive procedure in Reed's algorithm

Result: A weakly balanced separation (X, S, Y) of G of size $\leq k$
Procedure SEPARATION(G, k)
  •**if** $k > 0$ **then**
    | $(W', \{w(v) : v \in W'\}) =$ DFS-Trees(G, k)
  end
  •**for** *all* $v \in W'$ // Here v is placed into separator S.
  do
    | •$(X, S, Y) =$ SEPARATION$(G - v, k - 1)$
    | •**if** $\neg failure$ **then**
    | | •**return** $(X, S \cup \{v\}, Y)$
    | **end**
  end
      // The set of vertices W' is partitioned into L and $R = W' \setminus L$.
  •**for** *all* $X \subseteq W'$ // Here no vertex is put into S.
  do
    | •**if** $(\frac{1}{3} - \frac{1}{24})n \leq w(X) \leq (\frac{2}{3} + \frac{1}{24})n)$ **then**
    | | •Split$(G, X, W' \setminus X, k)$
    | | •**if** $\neg failure$ **then**
    | | | •**return** (X, S, Y)
    | | **end**
    | **end**
  end
  return *failure*
End Procedure

S, one vertex v at a time. This vertex v is removed from G, but otherwise, we still consider the same tree decomposition. $|B_x|$ has now decreased by 1, as v is removed from it.

We then consider the case that none of the remaining vertices in W' are in the separator. Now the weight of each part is at most $(\frac{2}{3} + \frac{1}{24})n$ as at most k small subtrees can have some of their vertices on the wrong side. And this is at most k times at most $\frac{n}{24k}$.

On the branch of the procedure Separation(G, k) which finds this partition of W', there is the separator B_x of size at most k between X and Y. Our algorithm cannot guarantee to find this separator B_x, but it will find some separator S of size at most k between X and Y. Again up to $\frac{1}{24}n$ vertices can be on the other side than their representatives. Now the larger side can contain at most $(\frac{2}{3} + 2 \cdot \frac{1}{24})n = \frac{3}{4}n$ vertices. Thus we have a somewhat balanced partition (a constant fraction on each side).

The overall algorithm can alternate between splitting W as in the $O(2^{3k}n^2)$ algorithm and splitting V. Now W can be of size at most $6k$. On each side, we have at most $(\frac{2}{3} \cdot 6k) = 4k$. Splitting by W as well as splitting by V adds k to the separator. Thus, we are back at $6k$. The constructed tree decomposition has then width at most $7k$. But we show that this can be actually a 5-approximation algorithm. We do not need to alternate between splitting W and V. Splitting

V is a costly procedure. We can do it only after every $\log_{\frac{3}{2}} k$ steps and we still spend $\mathcal{O}(n \log n)$ time.

We start with W of size at most $4k$ ($3k$ and k_{excess} as excess). Initially, $k_{excess} = k$. Each time we split W, we get $|W| \leq \frac{2}{3} \cdot 3k + \underbrace{k}_{\text{adding separator}} + \frac{2}{3} k_{excess} = 3k + \frac{2}{3} k_{excess}$, and then update $k_{excess} \leftarrow \lfloor \frac{2}{3} k_{excess} \rfloor$.

The excess drops by a factor of $\frac{2}{3}$. After $\log_{\frac{3}{2}} k$ many steps, the excess becomes zero and then we can split by V, where $|W|$ becomes $\leq 4k$ again ($3k$ was the size of W before this step, and when we split by V, we have to include the separator as well). In the end, we end up with $|W| \leq 4k$ and we add the separator to the root bag, which means the largest bag has size at most $5k$. Therefore, it is a 5-approximation algorithm. Reed mentions $5k$ in his paper but he does not bother himself giving the details. We think based on what it has been described in Reed's algorithm, it seems we should have $7k$. However, if splitting by V does not happen very often, we can achieve $5k$.

Lemma 5. *Assume $0 < \epsilon \leq \frac{1}{2}$, $0 < c' \leq c$, $2 \leq k$, and $n_1 + n_2 = n$. Then the recurrence*

$$f(n+k) \leq \begin{cases} c'(n+k) & \text{if } n \leq 4k \\ f(n_1 + k) + f(n_2 + k) + c(n+k) & \text{otherwise,} \end{cases}$$

where $\frac{1}{2} n \leq n_1 \leq (1-\epsilon)n$ has a solution with $f(n+k) \leq \frac{c}{\epsilon} n \ln n - ck$, for $n \geq 2k$.

Proof. Case 1: $2k \leq n \leq 4k$. Then $n \geq 4$ and

$$f(n+k) \leq c(n+k) < 2cn - ck < \frac{c}{\epsilon} n \ln n - ck.$$

Case 2: $n \geq 4k$ and $n_2 \geq 2k$.

$$\begin{aligned}
f(n+k) &\leq f(n_1+k) + f(n_2+k) + c(n+k) \\
&\leq \frac{c}{\epsilon} (n_1 \ln n_1 + (n-n_1)\ln(n-n_1)) - 2ck + c(n+k) \\
&\leq \frac{c}{\epsilon}(1-\epsilon)n(\underbrace{\ln(1-\epsilon)}_{< -\epsilon} + \ln n) + \epsilon n(\ln \epsilon + \ln n) + c(n-k) \\
&< \frac{c}{\epsilon} n \ln n - c(1-\epsilon)n + cn \ln \epsilon + c(n-k) \\
&\leq \frac{c}{\epsilon} n \ln n - cn + c\epsilon n + cn \ln \epsilon + c(n-k) \\
&\leq \frac{c}{\epsilon} n \ln n + (\underbrace{\epsilon + \ln \epsilon}_{<0 \text{ for } \epsilon \leq \frac{1}{2}}) cn - ck \\
&\leq \frac{c}{\epsilon} n \ln n - ck
\end{aligned}$$

Case 3: $n \geq 4k$ and $n_2 = n - n_1 < 2k$.

$$
\begin{aligned}
f(n + k) &\leq f(n_1 + k) + f(n_2 + k) + c(n + k) \\
&\leq \frac{c}{\epsilon} n_1 \ln n_1 - ck + c'(n_2 + k) + c(n + k) \\
&\leq \frac{c}{\epsilon} (1 - \epsilon) n (\ln(1 - \epsilon) + \ln n) - ck + c'(n_2 + k) + c(n + k) \\
&< \frac{c}{\epsilon} (1 - \epsilon) n (-\epsilon + \ln n) - ck + c'(n_2 + k) + c(n + k) \\
&\leq \frac{c}{\epsilon} n \ln n - ck - cn \ln n - c(1 - \epsilon) n + c'(n_2 + k) + c(n + k) \\
&< \frac{c}{\epsilon} n \ln n - ck
\end{aligned}
$$

The last inequality is true, because $n \geq 4k \geq 8$ implying $\ln n > 2$.

References

1. Amir, E.: Approximation algorithms for treewidth. Algorithmica **56**(4), 448–479 (2010)
2. Arnborg, S., Corneil, D.G., Proskurowski, A.: Complexity of finding embeddings in a k-tree. SIAM J. Algebraic Discret. Methods **8**, 277–284 (1987)
3. Bertele, U., Brioschi, F.: On non-serial dynamic programming. J. Comb. Theory Ser. A **14**(2), 137–148 (1973)
4. Bodlaender, H.L.: A linear-time algorithm for finding tree-decompositions of small treewidth. SIAM J. Comput. **25**(6), 1305–1317 (1996)
5. Bodlaender, H.L.: Discovering treewidth. In: Vojtáš, P., Bieliková, M., Charron-Bost, B., Sýkora, O. (eds.) SOFSEM 2005. LNCS, vol. 3381, pp. 1–16. Springer, Heidelberg (2005). https://doi.org/10.1007/978-3-540-30577-4_1
6. Bodlaender, H.L., Drange, P.G., Dregi, M.S., Fomin, F.V., Lokshtanov, D., Pilipczuk, M.: A $\mathcal{O}(c^k n)$ 5-approximation algorithm for treewidth. SIAM J. Comput. **45**(2), 317–378 (2016)
7. Courcelle, B.: The monadic second-order logic of graphs. I. Recognizable sets of finite graphs. Inf. Comput. **85**(1), 12–75 (1990)
8. Feige, U., Hajiaghayi, M.T., Lee, J.R.: Improved approximation algorithms for minimum weight vertex separators. SIAM J. Comput. **38**(2), 629–657 (2008)
9. Flum, J., Grohe, M.: Parameterized Complexity Theory. TTCS. Springer, Heidelberg (2006). https://doi.org/10.1007/3-540-29953-X
10. Halin, R.: S-functions for graphs. J. Geom. **8**(1–2), 171–186 (1976)
11. Lagergren, J.: Efficient parallel algorithms for graphs of bounded tree-width. J. Algorithms **20**(1), 20–44 (1996)
12. Reed, B.A.: Finding approximate separators and computing tree width quickly. In: Proceedings of the Twenty-Fourth Annual ACM Symposium on Theory of Computing, pp. 221–228. ACM (1992)
13. Robertson, N., Seymour, P.D.: Graph minors. III. Planar tree-width. J. Comb. Theory Ser. B **36**(1), 49–64 (1984)
14. Robertson, N., Seymour, P.D.: Graph minors. XIII. The disjoint paths problem. J. Comb. Theory Ser. B **63**(1), 65–110 (1995)

Homomorphisms to Digraphs with Large Girth and Oriented Colorings of Minimal Series-Parallel Digraphs

Frank Gurski$^{(\boxtimes)}$, Dominique Komander, and Marvin Lindemann

Institute of Computer Science, Algorithmics for Hard Problems Group,
Heinrich-Heine-University Düsseldorf, 40225 Düsseldorf, Germany
`frank.gurski@hhu.de`

Abstract. An oriented r-coloring of an oriented graph G corresponds to an oriented graph H on r vertices, such that there exists a homomorphism from G to H. The problem of deciding whether an acyclic digraph allows an oriented 4-coloring is already NP-hard. The oriented chromatic number of an oriented graph G is the smallest integer r such that G allows an oriented r-coloring.

In this paper we consider msp-digraphs (short for minimal series-parallel digraphs), which can be defined from the single vertex graph by applying the parallel composition and series composition. In order to show several results for coloring msp-digraphs, we introduce the concept of oriented colorings excluding homomorphisms to digraphs H with short cycles. A g-oriented r-coloring of an oriented graph G is a homomorphism from G to some digraph H on r vertices of girth at least $g+1$. The g-oriented chromatic number of G is the smallest integer r such that G allows a g-oriented r-coloring.

As our main result we show that for every msp-digraph the g-oriented chromatic number is at most $2^{g+1} - 1$. We use this bound together with the recursive structure of msp-digraphs to give a linear time solution for computing the g-oriented chromatic number of msp-digraphs. This implies that every msp-digraph has oriented chromatic number at most 7. Furthermore, we conclude that the chromatic number of the underlying undirected graphs of msp-digraphs can be bounded by 3. Both bounds are best possible and the exact chromatic numbers can be computed in linear time. Finally, we conclude that k-power digraphs of msp-digraphs have oriented chromatic number at most $2^{2k+1} - 1$.

Keywords: Oriented coloring · Graph homomorphisms · Minimal series-parallel digraphs · Linear time algorithms

1 Introduction

An oriented r-coloring of an oriented graph $G = (V, E)$ is a partition of the vertex set V into r independent sets, such that all arcs linking two of these

© Springer Nature Switzerland AG 2021
R. Uehara et al. (Eds.): WALCOM 2021, LNCS 12635, pp. 182–194, 2021.
https://doi.org/10.1007/978-3-030-68211-8_15

subsets have the same direction. The *oriented chromatic number* of an oriented graph G, denoted by $\chi_o(G)$, is the smallest integer r such that G allows an oriented r-coloring. Equivalently, there is an oriented r-coloring of an oriented graph G if and only if there is a homomorphism from G to some oriented graph H on r vertices and the oriented chromatic number of G is the minimum number of vertices in an oriented graph H, such that there is a homomorphism from G to H. Oriented coloring appears e.g. in scheduling models where incompatibilities are oriented, see [6]. Even deciding whether an acyclic digraph allows an oriented 4-coloring is NP-hard [6], which motivates to study the problem on special graph classes.

In [8], there was shown an FPT-algorithm for the oriented chromatic number problem (OCN) w.r.t. the parameter tree-width (of the underlying undirected graph) by Ganian. Additionally, he has shown that OCN is DET-hard[1] for classes of oriented graphs of bounded rank-width (of the underlying undirected class).

Presently, the definition of oriented coloring is mainly considered for undirected graphs, where the maximum value $\chi_o(G')$ of all possible orientations G' of an undirected graph G is considered. For various special undirected graph classes the oriented chromatic number has been bounded, e.g. for forests and cycles [18], for outerplanar graphs [17], and Halin graphs [7]. Moreover, the oriented chromatic number of planar graphs with large girth was intensively investigated e.g. in [14–16].

Until now, it seems that the subject of oriented coloring of special classes of oriented graphs is not well studied. In [10,11] we investigated the problem on oriented co-graphs, which can be defined from the single vertex graph by applying the disjoint union and order composition. In [11] we considered oriented coloring of msp-digraphs, which can be defined from the single vertex graph by applying the parallel composition and series composition. These graphs are useful for modeling flow diagrams and dependency charts, as well as they are helpful in applications for scheduling under constraints, see [2, Section 11.1].

In this paper we expand these outcomes with further interesting more general results about the oriented coloring problem restricted to msp-digraphs. Therefore, we generalize the concept of oriented colorings by excluding homomorphisms to digraphs H with short cycles. A *g-oriented r-coloring* of an oriented graph G is a homomorphism from G to some digraph H on r vertices of girth at least $g + 1$. The *g-oriented chromatic number* of G is the smallest integer r such that G allows a g-oriented r-coloring. It follows that for $g = 1$ we obtain the chromatic number of the underlying undirected graph and for $g = 2$ we obtain the oriented chromatic number.

Our main result is that for every msp-digraph the g-oriented chromatic number is at most $2^{g+1} - 1$. We use this bound together with the recursive structure of msp-digraphs to give a linear time solution for computing the g-oriented chromatic number of msp-digraphs. This implies that every msp-digraph has oriented chromatic number at most 7, which re-proves our result from [11]. Furthermore,

---

[1] DET is the class of decision problems which are reducible in logarithmic space to the problem of computing the determinant of an integer valued $n \times n$-matrix.

we conclude that the chromatic number of the underlying undirected graphs of msp-digraphs can be bounded by 3. Both bounds are sharp and the exact chromatic numbers can be computed in linear time. Finally, we conclude that k-power digraphs of msp-digraphs have oriented chromatic number at most $2^{2k+1} - 1$.

2   Preliminaries

2.1   Graphs and Digraphs

We use the notations of Bang-Jensen and Gutin [1] for (di)graphs. A *graph* is a pair $G = (V, E)$ with a finite set of *vertices* V and a finite set of *edges* $E \subseteq \{\{u, v\} \mid u, v \in V, u \neq v\}$. A *directed graph* or *digraph* is a pair $G = (V, E)$, in which V is a finite set of *vertices* and $E \subseteq \{(u, v) \mid u, v \in V, u \neq v\}$ is a finite set of ordered pairs of distinct vertices called *arcs* or *directed edges*. For a vertex $v \in V$, the sets $N^+(v) = \{u \in V \mid (v, u) \in E\}$ and $N^-(v) = \{u \in V \mid (u, v) \in E\}$ are called the *set of all successors* and the *set of all predecessors* of v. The *outdegree* of v, outdegree(v) for short, is the number of successors of v, while the *indegree* of v, indegree(v) for short, is the number of predecessors of v.

For digraph $G = (V, E)$, we define its underlying undirected graph by ignoring the directions of the arcs, i.e. $un(G) = (V, \{\{u, v\} \mid (u, v) \in E, u, v \in V\})$. A digraph $G' = (V', E')$ is a *subdigraph* of digraph $G = (V, E)$ if $V' \subseteq V$ and $E' \subseteq E$. Further, if every arc of E with both end vertices in V' is in E', we call G' an *induced subdigraph* of G and we write $G' = G[V']$.

An *oriented graph* is a digraph without loops or opposite arcs. A *tournament* is a digraph in which exists exactly one edge between every two distinct vertices. A *directed acyclic graph (DAG for short)* is a digraph with no directed cycles. The *girth* of digraph G is the length (number of arcs) of a shortest directed cycle in G. If G is a DAG its girth is defined to be infinity.

2.2   Coloring Oriented Graphs

In 1994 Courcelle [4] introduced oriented graph coloring, where we look at oriented graphs, i.e. digraphs with no loops and no opposite arcs.

Definition 1 (Oriented graph coloring [4]). *An* oriented r-coloring *of an oriented graph* $G = (V, E)$ *is a mapping* $c : V \rightarrow \{1, \ldots, r\}$ *such that:*

- $c(u) \neq c(v)$ *for every* $(u, v) \in E$,
- $c(u) \neq c(y)$ *for every two arcs* $(u, v) \in E$ *and* $(x, y) \in E$ *with* $c(v) = c(x)$.

The oriented chromatic number *of* G, *denoted with* $\chi_o(G)$, *is the smallest* r *such that* G *has an oriented* r-coloring. *Then* $V_i = \{v \in V \mid c(v) = i\}$, $1 \leq i \leq r$ *is a partition of* V, *which we call* color classes.

Within our generalization of oriented colorings in the next section we will use the following equivalent characterization of oriented graph coloring from [17]. For two digraphs $G_1 = (V_1, E_1)$ and $G_2 = (V_2, E_2)$ a *homomorphism* from

G_1 to G_2 is a mapping $h : V_1 \to V_2$, which preserves the edges, i.e., $(u, v) \in E_1$ implies $(h(u), h(v)) \in E_2$. A homomorphism from G_1 to G_2 can be regarded as an oriented coloring of G_1 that uses the vertices of G_2 as colors classes. Thus, we call G_2 the *color graph* of G_1. This leads to equivalent definitions for the oriented coloring and the oriented chromatic number. There is an oriented r-coloring of an oriented graph G_1 if and only if there is a homomorphism from G_1 to some oriented graph G_2 on r vertices. Then, the oriented chromatic number of G_1 is the minimum number of vertices in an oriented graph G_2 such that there is a homomorphism from G_1 to G_2. Clearly, it is possible to choose G_2 as a tournament.

We denote the (undirected) chromatic number of a (-n undirected) graph G by $\chi(G)$.

Observation 1 ([11]). *For every oriented graph G it holds that $\chi(un(G)) \leq \chi_o(G)$.*

Name: Oriented Chromatic Number (OCN)
Given: An oriented graph $G = (V, E)$ and a positive integer $r \leq |V|$.
Question: Is there an oriented r-coloring for G?

If r is not part of the input but a constant, we call the related problem the r-Oriented Chromatic Number (OCN_r). If $r \leq 3$, then we can decide OCN_r in polynomial time, but OCN_4 is NP-complete [13]. Moreover, OCN_4 is also NP-complete for several restricted classes of digraphs, e.g. for DAGs [6].

So far, the definition of oriented coloring was often used for undirected graphs, where the maximum value $\chi_o(G')$ of all possible orientations G' of a graph G is considered. This leads to the fact that every tree has oriented chromatic number at most 3. There are also bounds on the oriented chromatic number for other graph classes, e.g. for outerplanar graphs [17] and Halin graphs [7]. Moreover, the oriented chromatic number of planar graphs with large girth was intensively investigated e.g. in [14–16].

2.3   Generalization of Oriented Colorings

Next, we generalize the concept of oriented colorings by excluding homomorphisms to oriented graphs with short cycles. This allows us to show good bounds on the oriented chromatic number of msp-digraphs, k-power digraphs of msp-digraphs, and the chromatic number of the underlying undirected graphs.

Definition 2 (g-oriented graph coloring). *A g-oriented r-coloring of an oriented graph G is a homomorphism from G to some digraph H on r vertices of girth at least $g + 1$. The g-oriented chromatic number of G, denoted by $\chi_g(G)$, is the minimum number of vertices in a digraph H of girth at least $g + 1$, such that there is a homomorphism from G to H.[2]*

---

[2] While within oriented graph coloring only homomorphisms to oriented graphs are important, in Definition 2 we consider colorings of oriented graphs G where the color graph H is only oriented for $g \geq 2$.

Name: g-Oriented Chromatic Number (g-OCN)
Instance: An oriented graph $G = (V, E)$ and a positive integer $r \leq |V|$.
Question: Is there a g-oriented r-coloring for G?

If r is constant and not part of the input, the corresponding problem is denoted by r-Generalized Oriented Chromatic Number (g-OCN$_r$).

The following two observations state that g-oriented r-colorings of oriented graphs generalize oriented r-colorings as well as r-colorings of the underlying undirected graph.

Observation 2. *For every oriented graph G it holds $\chi_1(G) = \chi(un(G))$.*

Observation 3. *For every oriented graph G it holds $\chi_2(G) = \chi_o(G)$.*

Observation 4. *For every oriented graph G and every integer g it holds that $\chi_g(G) \leq \chi_{g+1}(G)$.*

By Observations 2 and 4 we obtain the next result.

Corollary 1. *For every oriented graph G and every integer g it holds that $\chi(un(G)) \leq \chi_g(G)$.*

Lemma 1. *Let G be an oriented graph and h be a homomorphism from G to some digraph H on r vertices of girth at least $g + 1$. Then, for every subdigraph G' of G there is a homomorphism h' from G' to an induced subdigraph H' of H (on at most r vertices) of girth at least $g + 1$.*

Corollary 2. *Let G be an oriented graph and G' be a subdigraph of G, then $\chi_g(G') \leq \chi_g(G)$.*

3 Coloring Msp-Digraphs

3.1 Msp-Digraphs

We recall definitions from [2] which are based on [20].

Definition 3 (Msp-digraphs [2]). *The class of minimal series-parallel digraphs, msp-digraphs for short, is recursively defined as follows.*

1. *Every digraph on a single vertex $(\{v\}, \emptyset)$, denoted by v, is a minimal series-parallel digraph.*
2. *If $G_1 = (V_1, E_1)$ and $G_2 = (V_2, E_2)$ are minimal series-parallel digraphs and O_1 is the set of vertex of outdegree 0 (set of sinks) in G_1 and I_2 is the set of vertices of indegree 0 (set of sources) in G_2, then*

   (a) *the parallel composition $G_1 \cup G_2 = (V_1 \cup V_2, E_1 \cup E_2)$ is a minimal series-parallel digraph and*
   (b) *the series composition $G_1 \times G_2 = (V_1 \cup V_2, E_1 \cup E_2 \cup (O_1 \times I_2))$ is a minimal series-parallel digraph.*

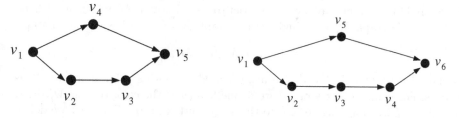

Fig. 1. Digraph(X_1) in Example 1.   **Fig. 2.** Digraph(X_2) in Example 1.

Every expression X using the operations of Definition 3 is called an *msp-expression*. The digraph defined by the expression X is denoted by digraph(X). We illustrate this by two expressions which we will refer to later.

Example 1. The following msp-expressions X_1 and X_2 define msp-digraphs on five and six vertices shown in Fig. 1 and Fig. 2.

$$X_1 = (v_1 \times ((v_2 \times v_3) \cup v_4)) \times v_5$$

$$X_2 = (v_1 \times ((v_2 \times (v_3 \times v_4)) \cup v_5)) \times v_6$$

If vertex v_3 is deleted from the msp-digraph digraph(X_2) in Example 1, we resulting oriented graph is not an msp-digraph. Consequently, the set of all msp-digraphs is not closed under taking induced subdigraphs.

For every msp-digraph we can define a tree structure, which is denoted as *msp-tree*. The leaves of an msp-tree correspond to the vertices of the digraph, while the inner vertices of the msp-tree represent the operations applied on the subexpressions defined by the subtrees. For every msp-digraph, an msp-tree is computable in linear time [20].

3.2   Generalized Oriented Coloring Msp-Digraphs

Next, we show a bound on the g-oriented chromatic number of msp-digraphs, which generalizes some of our results from [11].

Theorem 1. *Let G be an msp-digraph. Then, it holds that $\chi_g(G) \leq 2^{g+1} - 1$.*

Proof. We work with recursively defined msp-digraphs M_i, where digraph M_0 is the single vertex graph and for $i \geq 1$ it holds that

$$M_i = M_{i-1} \cup M_{i-1} \cup (M_{i-1} \times M_{i-1}).$$

By [11], every msp-digraph G is a (-n induced) subdigraph of a digraph M_i, such that every source in G is a source in M_i and every sink in G is a sink in M_i.

Since by Lemma 1 a g-oriented coloring of an oriented graph is also a g-oriented coloring for every subdigraph, we can show the theorem by giving a coloring for the digraphs M_i. Since the first two occurrences of M_{i-1} in M_i can be colored in the same way, we can restrict to msp-digraphs M_i' which are defined as follows. Digraph M_0' is a single vertex graph and for $i \geq 1$ we define digraph

$$M_i' = M_{i-1}' \cup (M_{i-1}' \times M_{i-1}').$$

Let $d = 2^{g+1} - 1$. We define a g-oriented d-coloring c for digraph M_i' as follows. For some vertex v of M_i' we define by $c(v, i)$ the color of v in M_i'. First we color M_0' by assigning color 0 to the single vertex in M_0'. For $i \geq 1$ we define the colors for the vertices v in $M_i' = M_{i-1}' \cup (M_{i-1}' \times M_{i-1}')$ according to the three copies of M_{i-1}' in M_i' (numbered from left to right). Therefore, we use the two functions $p(x) = (2^g \cdot x) \bmod d$ and $q(x) = (2^g \cdot x + 1) \bmod d$. We define

$$c(v, i) = \begin{cases} c(v, i-1) & \text{if } v \text{ is from the first copy,} \\ p(c(v, i-1)) & \text{if } v \text{ is from the second copy, and} \\ q(c(v, i-1)) & \text{if } v \text{ is from the third copy.} \end{cases}$$

Next we show that c leads to a g-oriented d-coloring for M_i'. For two integers a and b we defined $a \equiv_d b$ if and only if $a \bmod d = b \bmod d$. Let $C_i = (W_i, F_i)$ with $W_i = \{0, \ldots, d-1\}$ and $F_i = \{(c(u, i), c(v, i)) \mid (u, v) \in E_i\}$ be the color graph of $M_i' = (V_i, E_i)$. By the definition of digraph M_i' it holds that

$$\begin{aligned} F_i = F_{i-1} &\cup \{(p(x), p(y)) \mid (x, y) \in F_{i-1}\} \\ &\cup \{(q(x), q(y)) \mid (x, y) \in F_{i-1}\} \\ &\cup \{(p(c(v, i-1)), q(c(w, i-1))) \mid v \text{ sink of } M_{i-1}', w \text{ source of } M_{i-1}'\}. \end{aligned}$$

In Fig. 3 the color graph C_i for $g = 2$ $(d = 7)$ and $i \geq 5$ is given [11].

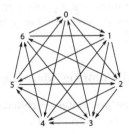

Fig. 3. The color graph C_i for $g = 2$ $(d = 7)$ and $i \geq 5$ used in the proof of Theorem 1 which corresponds to the Paley tournament of order 7.

In order to ensure a g-oriented d-coloring of M_i' we verify that C_i is a digraph of girth at least $g + 1$.

Every source in M_i' is colored by 0 since $p(0) = 0$. Every sink in M_i' is colored by some color from $T = \{\sum_{j=k}^{g+1} 2^j \bmod d \mid k \in \{1, \ldots, g+1\}\}$, since $0 \in T$:

$$0 \equiv_d 2^{g+1} - 1 = \sum_{j=0}^{g} 2^j \equiv_d \sum_{j=1}^{g+1} 2^j \qquad \text{(since } 2^{g+1} \equiv_d 1)$$

and for $x = \sum_{j=k}^{g+1} 2^j \bmod d \in T$ it also holds that $q(x) \in T$:

$$q(x) = q\left(\sum_{j=k}^{g+1} 2^j \bmod d\right) \equiv_d \sum_{j=k-1}^{g} 2^j + 1 \equiv_d \sum_{j=k-1}^{g} 2^j + 2^{g+1} \equiv_d \sum_{j=k-1}^{g+1} 2^j$$

Remember that $2^{g+1} \equiv_d 1$ and thus, $2^g \equiv_d \frac{1}{2}$ such that $q(x) = (\frac{x}{2}+1) \bmod d$. Since for $k = 1$ it holds that $\sum_{j=k-1}^{g+1} 2^j = \sum_{j=0}^{g+1} 2^j = d+2^{g+1} \equiv_d 2^{g+1}$, we know that $q(x) \equiv_d \sum_{j=k-1}^{g+1} 2^j \in T$.

In order to consider the girth of color graph $C_i = (W_i, F_i)$ of M_i' we color each edge $(c_1, c_2) \in F_i$, $c_1, c_2 \in \{0, \dots, d\}$ by the value $c_2 - c_1 \bmod d$. This implies that every edge in C_i is colored by a value in $T_2 = \{2^k \bmod d \mid k \in \{0, \dots, g\}\}$. This holds true, since every edge which is not from a copy of M_{i-1}' is of the type $(p(c(t, i-1)), q(c(s, i-1)))$ for some sink t and some source s in M_{i-1}', which implies for the coloring of this edge

$$q(c(s, i-1)) - p(c(t, i-1)) = q(0) - p(\sum_{j=k}^{g+1} 2^j) \equiv_d 1 - \sum_{j=k-1}^{g} 2^j \equiv_d 2^{g+1} - \sum_{j=k-1}^{g} 2^j$$
$$= 2^{k-1}$$

for some $k \in \{1, \dots, g+1\}$ (which implies that $2^{k-1} \in T_2$). Also the edges which come from a copy of M_{i-1}' fulfill this property, since for $c_2 - c_1 \equiv_d 2^k \in T_2$ it holds

$$p(c_2) - p(c_1) = q(c_2) - q(c_1) \equiv_d \frac{1}{2}(c_2 - c_1) = 2^{k-1}$$

and since $2^{-1} \equiv_d 2^g$, it holds $2^{k-1} \in T_2$. Thus, for all edges (c_1, c_2) from C_i it holds that $c_2 - c_1 \in T_2$. This can be used to show that the girth of C_i is at least $g + 1$. Every cycle in C_i fulfills that the sum of the colors of the edges modulo d is 0. Assume that there is a cycle of length $g' \le g$ in C_i, then it holds that

$$0 \equiv_d \sum_{j=1}^{g'} 2^{k_j}.$$

We consider the exponents $k_1, \dots, k_{g'}$ of this sum. We combine two equal exponents $2^k + 2^k = 2^{k+1}$ until there are no equal exponents and for $2^{k+1} = 2^{g+1}$ we replace 2^{g+1} by 1. Since in each of these combination steps two summands are replaced by one, this process terminates. This implies that $0 \equiv_d \sum_{j=1}^{g''} 2^{k_j'}$ for $0 < g'' \le g$ and all $k_j' \in T_2$ are mutually different. Thus, it holds that

$$\sum_{j=1}^{g''} 2^{k_j'} \le \sum_{j=1}^{g} 2^j < \sum_{j=0}^{g} 2^j = d.$$

But since all summands $2^{k_1'}, \dots, 2^{k_{g''}'}$ are positive and their sum is less than d their sum can not be 0 modular d. Thus, our assumption was false and all cycles in color graph C_i must have at least $g + 1$ directed edges.   □

Let $g \geq 1$ be some integer. In order to compute the g-oriented chromatic number of some msp-digraph G defined by some msp-expression X we compute the set $F(X)$ of all triples (H, L, R). In this context H is a color graph for G and L and R are the sets of colors of all sinks and all sources in G with respect to the coloring by H. The number of vertex labeled oriented graphs on n vertices is $3^{n(n-1)/2}$, where vertex labeled means that the vertices are distinguishable from each other. By Theorem 1 we can conclude that

$$|F(X)| \leq 3^{(2^{g+1}-1)(2^{g+1}-2)/2} \cdot 2^{2^{g+1}-1} \cdot 2^{2^{g+1}-1} \in \mathcal{O}(1)$$

which is independent of the size of G.

Given two color graphs $H_1 = (V_1, E_1)$ and $H_2 = (V_2, E_2)$ we define $H_1 + H_2 = (V_1 \cup V_2, E_1 \cup E_2)$.

Lemma 2. 1. Let $v \in V$, then it holds that $F(v) = \{((\{i\}, \emptyset), \{i\}, \{i\}) \mid 0 \leq i \leq 2^{g+1} - 2\}$.

2. Let X_1 and X_2 be two msp-expressions, then we obtain $F(X_1 \cup X_2)$ from $F(X_1)$ and $F(X_2)$ as follows. For every $(H_1, L_1, R_1) \in F(X_1)$ and every $(H_2, L_2, R_2) \in F(X_2)$ such that graph $H_1 + H_2$ has girth at least $g + 1$ we put $(H_1 + H_2, L_1 \cup L_2, R_1 \cup R_2)$ into $F(X_1 \cup X_2)$.

3. Let X_1 and X_2 be two msp-expressions, then we obtain $F(X_1 \times X_2)$ from $F(X_1)$ and $F(X_2)$ as follows. For every $(H_1, L_1, R_1) \in F(X_1)$ and every $(H_2, L_2, R_2) \in F(X_2)$ such that graph $H_1 + H_2$ together with the arcs in $R_1 \times L_2$ has girth at least $g + 1$ we put $((V_1 \cup V_2, E_1 \cup E_2 \cup R_1 \times L_2), L_1, R_2)$ into $F(X_1 \times X_2)$.

Proof. 1. $F(v)$ includes obviously all possible solutions to color every vertex on its own with the $2^{g+1} - 1$ given colors.

2. Let (H_1, L_1, R_1) be any possible solution for coloring digraph(X_1), which therefore is included in $F(X_1)$, as well as a possible solution (H_2, L_2, R_2) for coloring digraph(X_2) which is included in $F(X_2)$. Let further $H_1 + H_2$ be a digraph of girth at least $g + 1$. Since the operation $\cup$ creates no additional edges in digraph($X_1 \cup X_2$), the vertices of digraph(X_1) can still be colored with H_1 and the vertices of digraph(X_2) can still be colored with H_2 such that all vertices from digraph($X_1 \cup X_2$) are legally colored. Further, all sinks in digraph(X_1) and digraph(X_2) are also sinks in digraph($X_1 \cup X_2$). The same holds for the sources. For a digraph $H_1 + H_2$ of girth at least $g + 1$ this leads to $(H_1 + H_2, L_1 \cup L_2, R_1 \cup R_2) \in F(X_1 \cup X_2)$.

Let $(H, L, R) \in F(X_1 \cup X_2)$, then by Lemma 1 there is an induced subdigraph H_1 of the color graph H of girth at least $g + 1$ which colors digraph(X_1) which is an induced subdigraph of digraph($X_1 \cup X_2$). Let $L_1 \subseteq L$ be the sources with vertices in digraph(X_1) and $R_1 \subseteq R$ be the sinks for vertices in digraph(X_1). Then, it holds that $(H_1, L_1, R_1) \in F(X_1)$. The same arguments hold for X_2, such that $(H_2, L_2, R_2) \in F(X_2)$.

3. Let (H_1, L_1, R_1) be any possible solution for coloring digraph(X_1), which therefore is included in $F(X_1)$, as well as a possible solution (H_2, L_2, R_2) for coloring X_2 which is included in $F(X_2)$. Further, let $H_1 + H_2$ together

with edges from $R_1 \times L_2$ be a digraph of girth at least $g + 1$. Then, $H = (V_1 \cup V_2, E_1 \cup E_2 \cup R_1 \times L_2)$ is a legit coloring for $X = X_1 \times X_2$. Since the sinks of digraph(X_1) are connected with the sources of digraph(X_2) in digraph(X), the sources of L_1 are the only sources left in digraph(X) as well as the sinks in R_2 are the only sinks left in digraph(X). This leads to $(H, L_1, R_2) \in F(X)$.

Let $(H, L, R) \in F(X_1 \times X_2)$, then by Lemma 1 there is an induced subdigraph H_1 of the color graph H of girth at least $g + 1$ which colors digraph(X_1) which is an induced subdigraph of digraph$(X_1 \times X_2)$. Since all the sources of digraph$(X_1 \times X_2)$ are in digraph(X_1) it holds that $L_1 = L$ are also sources of digraph(X_1). Let R_1 be the vertices in digraph(X_1) which only have outgoing neighbors in digraph(X_2) but not in digraph(X_1), then R_1 are the sinks of digraph(X_1). Thus, it holds that $(H_1, L_1, R_1) \in F(X_1)$. Simultaneously, by Lemma 1 there is an induced subdigraph H_2 of the color graph H of girth at least $g + 1$ which colors digraph(X_2) which is an induced subdigraph of digraph$(X_1 \times X_2)$. Since all the sinks of digraph$(X_1 \times X_2)$ are in digraph(X_2) it holds that $R_2 = R$ are also sinks of digraph(X_2). Let L_2 be the vertices in digraph(X_2) which only have in-going neighbors in digraph(X_1) but not in digraph(X_2), then L_2 are the sources of digraph(X_2). Thus, it holds that $(H_2, L_2, R_2) \in F(X_2)$.

This shows the statements of the lemma.   □

Since every oriented coloring of G is considered of the set $F(X)$, where X is an msp-expression for G, we can find a minimum coloring for G as follows.

Corollary 3. *Let G be an msp-digraph which is given by some msp-expression X. Then, there is a g-oriented r-coloring for G if and only if there is some $(H, L, R) \in F(X)$, such that color graph H has r vertices. Therefore, $\chi_g(G) = \min\{|V| \mid ((V, E), L, R) \in F(X)\}$.*

Theorem 2. *Let G be an msp-digraph and g be some integer. Then, the g-oriented chromatic number of G can be computed in linear time.*

Proof. Let $G = (V, E)$ be an msp-digraph with $n = |V|$ vertices and $m = |E|$ edges and let T be an msp-tree for G with root r. For a vertex u of T we denote by T_u the subtree rooted at u and by X_u the msp-expression defined by T_u.

For computing the g-oriented chromatic number for some msp-digraph G, we traverse msp-tree T in bottom-up order. For every vertex u of T we can compute $F(X_u)$ by following the rules given in Lemma 2. By Corollary 3 we can solve our problem using $F(X_r) = F(X)$.

An msp-tree T can be computed in $\mathcal{O}(n + m)$ time from G, see [20]. By Lemma 2 we obtain the following running times.

- For every vertex $v \in V$ set $F(v)$ is computable in $\mathcal{O}(1)$ time.
- For every two msp-expressions X_1 and X_2 set $F(X_1 \cup X_2)$ can be computed in $\mathcal{O}(1)$ time from $F(X_1)$ and $F(X_2)$.
- For every two msp-expressions X_1 and X_2 set $F(X_1 \times X_2)$ can be computed in $\mathcal{O}(1)$ time from $F(X_1)$ and $F(X_2)$.

Since T consists of n leaves and $n-1$ inner vertices, the overall running time is in $\mathcal{O}(n+m)$.   □

3.3   Oriented Coloring Msp-Digraphs

By Theorem 1 and Observation 3 we obtain the following result, which re-proves our result from [11].

Corollary 4. *Let G be an msp-digraph. Then, it holds that $\chi_o(G) \leq 7$.*

Digraph G in Example 2 satisfies $\chi_o(G) = 7$, which was found in [11] by a computer program. This implies that the bound of Corollary 4 is best possible.

Example 2. In the subsequent msp-expression we assume that the series composition binds more strongly than the parallel composition.

$$X = v_1 \times (v_2 \cup v_3 \times (v_4 \cup v_5 \times v_6)) \times (v_7 \cup (v_8 \cup v_9 \times v_{10}) \times (v_{11} \cup v_{12} \times v_{13}))$$
$$\times (v_{14} \cup (v_{15} \cup (v_{16} \cup v_{17} \times v_{18}) \times (v_{19} \cup v_{20} \times v_{21}))$$
$$\times (v_{22} \cup (v_{23} \cup v_{24} \times v_{25}) \times v_{26})) \times v_{27}$$

By Theorem 2 and Observation 3 we obtain the following result.

Corollary 5. *Let G be an msp-digraph. Then, the oriented chromatic number of G can be computed in linear time.*

3.4   Coloring Underlying Undirected Graphs of Msp-Digraphs

Theorem 1 and Observation 2 imply the following result, which re-proves our result stated in [11].

Corollary 6. *Let G be an msp-digraph. Then, it holds that $\chi(un(G)) \leq 3$.*

For expression X_1 given in Example 1 we obtain by $un(\text{digraph}(X_1))$ a cycle on five vertices of chromatic number 3, which implies that the bound of Corollary 6 is sharp.

By Theorem 2 and Observation 2 we obtain the following result.

Corollary 7. *Let G be an msp-digraph. Then, the chromatic number of $un(G)$ can be computed in linear time.*

3.5   Coloring Powers of Msp-Digraphs

A well known concept in graph theory is the concept of graph powers, see [2] and [3].

Definition 4 (k-power digraph). *The k-power digraph G^k of a digraph G is a digraph with the same vertex set as G. There is an arc (u, v) in G^k if and only if $u \neq v$ and there is a directed path from u to v in G of length at most k.*

By this definition it holds that G^1 is G.

Theorem 3. *Let G be an oriented graph. Then, it holds that $\chi_g(G^k) \leq \chi_{g \cdot k}(G)$.*

Proof (idea). Every homomorphism from G to some graph H of girth at least $g \cdot k + 1$ is also a homomorphism from G^k to H^k and H^k has girth at least $g + 1$.   □

Corollary 8. *Let G be an msp-digraph. Then, it holds that $\chi_o(G^k) \leq 2^{2k+1} - 1$.*

Proof. By Observation 3, Theorem 3, and Theorem 1 it holds $\chi_o(G^k) = \chi_2(G^k) \leq \chi_{2k}(G) \leq 2^{2k+1} - 1$.   □

Digraph G in Example 2 satisfies $\chi_o(G) = 7$, which implies that the bound of Corollary 8 is sharp for $k = 1$.

4   Conclusions and Outlook

In this paper we generalized the concept of oriented r-colorings to g-oriented r-colorings by excluding homomorphisms to oriented graphs with cycles of length at most g. We have shown that every msp-digraph has g-oriented chromatic number at most $2^{g+1} - 1$. Using this bound together with the recursive structure of msp-digraphs we could give a linear time solution for computing the g-oriented chromatic number of msp-digraphs. This result immediately lead to sharp bounds on the oriented chromatic number of msp-digraphs and the chromatic number of the underlying undirected graphs.

For future work it could be interesting to apply the concept of g-oriented r-colorings to further recursively defined graph classes, such as oriented co-graphs [10]. Furthermore, it remains open to expand the solutions to superclasses e.g. series-parallel digraphs [20]. Moreover, the parameterized complexity of OCN and OCN$_r$ w.r.t. structural parameters has only been studied in [8] and [9]. For the parameters directed modular-width [19] and directed clique-width [5,12] the parameterized complexity of OCN is still open.

Acknowledgment. This work was funded by the Deutsche Forschungsgemeinschaft (DFG, German Research Foundation) – 388221852.

References

1. Bang-Jensen, J., Gutin, G.: Digraphs. Theory, Algorithms and Applications. Springer, Berlin (2009)
2. Bang-Jensen, J., Gutin, G.: Classes of Directed Graphs. Springer, Berlin (2018)
3. Brandstädt, A., Le, V., Spinrad, J.: Graph Classes: A Survey. SIAM Monographs on Discrete Mathematics and Applications. SIAM, Philadelphia (1999)
4. Courcelle, B.: The monadic second-order logic of graphs VI: on several representations of graphs by relational structures. Discret. Appl. Math. **54**, 117–149 (1994)
5. Courcelle, B., Olariu, S.: Upper bounds to the clique width of graphs. Discret. Appl. Math. **101**, 77–114 (2000)

6. Culus, J.-F., Demange, M.: Oriented coloring: complexity and approximation. In: Wiedermann, J., Tel, G., Pokorný, J., Bieliková, M., Štuller, J. (eds.) SOFSEM 2006. LNCS, vol. 3831, pp. 226–236. Springer, Heidelberg (2006). https://doi.org/10.1007/11611257_20

7. Dybizbański, J., Szepietowski, A.: The oriented chromatic number of Halin graphs. Inf. Process. Lett. **114**(1–2), 45–49 (2014)

8. Ganian, R.: The parameterized complexity of oriented colouring. In: Proceedings of Doctoral Workshop on Mathematical and Engineering Methods in Computer Science, MEMICS. OASICS, vol. 13. Schloss Dagstuhl - Leibniz-Zentrum fuer Informatik, Germany (2009)

9. Ganian, R., Hlinený, P., Kneis, J., Langer, A., Obdrzálek, J., Rossmanith, P.: Digraph width measures in parameterized algorithmics. Discret. Appl. Math. **168**, 88–107 (2014)

10. Gurski, F., Komander, D., Rehs, C.: Oriented coloring on recursively defined digraphs. Algorithms **12**(4), 87 (2019)

11. Gurski, F., Komander, D., Lindemann, M.: Oriented coloring of MSP-digraphs and oriented co-graphs (extended abstract). In: Wu, W., Zhang, Z. (eds.) COCOA 2020. LNCS, vol. 12577, pp. 743–758. Springer, Cham (2020). https://doi.org/10.1007/978-3-030-64843-5_50

12. Gurski, F., Wanke, E., Yilmaz, E.: Directed NLC-width. Theor. Comput. Sci. **616**, 1–17 (2016)

13. Klostermeyer, W., MacGillivray, G.: Homomorphisms and oriented colorings of equivalence classes of oriented graphs. Discret. Math. **274**, 161–172 (2004)

14. Marshall, T.: Homomorphism bounds for oriented planar graphs of given minimum girth. Graphs Combin. **29**, 1489–1499 (2013)

15. Marshall, T.: On oriented graphs with certain extension properties. Ars Combinatoria **120**, 223–236 (2015)

16. Ochem, P., Pinlou, A.: Oriented coloring of triangle-free planar graphs and 2-outerplanar graphs. Graphs Combin. **30**, 439–453 (2014)

17. Sopena, É.: The chromatic number of oriented graphs. J. Graph Theory **25**, 191–205 (1997)

18. Sopena, É.: Homomorphisms and colourings of oriented graphs: an updated survey. Discret. Math. **339**, 1993–2005 (2016)

19. Steiner, R., Wiederrecht, S.: Parameterized algorithms for directed modular width. In: Changat, M., Das, S. (eds.) CALDAM 2020. LNCS, vol. 12016, pp. 415–426. Springer, Cham (2020). https://doi.org/10.1007/978-3-030-39219-2_33

20. Valdes, J., Tarjan, R., Lawler, E.: The recognition of series-parallel digraphs. SIAM J. Comput. **11**, 298–313 (1982)

Overall and Delay Complexity of the CLIQUES and Bron-Kerbosch Algorithms

Alessio Conte[1(✉)] and Etsuji Tomita[2]

[1] University of Pisa, Pisa, Italy
conte@di.unipi.it
[2] The Advanced Algorithms Research Laboratory,
The University of Electro-Communications,
Chofugaoka 1-5-1, Chofu, Tokyo 182–8585, Japan
e.tomita@uec.ac.jp

Abstract. We revisit the maximal clique enumeration algorithm CLIQUES by Tomita et al. that appeared in Theoretical Computer Science 2006. It is known to work in $O(3^{n/3})$-time in the worst-case for an n-vertex graph. In this paper, we extend the time-complexity analysis with respect to the maximum size and the number of maximal cliques, and to its delay, solving issues that were left as open problems since the original paper. In particular, we prove that CLIQUES does *not* have polynomial delay, unless $P = NP$, and that this remains true for *any* possible pivoting strategy, for both CLIQUES and Bron-Kerbosch. As these algorithms are widely used and regarded as fast "in practice", we are interested in observing their practical behavior: we run an evaluation of CLIQUES and three Bron-Kerbosch variants on over 130 real-world and synthetic graphs, and observe how their performance seems far from its theoretical worst-case behavior in terms of both total time and delay.

Keywords: Maximal cliques · Graph enumeration · Output sensitive · Delay

1 Introduction

A *clique* is defined to be a subgraph in which all vertices are pairwise adjacent. In particular, it is *maximal* if it is not contained in a strictly larger clique. Given a graph, the enumeration of all its maximal cliques is a fundamental and important problem in graph theory [20] and has many practical applications in clustering, data mining, bioinformatics, social networks, and more, mostly related to community detection [10]. An *independent set* of a graph G is a clique of the complement graph $\bar{G}$.

Tsukiyama et al. [25] gave the first algorithm MIS for enumerating maximal independent sets with a theoretical time-complexity analysis. For a graph G with n vertices and m edges, MIS enumerates all maximal independent sets in time

© Springer Nature Switzerland AG 2021
R. Uehara et al. (Eds.): WALCOM 2021, LNCS 12635, pp. 195–207, 2021.
https://doi.org/10.1007/978-3-030-68211-8_16

$O(nm)$ per maximal independent set; this can be adapted to enumerate maximal cliques in the same complexity per solution [15].

Tomita et al. [23] and Bron and Kerbosch [2] independently presented different algorithms for the problem, although it was later understood that their pruning techniques were the same. These algorithms do not have output-sensitive guarantees but boast good practical performance.

Furthermore, while the complexity of Bron-Kerbosch is as of today still unknown, CLIQUES [23], based on *depth-first search* algorithms for finding a *maximum* clique [11,21], was the first maximal clique enumeration algorithm with proven worst-case optimal time (as a function of n): indeed its $O(3^{n/3})$-time worst-case time complexity matches the number of maximal cliques in Moon-Moser graphs [18].

The algorithms modeled after Tsukiyama et al. typically follow the *reverse-search* framework [1] and enumerate the cliques in an *output-sensitive* fashion: if a problem has size n and α solutions, an algorithm is output-sensitive if its complexity is $O(poly(\alpha, n))$-time, for some polynomial $poly(\cdot)$, and *amortized polynomial time* if it is just $O(\alpha\, poly(n))$.

In this line, steady improvements have been made by Chiba and Nishizeki [5], Johnson et al. [12], Makino and Uno [15], Chang et al. [4], Comin and Rizzi [6], and Conte et al. [8] and Manoussakis [16]. Most of these algorithms prove a stronger result than output-sensitivity, that is *polynomial delay*, *i.e.*, the time elapsed between two consecutive outputs of a solution is polynomial. Some rely on matrix multiplication, like the one by Comin and Rizzi [6], with $O(n^{2.094})$-time delay, while others on combinatorial techniques, such as the one by Conte et al. [8], with $O(qd\Delta)$-time delay, where q is the size of a maximum clique, d is the *degeneracy* of G (the smallest number such that every subgraph of G contains a vertex of degree at most d), and Δ the maximum degree. Based on CLIQUES, Eppstein et al. proposed an improved algorithm for sparse graphs that runs in $O(d(n - d)3^{d/3})$-time. In general, it is experimentally observed that algorithms based on CLIQUES and Bron-Kerbosch are fast in practice [7,9,19,24]. However, no theoretical time-complexity analysis with respect to the number of maximal cliques is made for CLIQUES in 2006 [24], where the problem is noted in [24] as an important open problem. This is in contrast to the time-complexity analysis in reverse-search approach.

It is natural to ask whether CLIQUES is output sensitive, and whether it has polynomial delay, under general or restricted conditions. In this paper, we are concerned with a new complexity analysis of CLIQUES and similar approaches with respect to the number and the maximum size of maximal cliques. A preliminary version of this paper appeared in [22].

2   Definitions and Notation

We consider a simple undirected *graph* $G = (V(G), E(G))$, or simply (V, E) when G is clear from the context, with a finite set V of *vertices* and a finite set E of *unordered* pairs (v, w) of distinct vertices, called *edges*. A pair of vertices v and w

are *adjacent* if $(v, w) \in E$. For a vertex $v \in V$, let $\Gamma(v)$ be the set of all vertices that are adjacent to v in $G = (V, E)$, i.e., $\Gamma(v) = \{w \in V \mid (v, w) \in E\}(\not\ni v)$.

For a subset $W \subseteq V$ of vertices, $G(W) = (W, E(W))$ with $E(W) = (W \times W) \cap E$ is called a *subgraph* of $G = (V, E)$ *induced* by W. For a set W of vertices, $|W|$ denotes the number of elements in W.

Given a subset $Q \subseteq V$ of vertices, if $(v, w) \in E$ for all $v, w \in Q$ with $v \neq w$ then the induced subgraph $G(Q)$ is called a *clique*. In this case, we may simply say that Q is a clique. If a clique is not a proper subgraph of another clique then it is called a *maximal* clique.

3 Maximal Clique Enumeration Algorithm CLIQUES

We briefly revisit a depth-first search algorithm, CLIQUES [23,24], which enumerates all maximal cliques of an undirected graph $G = (V, E)$, with $|V| = n$ vertices, giving the output in a tree-like form. A more detailed explanation is deferred to [22].

The algorithm (detailed in Algorithm 1) consists of a recursive call procedure based on two vertex sets *SUBG* and *CAND*. Initially, we set $SUBG \leftarrow V$ and $CAND \leftarrow V$, and the recursive task of CLIQUES(*SUBG*, *CAND*) is to enumerate all maximal cliques in $G(SUBG)$ which are fully contained in *CAND*.

We maintain a global variable $Q = \{p_1, p_2, ..., p_h\}$ with the vertices of a current clique, and $SUBG = V \cap \Gamma(p_1) \cap \Gamma(p_2) \cap \cdots \cap \Gamma(p_h)$. The cliques found in a recursive subtree correspond to extensions of Q, which is initially empty.

The algorithm employs two pruning methods to avoid unnecessary recursive calls, which happen to be the same as in the Bron-Kerbosch algorithms [2]:

Avoiding Duplication. The set *FINI* = *SUBG* \ *CAND* (short for _FINISHED_) contains vertices that were already processed and can be skipped as they can only lead to duplicate solutions.

Pivoting. The vertex u selected on Line 6, called the *pivot*, allows us to skip recurring on vertices in $\Gamma(u)$: any maximal clique Q' in $G(SUBG \cap \Gamma(u))$ is *not* maximal in $G(SUBG)$, as it is contained in $Q' \cup \{u\}$; therefore, any maximal clique either contains u or a vertex in $SUBG \setminus \Gamma(u)$.

CLIQUES selects $u \in SUBG$ that *maximizes* $|CAND \cap \Gamma(u)|$, *crucial* to the worst-case optimal complexity of $O(3^{n/3})$-time, and prints the output in a tree-like form (shown in Fig. 1).

3.1 Bron-Kerbosch Algorithms

Let us recall the antecedent algorithm by Bron and Kerbosch for enumerating maximal cliques [2], which use a backtracking strategy essentially similar to CLIQUES. However, we will highlight some key differences. We will call BK the version of the Bron-Kerbosch algorithm *without* pivoting, and BKP a variant of the Bron-Kerbosch algorithm with pivoting (where no specific pivot choice is mandated).

Algorithm 1: Algorithm CLIQUES in [24]

    Input : A graph $G = (V, E)$.
    Output: All maximal cliques in G.
    `/* Q ← ∅ is a global variable representing a clique */`

1   CLIQUES (V, V)
2   **Function** CLIQUES($SUBG$, $CAND$)
3 | **if** $SUBG = \emptyset$ **then**
4 | └ print (``clique,'') `/* Q is a maximal clique */`
5 | **else**
6 | | $u \leftarrow$ a vertex in $SUBG$ maximizing $|CAND \cap \Gamma(u)|$
          | | `/* FINI ← ∅ */`
7 | | **while** $CAND \setminus \Gamma(u) \neq \emptyset$ **do**
8 | | | $p \leftarrow$ a vertex in $CAND \setminus \Gamma(u)$
9 | | | print $(p, \text{``,''})$ `/* Q ← Q ∪ {p} */`
10 | | | $SUBG_p \leftarrow SUBG \cap \Gamma(p)$
11 | | | $CAND_p \leftarrow CAND \cap \Gamma(p)$
12 | | | CLIQUES($SUBG_p$, $CAND_p$)
13 | | | $CAND \leftarrow CAND \setminus \{p\}$
          | | | `/* FINI ← FINI ∪ {p} */`
14 | | └ print (``back,'') `/* Q ← Q \ {p} */`

One difference is that CLIQUES outputs all maximal cliques in a *tree*-like format (Lines 4, 9, 14), to pay just $O(1)$ time per output; another is the choice of pivot u maximizing $|CAND \cap \Gamma(u)|$. Both of these factors are crucial to prove the worst-case optimal $O(3^{n/3})$-time complexity of CLIQUES, and indeed the worst-case running time of BKP is yet unknown.

While BK and CLIQUES handle vertices differently (see the pseudo code in [2]), the recursion tree and pivoting step of BKP are inherently similar to that of CLIQUES, thus the results proven in this paper for the search tree of CLIQUES extend to BKP as well.

3.2 Search Tree

We use the *search tree* (illustrated in Fig. 1) to represent the enumeration process of CLIQUES.

- The root of the search tree is a newly introduced *dummy* root p_0 $(\notin V)$ to form a tree.
- Every vertex in V is a child of the dummy root p_0, and every node of the search tree except the root corresponds to a vertex in V.
- Assume we have a path from the dummy root p_0 to a certain node p_h in the search tree as a sequence of nodes $p_0, p_1, p_2, ..., p_h$, and let $SUBG_h = V \cap \Gamma(p_1) \cap \Gamma(p_2) \cap \cdots \cap \Gamma(p_h)$. Then, every vertex in $SUBG_h$ is a child of p_h in the search tree.

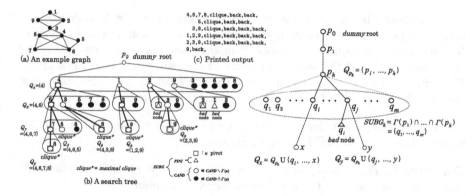

Fig. 1. Left: An example run of CLIQUES from [24]. Right: part of a search tree

Furthermore, every vertex in *FINI* corresponds to a *bad node*, i.e., a recursive node whose execution is prevented and, similarly, every neighbor of the pivot (that also is not expanded by the algorithm) corresponds to a *black node*.

When the above $SUBG_h$ is a singleton $\{q_1\}$, q_1 is a leaf in the search tree.

Let q_i be a child of p_h, then the set $\{p_1, p_2, ..., p_h, q_i\}$ constitutes a *clique*, called an *accompanying clique* and is denoted by $Q_{p_1,...,q_i}$, or simply Q_{q_i}, or Q when it is clear. Figure 1 shows an example run of CLIQUES [24] (left), and a part of a general search tree (right).

4   Overall Complexity of CLIQUES

The time-complexity of CLIQUES directly depends on the size of the search tree. Suppose we have a path from the dummy root p_0 to a certain node p_h in the search tree as a sequence of nodes $p_0, p_1, p_2, ..., p_h$. Then the set $\{p_1, p_2, ..., p_h\}$ is an *accompanying clique*.

We are interested in the accompanying clique Q across different search tree nodes, and in particular, let us observe the following:

Lemma 1. *The accompanying clique Q is distinct in any internal (non-leaf) node of a search tree of* CLIQUES.

Proof. Proof sketch (full proof in [22]). Tracing any two internal nodes x and y (assuming x is executed before y) to their nearest common ancestor in the search tree we can observe how a vertex in the accompanying clique of x must be in *FINI* when y is executed. $\square$

Now let q be the size of a maximum clique and α the number of maximal cliques; we can prove the following on the complexity of CLIQUES:

Theorem 1. *The search tree of* CLIQUES *has at most* $(1 + \Delta)\alpha 2^q$ *nodes. Consequently, the running time of* CLIQUES *is* $O(\alpha 2^q n^2 \Delta)$.

Proof. Lemma 1 implies that the number of internal nodes is bounded by the number of possible accompanying cliques, *i.e.*, distinct non-maximal cliques of G. Each maximal clique has at most 2^q distinct subsets, so the internal nodes are at most $\alpha 2^q$, and the number of leaves of the search tree is at most $\Delta\alpha 2^q$. The number of nodes of a search tree is thus at most $(1 + \Delta)\alpha 2^q$. Since each node can be executed in $O(n^2)$-time [24], the statement follows. □

Theorem 1 has the following noteworthy consequence (the proof is straight-forward, but reported in [22] for completeness):

Corollary 1. *The running time of* CLIQUES *on a graph G with n vertices and maximum clique size $q = O(\log n)$ is amortized polynomial.*

This condition immediately applies to many sparse graphs, where Δ is assumed to be small, and even in graphs with large Δ but small degeneracy d, as $q \leq d + 1 \leq \Delta + 1$. Corollary 1 however claims more: even dense graphs may satisfy this property. A simple example is the complete bipartite graph $K_{\frac{n}{2},\frac{n}{2}}$, which is by no means sparse as all vertices have degree $n/2$, but the size of a maximum clique is 2. It is often observed that the size q of a maximum clique is $O(\log n)$ in real-world graphs (with n vertices). To support this claim, and the scope of Corollary 1, we analyzed over a hundred real-world graphs, reporting our findings in Sect. 6. Finally, we recall that this corollary also holds for BKP.

5   There Is No Polynomial Delay Strategy for CLIQUES and Bron-Kerbosch Unless P = NP

Known pivoting strategies can have a large impact on the number of recursive nodes, this can be observed e.g., in [3] and our experiments in Sect. 6.

It is natural to ask: is there an ideal pivoting strategy with polynomial delay? That is, is there a strategy that guarantees a new clique —or the end of the algorithm— is always reached within polynomial time? We complete our results by showing that no such strategy can exist unless $P = NP$.[1]

To do so, we define the *extension problem* for maximal cliques, showing that it is NP-complete. Then, we show how a pivoting strategy for Algorithm 1 that guarantees polynomial delay could be used to solve this problem in polynomial time.

Hardness of the Extension Problem

Problem 1 (Extension Problem, EXT-P $(G(V, E), X)$). Given a graph $G(V, E)$ and $X \subset V$, does G have a maximal clique Q that does not intersect X?

---

[1] Polynomial delay might be achieved by other means, what we prove here is that CLIQUES and BKP cannot guarantee polynomial delay even changing the pivot selection strategy, unless $P = NP$.

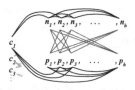

Fig. 2. Example construction of G from the $\mathcal{F}$ formula for a clause $c_1 = (v_1, \neg v_3, \neg v_h)$. Note how c_1 is connected to *all* literals that do not satisfy it, including, e.g., p_2 and n_2. The set X is composed of all c_i vertices.

Looking at a recursive node of CLIQUES, with its sets Q, $CAND$, and $SUBG$, we can observe how a maximal clique will be output in its recursive subtree iff there exists a maximal clique in $G(SUBG)$ that does not intersect $SUBG \setminus CAND$. In other words, the problem answers the question "will a maximal clique be output in this recursive subtree?". This problem is, however, NP-complete, as we show by a reduction from CNFSAT.

Theorem 2. *The extension problem for maximal cliques is NP-complete.*

Proof. Let $\mathcal{F}$ be a CNF Boolean formula on h variables $v_1, \ldots, v_h$ and l clauses $c_1, \ldots, c_l$, and let the positive and negative literals of the variable v_i be represented by v_i and $\neg v_i$.

We build a graph G, with a suitable vertex set X, such that $\mathcal{F}$ can be satisfied iff G has a maximal clique not intersecting X. Let $V(G)$ and $E(G)$ be as follows:

- $V(G)$ contains a vertex p_i for each positive literal v_i in $\mathcal{F}$.
- $V(G)$ contains a vertex n_i for each negative literal $\neg v_i$ in $\mathcal{F}$.
- $V(G)$ contains a vertex c_i for each clause of $\mathcal{F}$.
- $E(G)$ contains, for all distinct i and j, (p_i, p_j), (p_i, n_j), (n_i, p_j), and (n_i, n_j), *i.e.*, all literals are connected to all others (positive and negative) except their own negation.
- $E(G)$ contains (c_i, p_i) if literal v_i does *not* appear in c_i. Similarly, $(c_i, n_i) \in E(G)$ if $\neg v_i$ does *not* appear in c_i, *i.e.*, clauses are connected to all literals that do not satisfy them.
- Finally, let X be the set of all vertices c_i corresponding to clauses. Note that $V \setminus X$ is the set of all vertices corresponding to literals.

An example is shown in Fig. 2.

Observe how a clique cannot contain both p_i and n_i, and any set $S \subseteq (V \setminus X)$ is a clique iff it does *not* contain both a literal and its negation: thus any clique in $V \setminus X$ corresponds to valid truth assignments of the variables of $\mathcal{F}$. We can now prove that a maximal clique $S \subseteq (V \setminus X)$ exists iff $\mathcal{F}$ can be satisfied.

Firstly, if S is a maximal clique, each c_i is *not* adjacent to some vertex $v \in S$: from the construction of the graph, this means c_i is satisfied by the literal corresponding to v, meaning the literals in S satisfy all clauses in $\mathcal{F}$.

To prove the converse, assume S does *not* contain any literal satisfying a clause c_i: then (again from how G is built) c_i is adjacent to all literals in S and $S \cup \{c_i\}$ is a clique, so S is not maximal.

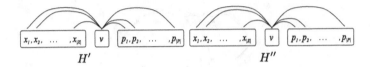

Fig. 3. Graph G', obtained as two copies of the graph H

It follows that EXT-P $(G(V, E), X)$ has a positive answer iff $\mathcal{F}$ can be satisfied. As EXT-P $(G(V, E), X)$ is in NP, because we can test a solution by verifying the maximality of a clique, the proof is complete. $\square$

A Polynomial Delay Pivoting Strategy Implies P = NP

Now, given $G(V, E)$ and X, we build a graph $G'(V', E')$ such that, if Algorithm 1 runs on $G'(V', E')$ with polynomial delay, then we can solve the NP-complete problem EXT-P $(G(V, E), X)$ in polynomial time.

$G'(V', E')$ consists of two identical disjoint copies of a graph $H(V(H), E(H))$, built as follows. $V(H) = X \cup \{v\} \cup P$, where:

- $X = \{x_1, \ldots, x_{|X|}\}$ is the set of vertices of X from EXT-P $(G(V, E), X)$.
- $P = \{p_1, \ldots, p_{|P|}\}$ is the set of vertices of $V \setminus X$ from EXT-P $(G(V, E), X)$.

$E(H)$ is obtained connecting all vertices from $G(V, E)$ as they are connected in G, and the vertex v to all of P and X. Figure 3 shows a graphical example.

Let the two copies of H be H' and H''. If we run Algorithm 1 on G', it will first choose a pivot vertex from either H' or H'': assume wlog it is H'' (the other case is identical; this means that no vertex of H' is adjacent to the pivot, so at least every vertex of H' is processed by Algorithm 1. By processing we mean it is considered in the **foreach** loop of the root recursive call of the algorithm.

As the algorithm does *not* specify in which order vertices are processed, we assume the order is as in Fig. 3, *i.e.*, first $x_1, \ldots, x_{|X|}$, then v, then $p_1, \ldots, p_{|P|}$ (vertices of H'' are not relevant and can be disregarded). Now, take the moment when v is processed: We have that $CAND \cap \Gamma(v)$ is exactly P, and $(SUBG \setminus CAND) \cap \Gamma(v) = FINI \cap \Gamma(v)$ is X as we processed all vertices of X.

If Algorithm 1 —with any arbitrary pivoting strategy— has polynomial delay, it must either find a new maximal clique or terminate, in polynomial time: as any maximal clique containing vertices of X has already been found, this process will output a new maximal clique iff there is a maximal clique in P that cannot be extended with vertices of X, *i.e.*, since P corresponds to $V \setminus X$, there is a maximal clique in $G(V, E)$ that does not intersect X. Finally, since Algorithm 1 may spend exponential time before processing v, we want to skip this time: we do so by simply running the algorithm with $Q = \{v\}$, $CAND = P$ and $SUBG = X \cup P$. We can thus state:

Theorem 3. *No pivoting strategy for the* CLIQUES *[24] and* BKP *[2] can guarantee polynomial delay unless* $P = NP$. $\square$

We finally remark that a class of graphs where CLIQUES has exponential delay can be constructed, although details are omitted for space reasons.

6   Experimental Results

While CLIQUES and Bron-Kerbosch may not have worst-case polynomial delay, we are also interested in their delay and overall performance in practice. To give a complete picture, we present an experimental evaluation of CLIQUES and Bron-Kerbosch variants on real-world networks, showing how their behavior *appears* to be output-sensitive and to have small delay on real-world networks. The worst-case amortized cost remains an open question. Aiming to get substantial experimental evidence we ran our experiments on 138 real-world and synthetic graphs taken from the SNAP [14] and LASAGNE [13] repositories, with up to 3 million edges. For space reasons, we report only a subset in Table 1.

6.1   Maximum Clique Size in Real-World Networks

Firstly, to gauge the scope of Corollary 1, we computed the maximum clique size q of real-world networks in our dataset (128 out of 138). We report this in Fig. 4 against $\log n$: indeed q is below $\log n$ on the majority of networks, and below $10 \log n$ in almost all cases. The significant outlier is the co-authorship network ca-HepPh concerning *High Energy Physics* papers on ArXiv, with 12 006 vertices and a maximum clique of size 239; this is perhaps not surprising as each paper makes a clique of its co-authors, and hyperauthorship is not uncommon in the Physics literature.

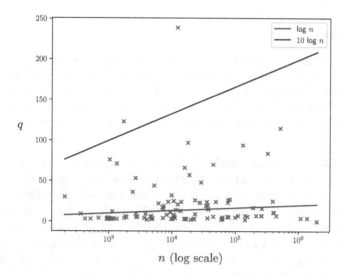

Fig. 4. Maximum clique size q against number of vertices n in 128 real-world graphs.

6.2   Experimental Setup

We consider the following algorithms.

- CLIQUES: the algorithm in [24], described in Algorithm 1 (with pivot u chosen as the vertex in $SUBG$ maximizing $|CAND \cap \Gamma(u)|$).
- BKP_M: BKP [2] with pivot u chosen as the highest-degree vertex in $SUBG$.
- BKP_R: BKP Randomized, i.e., with pivot u chosen randomly in $SUBG$.
- BK: Bron-Kerbosch, without pivoting.

Other efficient algorithms exist, but their inclusion is not meaningful, as we aim to judge pruning effectiveness of pivoting strategies and not the running time. Notable examples are the algorithm by Eppstein et al. [9], effective on sparse graphs, that uses CLIQUES as a subroutine, and the algorithms by San Segundo et al. [19], which aim to quickly find a good (but not optimal) candidate according to CLIQUES's metric.

Metrics. We are not strictly interested in the running time, as a recursive node has polynomial cost. As solutions are output in leaves, we are interested in what portion of the leaves outputs a solution: the total running time is $O(poly(n))$ times the number of leaves, so the ratio $\frac{CLIQUES}{LEAVES}$ (number of maximal cliques divided by the number of leaves of the recursion tree) gives an idea of "how output-sensitive" the execution is. We also show the *delay* in terms of nodes and leaves, i.e., the longest sequence of nodes/leaves between two consecutive outputs. For completeness, we also report the total time and delay in milliseconds.

6.3   Results

For each algorithm, on each graph, we computed the number of nodes and leaves in the recursion tree, and the metrics discussed above. Due to the large number of experiments (and the tendency of BK to time out even on small graphs) to provide a fair comparison, we set a 30 minutes time limit on all reported executions. For space reasons we omit the raw data and only comment on the result obtained.

Nodes and Leaves Generated. We first observed how CLIQUES, thanks to its pivoting strategy, is more effective in pruning than BKP_M and BKP_R: CLIQUES often produces less half the recursive nodes of the next best algorithm, and sometimes orders of magnitude less (e.g., bccstk30, ca-AstroPh, ca-HepPh). Same goes for the delay: CLIQUES has typically lower delay, both in terms of time (see DELAY, column ms) and in terms of nodes and leaves. On the other hand, BK produces a far larger amount of recursive nodes, and frequently times out.

The most relevant value to observe is the $\frac{CLIQUES}{LEAVES}$ ratio: a high value shows that the algorithm is performing in an output-sensitive way. Again we observed how CLIQUES consistently has the highest ratio, sometimes by an order of magnitude (e.g., bcsstk30, Slashdot090221, soc-sign-epinions). For completeness, it is worth observing that the $\frac{CLIQUES}{LEAVES}$ ratios of BKP_M and BKP_R, while worse than CLIQUES, are still often high. Furthermore, $\frac{CLIQUES}{LEAVES}$ for CLIQUES, BKP_M and BKP_R

Table 1. Excerpt of the graphs used, with number of vertices (n), edges (m), maximum degree (Δ), degeneracy (d) and the number of maximal cliques (#cliques).

GRAPH	n	m	Δ	d	q	#cliques
GoogleNw	15 763	148 585	11 401	102	66	75 258
Meth	956	1 157	31	3	3	1 046
add32	4 960	9 462	31	3	4	4 519
amazon0601	403 394	2 443 408	2 752	10	11	1 023 572
auto	448 695	3 314 611	37	9	7	2 164 046
bcsstk30	28 924	1 007 284	218	58	48	6 706
brack2	62 631	366 559	32	7	5	282 557
ca-AstroPh	18 771	198 050	504	56	57	36 427
ca-HepPh	12 006	118 489	491	238	239	14 937

GRAPH	n	m	Δ	d	q	#cliques
darwinBookInter	7 381	45 229	2 686	306	16	127 055
fe_ocean	143 437	409 593	6	4	2	409 593
forest1e4_2	10 000	153 925	1 124	101	29	96 861 484
interdom	1 706	78 983	728	129	123	3 351
Slashdot090221	82 140	500 480	2 548	54	27	854 407
soc-sign-epinions	131 827	711 209	3 558	121	94	22 226 172
spanishBookInter	11 586	44 214	3 327	342	14	66 505
ud_1e4	10 000	313 726	523	285	258	132 557
yeast_bo	1 846	2 203	56	5	6	1 940

is seemingly independent of the size of the graph, and in most cases even close to 1. This supports the idea that $\frac{\text{CLIQUES}}{\text{LEAVES}}$ is in practice $\Omega(1/poly(n))$ for CLIQUES, BKP$_M$ and BKP$_R$, and that the algorithms behave in an output-sensitive way in practice.

Running Time. While a running time comparison is not the goal of this paper (and the implementations are not optimized for this purpose), it is worth observing that CLIQUES seems to perform best on graphs with highest degeneracy, denser and with more solutions; in some cases it is the only one to terminate (e.g., forest1e4_2, soc-sign-epinions, ud_1e4). When BKP$_M$ and BKP$_R$ terminate, in some cases CLIQUES is still significantly faster (e.g., ca-HepPh, interdom, bcsstk30). In others, BKP$_M$ is competitive (e.g., GoogleNW, spanishBookInter) or faster (e.g., darwinBookInter) despite generating more recursive nodes, probably due to CLIQUES having an expensive pivot computation. On small graphs, with low degeneracy, few maximal cliques (e.g., Meth, add32, brack2, fe_ocean, yeast_bo) the differences flatten out, and performance become comparable.

7   Concluding Remarks

We presented a study of the CLIQUES and Bron-Kerbosch algorithms, showing how their delay is exponential in the worst case, unless $P = NP$, settling a question unsolved for a long time. Furthermore, we have shown that the claim remains true for any pivoting strategy that can be computed in polynomial time. On the other hand, we proved that their time complexity is amortized polynomial on graphs whose largest clique has logarithmic size; we showed this condition can hold in both sparse and dense graphs, and observed experimentally that it is generally true in real-world graphs. Our experiments further support this claim as both algorithms perform well in practice on over a hundred real-world graphs. This result partially fills the long-standing gap between the theoretical worst-case exponential time complexity of CLIQUES and its practical efficiency.

While writing this manuscript, we noticed a preprint [17] claiming "The Bron-Kerbosch algorithm with vertex ordering is output-sensitive." However, a bug in the approach was pointed out and later confirmed by the author. The worst-case amortized cost per solution of CLIQUES and Bron-Kerbosch, and the worst-case time of Bron-Kerbosch remain open.

Acknowledgement. The authors would like to thank G. Manoussakis, E. Harley and L. Versari for useful discussions. This work was supported in part by JSPS KAKENHI, Grant JP17K00006 and MIUR, Grant 20174LF3T8 AHeAD.

References

1. Avis, D., Fukuda, K.: Reverse search for enumeration. Discret. Appl. Math. **65**(1–3), 21–46 (1996)
2. Bron, C., Kerbosch, J.: Algorithm 457: finding all cliques of an undirected graph. Commun. ACM **16**(9), 575–577 (1973)
3. Cazals, F., Karande, C.: Reporting maximal cliques: new insights into an old problem. Research report RR-5615, INRIA (2006)
4. Chang, L., Yu, J.X., Qin, L.: Fast maximal cliques enumeration in sparse graphs. Algorithmica **66**(1), 173–186 (2013)
5. Chiba, N., Nishizeki, T.: Arboricity and subgraph listing algorithms. SIAM J. Comput. **14**(1), 210–223 (1985)
6. Comin, C., Rizzi, R.: An improved upper bound on maximal clique listing via rectangular fast matrix multiplication. Algorithmica **80**(12), 3525–3562 (2018)
7. Conte, A., Grossi, R., Marino, A., Versari, L.: Sublinear-space bounded-delay enumeration for massive network analytics: maximal cliques. In: ICALP 2016, vol. 55, pp. 148:1–148:15 (2016)
8. Conte, A., Grossi, R., Marino, A., Versari, L.: Sublinear-space and bounded-delay algorithms for maximal clique enumeration in graphs. Algorithmica **82**(6), 1547–1573 (2020)
9. Eppstein, D., Löffler, M., Strash, D.: Listing all maximal cliques in large sparse real-world graphs. ACM J. Exp. Algorithmics **18**, 3.1:1–3.1:21 (2013)
10. Fortunato, S.: Community detection in graphs. Phys. Rep. **486**(3), 75–174 (2010)
11. Fujii, T., Tomita, E.: On efficient algorithms for finding a maximum clique. Technical report of IECE, AL81-113, pp. 25–34 (1982)
12. Johnson, D.S., Yannakakis, M., Papadimitriou, C.H.: On generating all maximal independent sets. Inf. Process. Lett. **27**(3), 119–123 (1988)
13. Laboratory of Algorithms, models, and Analysis of Graphs and NEtworks. https://www.pilucrescenzi.it/wp/software/lasagne/. Accessed September 2020
14. Leskovec, J., Krevl, A.: SNAP datasets: stanford large network dataset collection (2015). https://snap.stanford.edu/data/
15. Makino, K., Uno, T.: New algorithms for enumerating all maximal cliques. In: Hagerup, T., Katajainen, J. (eds.) SWAT 2004. LNCS, vol. 3111, pp. 260–272. Springer, Heidelberg (2004). https://doi.org/10.1007/978-3-540-27810-8_23
16. Manoussakis, G.: A new decomposition technique for maximal clique enumeration for sparse graphs. Theor. Comput. Sci. **770**, 25–33 (2019)
17. Manoussakis, G.: The Bron-Kerbosch algorithm with vertex ordering is output-sensitive. arXiv:1911.01951v2 (2019). (Pdf not served as of Sept. 2020)
18. Moon, J.W., Moser, L.: On cliques in graphs. Isr. J. Math **3**(1), 23–28 (1965)
19. San Segundo, P., Artieda, J., Strash, D.: Efficiently enumerating all maximal cliques with bit-parallelism. Comput. Oper. Res. **92**, 37–46 (2018)
20. Tomita, E.: Clique enumeration. In: Kao, M.-Y. (ed.) Encyclopedia of Algorithms, 2nd edn, pp. 1–6. Springer, Boston (2016). https://doi.org/10.1007/978-3-642-27848-8_725-2

21. Tomita, E.: Efficient algorithms for finding maximum and maximal cliques and their applications. In: Poon, S.-H., Rahman, M.S., Yen, H.-C. (eds.) WALCOM 2017. LNCS, vol. 10167, pp. 3–15. Springer, Cham (2017). https://doi.org/10.1007/978-3-319-53925-6_1

22. Tomita, E., Conte, A.: Another time-complexity analysis for maximal clique enumeration algorithm CLIQUES. Technical report of IEICE COMP, (1), pp. 1–8 (2020). http://id.nii.ac.jp/1438/00009571/

23. Tomita, E., Tanaka, A., Takahashi, H.: The worst-case time complexity for finding all the cliques. Techncial report of the University of Electro-Communications, UEC-TR-C 5(2), pp. 1–19 (1988). http://id.nii.ac.jp/1438/00001898/

24. Tomita, E., Tanaka, A., Takahashi, H.: The worst-case time complexity for generating all maximal cliques and computational experiments. Theor. Comput. Sci. **363**(1), 28–42 (2006)

25. Tsukiyama, S., Ide, M., Ariyoshi, H., Shirakawa, I.: A new algorithm for generating all the maximal independent sets. SIAM J. Comput. **6**(3), 505–517 (1977)

Computing $L(p, 1)$-Labeling
with Combined Parameters

Tesshu Hanaka[1]([✉]) [ID], Kazuma Kawai[2], and Hirotaka Ono[2] [ID]

[1] Chuo University, Tokyo 112-8551, Japan
hanaka.91t@g.chuo-u.ac.jp
[2] Nagoya University, Nagoya 464-8601, Japan
kawai.kazuma@g.mbox.nagoya-u.ac.jp, ono@i.nagoya-u.ac.jp

Abstract. Given a graph, an $L(p, 1)$-labeling of the graph is an assignment f from the vertex set to the set of nonnegative integers such that for any pair of vertices (u, v), $|f(u) - f(v)| \geq p$ if u and v are adjacent, and $f(u) \neq f(v)$ if u and v are at distance 2. The $L(p, 1)$-LABELING problem is to minimize the span of f (i.e., $\max_{u \in V}(f(u)) - \min_{u \in V}(f(u)) + 1$). It is known to be NP-hard even for graphs of maximum degree 3 or graphs with tree-width 2, whereas it is fixed-parameter tractable with respect to vertex cover number. Since the vertex cover number is a kind of the strongest parameter, there is a large gap between tractability and intractability from the viewpoint of parameterization. To fill up the gap, in this paper, we propose new fixed-parameter algorithms for $L(p, 1)$-LABELING by the twin cover number plus the maximum clique size and by the tree-width plus the maximum degree. These algorithms reduce the gap in terms of several combinations of parameters.

Keywords: Distance constrained labeling · $L(p, 1)$-labeling · Fixed Parameter Algorithm · Treewidth · Twin Cover

1 Introduction

Let G be an undirected graph, and p and q be constant positive integers. An $L(p, q)$-labeling of a graph G is an assignment f from the vertex set $V(G)$ to the set of nonnegative integers such that $|f(x) - f(y)| \geq p$ if x and y are adjacent and $|f(x) - f(y)| \geq q$ if x and y are at distance 2, for all x and y in $V(G)$. We call the former *distance-1 condition* and the latter *distance-2 condition*. A k-$L(p, q)$-labeling is an $L(p, q)$-labeling $f : V(G) \to \{0, \ldots, k\}$, where the labels start from 0 for conventional reasons. The k-$L(p, q)$-LABELING problem determines whether given G has a k-$L(p, q)$-labeling, or not, and the $L(p, q)$-LABELING problem asks the minimum k among all possible assignments. The minimum value k is called the $L(p, q)$-labeling number, and we denote it by $\lambda_{p,q}(G)$, or simply $\lambda_{p,q}$. Notice that we can use $k + 1$ different labels when $\lambda_{p,q}(G) = k$.

This work is partially supported by JSPS KAKENHI Grant Numbers JP17K19960, JP17H01698, JP19K21537 and JP20H05967. A full version is available in [21].

© Springer Nature Switzerland AG 2021
R. Uehara et al. (Eds.): WALCOM 2021, LNCS 12635, pp. 208–220, 2021.
https://doi.org/10.1007/978-3-030-68211-8_17

The original notion of $L(p, q)$-labeling can be seen in the context of frequency assignment. Suppose that vertices in a graph represent wireless devices. The presence/absence of edges indicates the presence/absence of direct communication between the devices. If two devices are very close, that is, they are connected in the graph, they need to use sufficiently different frequencies, that is, their frequencies should be apart at least p. If two devices are not very but still close, that is, they are at distance 2 in the graph, their frequencies should be apart at least q ($\leq p$). Thus, the setting of $q = 1$ as one unit and $p \geq q = 1$ is considered natural and interesting, and the minimization of used range becomes the issue. Note that $L(1, 1)$-labeling on G is equivalent to the ordinary coloring on the square of G. From these, $L(p, 1)$-LABELING for $p > 1$ is intensively and extensively studied among several possible settings of p. In particular, $L(2, 1)$-LABELING is considered the most important. A reason is that it is natural and suitable as a basic step to consider, and another reason is that the computational complexity (e.g., hardness or polynomial-time solvability) tends to be inherited from $L(2, 1)$ to $L(p, 1)$ of $p > 2$; for example, if $L(2, 1)$-LABELING is NP-hard in a setting, the hardness proof could be modified to $L(p, 1)$-LABELING in the same setting. Also many polynomial-time algorithms of $L(2, 1)$-labeling for specific graph classes can be easily extended to $L(p, 1)$. We can find various related results in surveys by Calamoneri [6]. See also [23] for algorithmic results.

The notion of $L(p, q)$-LABELING firstly appeared in [19] and [28]. Griggs and Yeh formally introduced the $L(2, 1)$-LABELING problem [17]. They also show that $L(2, 1)$-LABELING is NP-hard in general. Furthermore, $L(2, 1)$-LABELING is shown to be NP-hard even for planar graphs, bipartite graphs, chordal graphs [4], graphs with diameter of 2 [17] and graphs with tree-width 2 [11]. Moreover, for every $k \geq 4$, k-$L(2, 1)$-LABELING, that is the decision version of $L(2, 1)$-LABELING is NP-complete for general graphs [13] and even for planar graphs [8]. These results imply that k-$L(2, 1)$-LABELING is NP-complete for every $\Delta \geq 3$, where Δ denotes the maximum degree. On the other hand, $L(2, 1)$-LABELING can be solved in polynomial time for paths, cycles, wheels [17], but these are rather trivial. For non-trivial graph classes, only a few graph classes (e.g., cographs [7]) are known to be solvable in polynomial time. In particular, Griggs and Yeh conjectured that $L(2, 1)$-LABELING on trees was NP-hard, which was later disproved (under P$\neq$NP) by the existence of an $O(n^{5.5})$-time algorithm [7]. It is now known that $L(p, 1)$-LABELING on trees can be solved in linear time [22].

From these results, we roughly understand the boundary between polynomial time solvability and NP-hardness concerning graph classes, and studies are going to fixed-parameter (in)tractability. For a problem A with input size n and parameter t, A is called *fixed-parameter tractable* with respect to t if there is an algorithm whose running time is $g(t)n^{O(1)}$, where g is a certain function. Such an algorithm is called a *fixed-parameter* algorithm. If problem A is NP-hard for a constant value of t, there is no fixed-parameter algorithm unless P=NP; we say A is paraNP-hard. Unfortunately, $L(2, 1)$-LABELING is already shown to be paraNP-hard for several parameters such as $\lambda_{2,1}$, maximum degree and tree-width as seen above. For positive results, it is fixed-parameter tractable with respect to vertex cover number [12] or neighborhood diversity [10]. Note that vertex cover number is a stronger parameter than tree-width, which means that if the vertex cover number is bounded, the tree-width is also. There is still a gap on fixed-parameter (in)tractability between them. For such a situation, two approaches can be taken. One is to finely classify intermediate parameters and see fixed-parameter (in)tractability for them, and the other is to combine two or more parameters and see fixed-parameter (in)tractability under the combinations. In this paper, we take the latter approach.

1.1 Our Contribution

In this paper, we present algorithms with combined parameters. The parameters that we focus on are clique-width (cw), tree-width (tw), maximum clique size (ω), maximum degree (Δ) and twin cover number (tc). These are selected in connection with aforementioned parameters, $\lambda_{p,1}$, maximum degree and tree-width. Maximum clique size and clique-width are well used parameters weaker than tree-width. Maximum degree itself is a considered parameter, which is strongly related to $\lambda_{p,q}(G)$. In fact, it is easy to see that $\lambda_{p,1} \geq \Delta + p - 1$, and $\lambda_{p,1} \leq \Delta^2 + (p-1)\Delta - 2$ [16]. Thus, $\lambda_{p,1}$ and Δ are parameters equivalent in terms of fixed-parameter (in)tractability. Twin cover number is picked up as a parameter that is moderately weaker than vertex cover number but stronger than clique-width and is also incomparable to neighborhood diversity.

These parameters are ordered in the following two ways: (1) (vc $\succeq$){tw, tc} $\succeq$ cw and (2) ($\lambda_{p,1} \simeq)\Delta \succeq \omega$. Here, for graph parameters α and β, $\alpha \succeq \beta$ represents that there is a positive function g such that $g(\alpha(G)) \geq \beta(G)$ holds for any G, and we denote $\alpha \simeq \beta$ if $\alpha \succeq \beta$ and $\beta \succeq \alpha$. For combined parameters of one from (1) and another from (2), we design fixed-parameter algorithms. Note that some combination yields essentially one parameter. For example, tw + ω is equivalent to tw, because tw $\geq \omega - 1$ holds. The obtained results are listed below:

- $L(p, 1)$-LABELING can be solved in time $\Delta^{O(\text{tw}\Delta)}n$ for $p \geq 1$. Since it is known that tw ≤ 3cw$\Delta - 1$ ([18]), it is also a $\Delta^{O(\text{cw}\Delta^2)}n$-time algorithm, which implies $L(p, 1)$-LABELING is actually FPT with respect to cw + Δ. This result also implies that $L(p, 1)$-LABELING is FPT when parameterized by band-width.
- $L(p, 1)$-LABELING is FPT when parameterized by tc+ω. Since tc+$\omega \leq$ vc+1 for any graph, it generalizes the fixed-parameter tractability with respect to vertex cover number in [12]. Since tc + $\omega \geq$ tw, tc + ω is located between tw and vc.
- $L(1, 1)$-LABELING is FPT when parameterized by *only* twin cover number. This also yields a fixed-parameter p-approximation algorithm for $L(p, 1)$-LABELING with respect to twin cover number.

Figure 1 illustrates the detailed relationship between graph parameters and the parameterized complexity of $L(p, 1)$-LABELING.

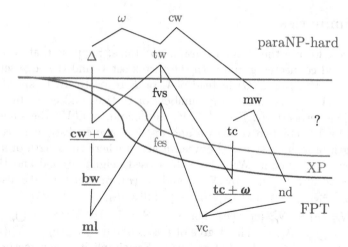

Fig. 1. The relationship between graph parameters and the parameterized complexity of $L(p, 1)$-LABELING. Let $\omega, \Delta, \mathrm{cw}, \mathrm{mw}, \mathrm{nd}, \mathrm{tc}, \mathrm{tw}, \mathrm{fvs}, \mathrm{fes}, \mathrm{bw}, \mathrm{ml}$, and vc denote maximum clique size, maximum degree, clique-width, modular-width, neighborhood diversity, twin cover number, tree-width, feedback vertex set number, feedback edge set number, band-width, max leaf number, and vertex cover number, respectively. Connections between two parameters imply that the upper is bounded by a function of the lower. The underlines for parameters indicate that they are obtained in this paper.

1.2 Related Work

As mentioned above, $L(p, 1)$-LABELING is NP-hard even on graphs of tree-width 2 [11]. Using stronger parameters than tree-width, Fiala et al. showed that $L(p, 1)$-LABELING is fixed-parameter tractable when parameterized by vertex cover [12] and neighborhood diversity [10]. Moreover, Fiala, Kloks and Kratochvíl showed that the problem is XP when parameterized by feedback edge set number [13]. For approximation, it is NP-hard to approximate $L(p, 1)$-LABELING within a factor of $n^{0.5-\varepsilon}$ for any $\varepsilon > 0$, whereas it can be approximated within $O(n(\log \log n)^2/ \log^3 n)$ [20]. For $L(1, 1)$-LABELING, it can be solved in time $O(\Delta^{2^{8(\mathrm{tw}+1)+1}} n + n^3)$, and hence it is XP by tree-width [30]. This result is tight in the sense of fixed-parameter (in)tractability, because it is W [1]-hard with respect to tree-width [12]. Moreover, it can be solved in time $O(\mathrm{cw}^3 2^{6\mathrm{cw}} n^{2^{4\mathrm{cw}} + 2^{2\mathrm{cw}} + 1})$ [29].

Apart from $L(p, 1)$-LABELING, twin cover number is a relatively new graph parameter, which is introduced in [14] as a stronger parameter than vertex cover number. In the same paper, many problems are shown to be FPT when parameterized by twin cover number, and it is getting to be a standard parameter (e.g., [1,9,15,24,25]). Recently, for IMBALANCE, which is one of graph layout problems, a parameterized algorithm is presented [27]. It is interesting that they also adopt twin cover number plus maximum clique size as the parameters.

2   Preliminaries

In this paper, we use the standard graph notations. Suppose that $G = (V, E)$ is a simple and connected graph with the vertex set V and the edge set E. We sometimes use $V(G)$ or $E(G)$ instead of V or E respectively, to specify graph G. For $G = (V, E)$, we denote the numbers of vertices and edges by $n = |V|$ and $m = |E|$, respectively. For $V' \subseteq V$, we denote by $G[V']$ the subgraph of G induced by V'. For two vertices u and v, the *distance* $\mathrm{dist}_G(u, v)$ is defined by the length of a shortest path between u and v where the length of a path is the number of edges of it. We denote the closed neighbourhood and the open neighbourhood of a vertex v by $N_G[v]$ and $N_G(v)$, respectively. We also define $N_G^\ell[v] = \{u \mid \mathrm{dist}_G(u, v) = \ell\}$, $N_G^{\leq \ell}[v] = \{u \mid \mathrm{dist}_G(u, v) \leq \ell\}$, $N_G^\ell(v) = N_G^\ell[v] \setminus \{v\}$, and $N_G^{\leq \ell}(v) = N_G^{\leq \ell}[v] \setminus \{v\}$. For a set $S \subseteq V$, let $N_G(S) = \bigcup_{v \in S} N_G(v)$ and $N_G[S] = \bigcup_{v \in S} N_G[v]$. The degree of v is denoted by $d_G(v) = |N_G(v)|$. The maximum degree of G is denoted by $\Delta(G)$. For simplicity, we sometimes omit the subscript G.

The k-th power $G^k = (V, E^k)$ of a graph $G = (V, E)$ is a graph such that the set of vertices is V and there is an edge (u, v) in E^k if and only if there is a path of length at most k between u and v in G [5]. In particular, G^2 is called the *square* of G.

Graph parameters

In the following, we introduce several graph parameters.

Definition 1 (Tree Decomposition). *A tree decomposition of a graph $G = (V, E)$ is defined as a pair $\langle \mathcal{X}, T \rangle$, where T is a tree with node set $I(T)$ and $\mathcal{X} = \{X_i \mid i \in I(T)\}$ is a collection of subsets, called bags, of V such that:*

1. *(vertex condition)* $\bigcup_{i \in I(T)} X_i = V$;
2. *(edge condition) For every $\{u, v\} \in E$, there exists an $i \in I(T)$ such that $\{u, v\} \subseteq X_i$;*
3. *(coherence property) For every $u \in V$, $I_u = \{i \in I(T) \mid u \in X_i\}$ induces a connected subtree of T.*

The width of a tree decomposition is defined as $\max_{i \in I} |X_i| - 1$ and the tree-width of G, denoted by $\mathrm{tw}(G)$, is defined as the minimum width among all possible tree decompositions of G.

Two vertices u, v are called *twins* if $N(u) = N(v)$. Moreover, if twins u, v have edge $\{u, v\}$, they are called *true twins* and the edge is called a *twin edge*. Then a *twin cover* of G is defined as follows.

Definition 2 (Twin Cover, [14]). *A set of vertices X is a twin cover of G if every edge $\{u, v\} \in E$ satisfies either (1) $u \in X$ or $v \in X$, or (2) u, v are true twins. The twin cover number of G, denoted by $\mathrm{tc}(G)$, is defined as the minimum size of twin covers in G.*

An important observation is that the complement $V \setminus X$ of a twin cover X induces disjoint cliques. Moreover, for each clique Z of $G[V \setminus X]$, $N(u) \cap X = N(v) \cap X$ for every $u, v \in Z$ [14].

A *vertex cover* X is the set of vertices such that for every edge, at least one endpoint is in X. The *vertex cover number* of G, denoted by $\mathsf{vc}(G)$, is defined as the minimum size of vertex covers in G. Since every vertex cover of G is also a twin cover of G, $\mathsf{tc}(G) \leq \mathsf{vc}(G)$ holds. Also, for any graph G, we have $\mathsf{tc}(G) + \omega(G) \leq \mathsf{vc}(G) + 1$.

Integer Linear Programming

INTEGER LINEAR PROGRAMMING FEASIBILITY is formulated as follows.

Input: An $q \times p$ matrix A with integer elements, an integer vector $b \in \mathbb{Z}^q$
Question: Is there a vector $x \in \mathbb{Z}^p$ such that $A \cdot x \leq b$?

Lenstra [26] proved that INTEGER LINEAR PROGRAMMING FEASIBILITY is FPT when parameterized by the number of variables.

3   Parameterization by $\mathsf{cw} + \Delta$ and $\mathsf{tw} + \Delta$

As $L(p, 1)$-LABELING is paraNP-hard for tree-width, so is for clique-width. In this section, as a complement, we show that $L(p, 1)$-LABELING (actually, $L(p, q)$-LABELING for any constant p and q) is fixed-parameter tractable when parameterized by $\mathsf{cw} + \Delta$.

To this end, we give a fixed-parameter algorithm for $L(p, 1)$-LABELING parameterized by not $\mathsf{cw} + \Delta$ but $\mathsf{tw} + \Delta$, which actually implies that the problem is FPT with respect to $\mathsf{cw} + \Delta$, because it is known that $\mathsf{tw} \leq 3\mathsf{cw}\Delta - 1$ [18]. The running time of the algorithm is $\Delta^{O(\mathsf{tw}\Delta)}n$, and so it is $\Delta^{O(\mathsf{cw}\Delta^2)}n$.

In the algorithm, we first construct the square G^2 of G and then compute $L(p, 1)$-LABELING of G by dynamic programming on a tree decomposition $\langle \mathcal{X}', T \rangle$ of G^2. Actually, the algorithm runs for $L(p, q)$-LABELING though the running time depends on $\lambda_{p,q}(G)$. One can obtain the square of G^2 in time $O(m\Delta(G)) = O(\Delta(G)^2 n)$. We then prove the following lemma.

Lemma 1. *Given a tree decomposition of a graph G of width t with ℓ bags, one can construct a tree decomposition of G^2 of width at most $(t + 1)\Delta(G) + t$ with ℓ bags in time $O(t\Delta(G)\ell)$.*

Proof. We are given a tree decomposition $\langle \mathcal{X}, T \rangle$ of G of width t. Let $X_i' = X_i \cup N(X_i)$ and $\mathcal{X}' = \{X_i' \mid i \in I(T)\}$ be the set of bags. We here define $\langle \mathcal{X}', T' \rangle$ as a tree decomposition of G^2, where T' and T are identical; T and T' has the same node set and the same structure, where each $i \in I(T')$ corresponds to $i \in I(T)$. In the following, we denote $\langle \mathcal{X}', T \rangle$ instead of $\langle \mathcal{X}', T' \rangle$.

We can see that $\langle \mathcal{X}', T \rangle$ is really a tree decomposition of G^2 with width $(t + 1)\Delta(G) + t$. It satisfies the properties of tree decomposition indeed: Since

$\bigcup_{i \in I(T)} X_i' = \bigcup_{i \in I(T)} (X_i \cup N(X_i)) = V(G) = V(G^2)$, the vertex condition is satisfied. We next see the edge condition. For each $e \in E$, there is X_i containing e, so $e \in X_i'$. For each $\{u, v\} \in E^2 \setminus E$, there is a vertex $v'(\neq u, v)$ such that $\{u, v'\} \in E$ and $\{v', v\} \in E$. Thus there is X_i satisfying $\{u, v'\} \subseteq X_i$, which implies $\{u, v\} \subseteq X_i \cup \{v\} \subseteq X_i \cup N(\{v'\}) \subseteq X_i'$. These show that the edge condition is satisfied.

Finally, we check coherence property: we show that for every $u \in V$, $I_u' = \{i \in I(T) \mid u \in X_i'\}$ induces a connected subtree of T. Note that

$$I_u' = \{i \in I(T) \mid u \in X_i'\} = \{i \in I(T) \mid u \in X_i\} \cup \bigcup_{v \in N(u)} \{i \in I(T) \mid v \in X_i\}.$$

Here, the subgraph T_v of T induced by $\{i \in I(T) \mid u \in X_i\}$ is connected by the coherent property of $\langle \mathcal{X}, T \rangle$. Also for each $v \in N(u)$, the subgraph T_v of T induced by $\{i \in I(T) \mid v \in X_i\}$ is connected. By $\{u, v\} \in E$, the edge condition of $\langle \mathcal{X}, T \rangle$ implies that there exists a bag X_j containing both u and v. Since T_u and T_v has a common node j, the subgraph of T induced by $\{i \in I(T) \mid u \in X_i\} \cup \{i \in I(T) \mid v \in X_i\}$ is also connected, which leads that the subgraph of T induced by I_u' is also connected.

Hence, $\langle \mathcal{X}', T \rangle$ is a tree decomposition of G^2. Since the size of bag X_i' is $|X_i'| = |X_i \cup N(X_i)| = |\bigcup_{u \in X_i} N[u]| \leq (t+1)(\Delta(G) + 1)$, the width is at most $(t+1)(\Delta(G) + 1) - 1 = (t+1)\Delta(G) + t$. The construction of $\langle \mathcal{X}', T \rangle$ is done by preparing each X_i', which takes $O(t\Delta(G))$ steps for each i. Thus it can be done in time $O(t\Delta(G)\ell)$ in total.   □

Corollary 1. $\mathtt{tw}(G^2) \leq (\mathtt{tw}(G) + 1)\Delta(G) + \mathtt{tw}(G)$ holds.

By the above lemma, the tree-width of G^2 is bounded if $\mathtt{tw}(G)$ and $\Delta(G)$ are bounded. Thus we can design a dynamic programming algorithm on a tree decomposition of G^2, although we omit the detail.

Lemma 2. Given a tree decomposition of G^2 of width at most t, one can compute k-$L(p, q)$-LABELING on G in time $O((k+1)^{t+1}t^2n)$.

Here, one can construct a tree decomposition $\langle \mathcal{X}, T \rangle$ of G of width $5\mathtt{tw}(G) + 4$ with $O(n)$ bags in time $2^{O(\mathtt{tw}(G))}n$ [3]. By Lemma 1, we can obtain a tree decomposition $\langle \mathcal{X}', T \rangle$ of G^2 of width $(5\mathtt{tw}(G) + 4 + 1)\Delta(G) + 5\mathtt{tw}(G) + 4 = O(\mathtt{tw}(G)\Delta(G))$ from $\langle \mathcal{X}, T \rangle$ in time $O(\mathtt{tw}(G)\Delta(G)n)$. By Lemma 2 and $\lambda_{p,q} \leq \max\{p, q\}\Delta^2$, we have the following theorem.

Theorem 1. For any positive constant p and q, there is an algorithm to solve $L(p, q)$-LABELING in time $\Delta^{O(\mathtt{tw}\Delta)}n$, which is also bounded by $\Delta^{O(\mathtt{cw}\Delta^2)}n$.

For the band-width $\mathtt{bw}(G)$ and the max leaf number $\mathtt{ml}(G)$ of G, we have $\mathtt{tw}(G) \leq \mathtt{bw}(G) \leq \mathtt{ml}(G)$ and $\Delta(G) \leq 2\mathtt{bw}(G)$ [2]. Thus, the following corollary holds.

Corollary 2. For any positive constant p and q, $L(p, q)$-LABELING is fixed-parameter tractable when parameterized by max leaf number, and even bandwidth.

4   Parameterization by Twin Cover Number

4.1 $L(p,1)$-LABELING parameterized by $\mathsf{tc} + \omega$

We design a fixed-parameter algorithm for $L(p,1)$-LABELING with respect to $\mathsf{tc} + \omega$. Notice that for a twin cover X of $G = (V,E)$, each of the connected components of $G[V \setminus X]$ forms a clique. We categorize vertices in $V \setminus X$ with respect to the neighbors in X. Let $T_1, T_2, \ldots, T_s$ be the sets of vertices having common neighbors in X, called *types* of vertices in $V \setminus X$, where s is the number of types. Moreover, we say that a clique $C \subseteq V \setminus X$ is of type T_i if $C \subseteq T_i$. Note that $V \setminus X = \bigcup_{i=1}^{s} T_i$. Let $n_i = |T_i|$ and ω_i be the maximum clique size in T_i.

We first see a general property about cliques with the common neighbors: Suppose that a graph G consists of only cliques $C_1, C_2, \ldots, C_h$ and the common neighbors Y of all the vertices in the cliques. That is, all the vertices are within distance 2. Note that a twin cover focuses on such a substructure in a graph. Then the following lemma holds.

Lemma 3. *Suppose that a graph G is above defined by cliques $C_1, C_2, \ldots, C_h$, in the descending order of the size and their common neighbors Y, where the vertices in Y are labeled by $a_1, a_2, \ldots, a_{|Y|}$. For an arbitrary set L of labels that are at least p apart from $a_1, a_2, \ldots, a_{|Y|}$, if $|L| \geq \sum_j |C_j|$ and $\sum_j |C_j| \geq p|C_1|$ hold, there exists an $L(p,1)$-labeling of $C_1, \ldots, C_h$ using only labels in L.*

Proof. Let $n' = \sum_j |C_j|$ and $\omega = |C_1|$. Let us assume $L = \{l_1, l_2, \ldots, l_{n'}\}$. Since we can use distinct labels for vertices in $C_1, C_2, \ldots, C_h$, only the distance-1 condition inside of a same clique matters. If $n' \equiv 1 \pmod{p}$, we label the vertices in $C_1, C_2, \ldots, C_{n'}$ in this order by using labels in order of $l_1, l_{p+1}, l_{2p+1}, \ldots, l_{n'}, l_2,$ $l_{p+2}, l_{2p+2} \ldots, l_{n'-p+2}, l_3 \ldots, l_p, l_{2p}, \ldots, l_{n'-1}$. Note that the vertices in C_1 are labeled by $l_1, l_{p+1}, \ldots, l_{p(\omega-1)+1}$ (note that $p\omega \leq n'$). Since the difference between $l_{\alpha p + i}$ and $l_{(\alpha+1)p+i}$ for each i and α is at least p, the labeling for cliques does not violate the distance-1 condition. We can choose similar orderings for the other residuals. $\square$

Now we go back to the algorithm parameterized by $\mathsf{tc} + \omega$. Given a twin cover X, we say that a k-$L(p,1)$-labeling is *good* for X if it uses only labels in $\{0, 1, \ldots, (2p-1)|X|-p\} \cup \{k-(2p-1)|X|+p, \ldots, k\}$ for X. The following lemma is also important. It can be shown by repeatedly applying Lemma 3 though we omit the detail.

Lemma 4. *Let X be a twin cover in G such that each T_i satisfies $\omega_i \leq n_i/p$. If G has a k-$L(p,1)$-labeling, then G also has a good k-$L(p,1)$-labeling for X.*

Thus, we consider to find a good $L(p,1)$-labeling. Using the lemma, we show that $L(p,1)$-LABELING is fixed-parameter tractable with respect to $\mathsf{tc} + \omega$.

Theorem 2. *$L(p,1)$-LABELING is fixed-parameter tractable when parameterized by $\mathsf{tc} + \omega$.*

Proof. We present an algorithm to solve k-$L(p, 1)$-LABELING instead of $L(p, 1)$-LABELING. We first compute a minimum twin cover X in time $O(1.2738^{\text{tc}} + \text{tc}n + m)$ [14]. For twin cover X, we define T_i's. Then, we define another twin cover of $X' = X \cup \bigcup_{i:\omega_i > n_i/p} T_i$, where X' is obtained by adding every T_i breaking the condition of Lemma 4 to X. By this modification, our algorithm can utilize a twin cover that satisfies the condition of Lemma 4. The size of X' is at most $\text{tc} + 2^{\text{tc}} \cdot p \cdot \omega$, because the number of types is at most 2^{tc} and the size of T_i joining X is at most $p \cdot \omega$. Let $\text{tc}' = |X'|$.

We are now ready to present the core of the algorithm. We classify an instance into two cases. If k is small enough, we can apply a brute-force type algorithm. Otherwise, we try to find a good k-$L(p, 1)$-labeling.

(**Case: $k < 4ptc'$**) For each type T_i, the distance between two vertices in T_i is at most 2. Thus, the labels of vertices in T_i must be different each other. Due to $k < 4ptc'$, if $|T_i| \geq 4ptc'$, we conclude that the input is a no-instance. Otherwise, $n = |X'| + \sum |T_i| \leq \text{tc}' + 4ptc'2^{\text{tc}}$ holds, because the number of T_i's is at most 2^{tc}. Thus we check all the possible labelings in time $O((4ptc')^{\text{tc}'(4p2^{\text{tc}}+1)})$.

(**Case: $k \geq 4ptc'$**) Let $\mathcal{C}_0, \mathcal{C}_1, \ldots \mathcal{C}_t$ be the family of all possible set systems on $\{T_1, \ldots, T_s\}$ such that whenever two distinct T_j and $T_{j'}$ are in $\mathcal{C}_i$ then $N(T_j) \cap N(T_{j'}) = \emptyset$. Here, $\mathcal{C}_0$ is the empty set. These are introduced to describe a set of T_j's that can use a same label. For each $\mathcal{C}_i$, we prepare a set L_i of labels, which will be used during the execution of the algorithm to represent the set of labels that could be used for vertices in $T_j \in \mathcal{C}_i$. Note that $L_0, L_1, \ldots, L_t$ must be disjoint each other, and a label in L_i is used exactly once per T_j. We also define L_0 as the set of labels not used in $V \setminus X'$. Each L_i can be empty.

By Lemma 4, there is a good k-$L(p, 1)$-labeling for X such that vertices in X only use labels in $\{0, 1, \ldots, 2p(\text{tc}' - 1) - p\} \cup \{k - 2p(\text{tc}' - 1) + p, \ldots, k\}$ if the input is an yes-instance. Thus we try all the possible partial labelings for X, each of which uses only labels in $\{0, 1, \ldots, 2p(\text{tc}'-1)-p\} \cup \{k-2p(\text{tc}'-1)+p, \ldots, k\}$. Since the number of labels is $2(2p(\text{tc}' - 1) - p + 1) \leq 4ptc'$, there are at most $(4ptc')^{\text{tc}'}$ possible labelings of X. For each of them we further try all the possible placement of labels in $\{0, 1, \ldots, 2p(\text{tc}' - 1) - 1\} \cup \{k - 2p(\text{tc}' - 1) + 1, \ldots, k\}$ into $L_0, L_1, \ldots, L_t$, which is a little wider than above. The number of possible placements is at most $t^{4ptc'}$ due to the disjointness of L_i's. Therefore, the total possible nonisomorphic partial labelings is at most $(4ptc')^{\text{tc}} \cdot t^{4ptc'}$. Note that no vertex will be labeled by a label in $\{0, 1, \ldots, 2p(\text{tc}' - 1) - 1\} \cup \{k - 2p(\text{tc}' - 1) + 1, \ldots, k\}$ hereafter. Thus we consider how we use labels in $\{2p(\text{tc}' - 1), \ldots, k - 2p(\text{tc}' - 1)\}$ for $V \setminus X$, which does not yield any conflict with X.

We then formulate how many labels should be placed in $L_0, L_1, \ldots, L_t$ for one partial labeling using $\{0, 1, \ldots, 2p(\text{tc}' - 1) - p\} \cup \{k - 2p(\text{tc}' - 1) + p, \ldots, k\}$ as Integer Linear Programming. For a fixed partial labeling, let a_i be the number of labels that have been already assigned to L_i, and x_i be a variable representing the number of labels used in L_i in the desired labeling.

The following is the ILP formulation.

$$\begin{cases} x_0 + \cdots + x_t \le k + 1 \\ x_i \ge a_i, & \text{for } i \in \{0, \ldots, t\} \\ \sum_{i:T_j \in \mathcal{C}_i} x_i = |T_j|, & \text{for } j \in \{1, \ldots, s\} \end{cases}$$

The first constraint shows that the total number of labels is at most $k + 1$. Note that the number of unused labels is x_0. The second one is for consistency to the partial labeling. The last one, which is the most important, guarantees that every vertex in T_j can receive a label; the number of usable labels is $|\{i \mid T_j \in \mathcal{C}_i\}|$, because a label in L_i is used exactly once per T_j.

If the above ILP has a feasible solution, it is possible to assign labels to all the vertices in $V \setminus X$ if we ignore the distance-1 condition inside of each clique. Actually, we can see that the information is sufficient to give a proper k-$L(p, 1)$-labeling. At the beginning of the algorithm, we take twin cover X', which means that for every $T_i \subseteq V \setminus X$, $n_i \ge p\omega_i$ holds. Since cliques in $G[T_i]$ have common neighbors and $n_i \ge p\omega_i$, only the number of available labels matters by Lemma 3. Since the existence of an ILP solution guarantees this, we can decide whether a partial labeling can be extended to a proper k-$L(p, 1)$-labeling, or not.

Because $t \le 2^s \le 2^{2^{tc}}$, the number of variables of the ILP is at most $2^{2^{tc}}$; it can be solved in FPT time with respect to $\mathtt{tc}$ [26]. Since $\mathtt{tc}' \le \mathtt{tc} + 2^{\mathtt{tc}} \cdot p \cdot \omega$, the total running time is FPT time with respect to $\mathtt{tc} + \omega$.   □

4.2 $L(1, 1)$-Labeling parameterized by tc

Unlike $L(p, 1)$-labeling with $p \ge 2$, the distance-1 condition of $L(1, 1)$-labeling requires just that the labels between adjacent vertices are different. Thus, $L(1, 1)$-Labeling seems to be easier than $L(p, 1)$-Labeling with $p \ge 2$. Actually, we can show that $L(1, 1)$-Labeling is fixed-parameter tractable parameterized only by twin cover number.

Lemma 5. *For a graph G, let u and v be twins with edge $\{u, v\}$. For $G' = (V, E')$ with $E' = E \setminus \{\{u, v\}\}$, any $L(1, 1)$-labeling on G' is also an $L(1, 1)$-labeling on G and vice versa.*

Proof. The statement is true, if $N_G^{\le 2}[y] = N_{G'}^{\le 2}[y]$ holds for any vertex $y \in V$, and we show this below. Since $N_G^{\le 2}[y] \supseteq N_{G'}^{\le 2}[y]$ is obvious, we show $N_G^{\le 2}[y] \subseteq N_{G'}^{\le 2}[y]$, that is, for any $y \in V$, if $w \in N_G^{\le 2}[y]$, w also belongs to $N_{G'}^{\le 2}[y]$. Note that $w \in N_G^{\le 2}[y]$ means there is a path with length at most 2 between y and w. If G has such a path between y and w not containing $\{u, v\}$, G' also does, and $w \in N_{G'}^{\le 2}[y]$ holds. Otherwise, every path with length at most 2 between y and w in G contains $\{u, v\}$, which implies that either y or w is u or v. We just see the case when $y = u$ for symmetry, and take such a path between $y(= u)$ and w. If the path length is 1 (that is, $w = v$) in G, $y(= u)$ and $w(= v)$ has a common neighbor because u and v are twins, which implies that w and y are within distance 2 in G'. If the path length is 2, the path forms (y, v, w). Namely, w is a neighbor of v and also of $u(= y)$ in G'. This completes the proof.   □

Corollary 3. *For G' defined as above, $\lambda_{1,1}(G') = \lambda_{1,1}(G)$ holds.*

Let X be a twin cover again, and then each connected component in $G[V \setminus X]$ forms a clique, each of the edges in which are twin edges. Lemma 5 implies that graph G' obtained by removing all the edges in $G[V \setminus X]$ has the same $L(1,1)$-labeling number of G. The above deletion shows that X is also a vertex cover of G'. Since $L(1,1)$-LABELING is fixed-parameter tractable when parameterized by vertex cover number [12], we have the following theorem.

Theorem 3. $L(1,1)$-LABELING *is fixed-parameter tractable when parameterized by twin cover number.*

Since $\lambda_{1,1}(G) \leq \lambda_{p,1}(G) \leq \lambda_{p,p}(G) = p\lambda_{1,1}(G)$ holds, an $L(1,1)$-labeling gives an approximation for $L(p,1)$-LABELING. In fact, by replacing the labels of an optimal $L(1,1)$-labeling of G with multiples of p, we obtain an $L(p,1)$-labeling whose approximation factor is at most p.

Corollary 4. *For $L(p,1)$-LABELING, there is a fixed-parameter p-approximation algorithm with respect to twin cover number.*

Acknowledgements. We are grateful to Dr. Yota Otachi for his insightful comments.

References

1. Asahiro, Y., Eto, H., Hanaka, T., Lin, G., Miyano, E., Terabaru, I.: Parameterized algorithms for the happy set problem. In: Rahman, M.S., Sadakane, K., Sung, W.-K. (eds.) WALCOM 2020. LNCS, vol. 12049, pp. 323–328. Springer, Cham (2020). https://doi.org/10.1007/978-3-030-39881-1_27

2. Blum, J.: Hierarchy of transportation network parameters and hardness results. In: International Symposium on Parameterized and Exact Computation (IPEC 2019), vol. 148, pp. 4:1–4:15 (2019)

3. Bodlaender, H.L., Drange, P.G., Dregi, M.S., Fomin, F.V., Lokshtanov, D., Pilipczuk, M.: A $c^k n$ 5-approximation algorithm for treewidth. SIAM J. Comput. **45**(2), 317–378 (2016)

4. Bodlaender, H.L., Kloks, T., Tan, R.B., Van Leeuwen, J.: Approximations for λ-colorings of graphs. Comput. J. **47**(2), 193–204 (2004)

5. Bondy, J.A., Murty, U.S.R.: Graph Theory. Springer, New york (2008)

6. Calamoneri, T.: The $L(h,k)$-labelling problem: an updated survey and annotated bibliography. Comput. J. **54**(8), 1344–1371 (2011)

7. Chang, G.J., Kuo, D.: The L(2,1)-labeling problem on graphs. SIAM J. Disc. Math. **9**(2), 309–316 (1996)

8. Eggemann, N., Havet, F., Noble, S.D.: k-$L(2,1)$-labelling for planar graphs is NP-complete for $k \geq 4$. Disc. Appl. Math. **158**(16), 1777–1788 (2010)

9. Eto, H., Hanaka, T., Kobayashi, Y., Kobayashi, Y.: Parameterized algorithms for maximum cut with connectivity constraints. In: International Symposium on Parameterized and Exact Computation (IPEC 2019), vol. 148, pp. 13:1–13:15 (2019)

10. Fiala, J., Gavenčiak, T., Knop, D., Koutecký, M., Kratochvíl, J.: Parameterized complexity of distance labeling and uniform channel assignment problems. Disc. Appl. Math. **248**, 46–55 (2018)
11. Fiala, J., Golovach, P.A., Kratochvíl, J.: Distance constrained labelings of graphs of bounded treewidth. In: Caires, L., Italiano, G.F., Monteiro, L., Palamidessi, C., Yung, M. (eds.) ICALP 2005. LNCS, vol. 3580, pp. 360–372. Springer, Heidelberg (2005). https://doi.org/10.1007/11523468_30
12. Fiala, J., Golovach, P.A., Kratochvíl, J.: Parameterized complexity of coloring problems: treewidth versus vertex cover. Theor. Comput. Sci. **412**(23), 2513–2523 (2011)
13. Fiala, J., Kloks, T., Kratochvíl, J.: Fixed-parameter complexity of λ-labelings. Disc. Appl. Math. **113**(1), 59–72 (2001)
14. Ganian, R.: Improving vertex cover as a graph parameter. Disc. Math. Theor. Comput. Sci. **17**(2), 77–100 (2015)
15. Gaspers, S., Najeebullah, K.: Optimal surveillance of covert networks by minimizing inverse geodesic length. In: AAAI Conference on Artificial Intelligence (AAAI 2019), pp. 533–540 (2019)
16. Gonçalves, D.: On the $L(p, 1)$-labelling of graphs. Disc. Math. **308**(8), 1405–1414 (2008)
17. Griggs, J.R., Yeh, R.K.: Labelling graphs with a condition at distance 2. SIAM J. Disc. Math. **5**(4), 586–595 (1992)
18. Gurski, F., Wanke, E.: The tree-width of clique-width bounded graphs without Kn,n. In: Brandes, U., Wagner, D. (eds.) WG 2000. LNCS, vol. 1928, pp. 196–205. Springer, Heidelberg (2000). https://doi.org/10.1007/3-540-40064-8_19
19. Hale, W.K.: Frequency assignment: theory and applications. Proc. IEEE **68**(12), 1497–1514 (1980)
20. Halldórsson, M.M.: Approximating the $L(h, k)$-labelling problem. Int. J. Mobile Netw. Des. Innov. **1**(2), 113–117 (2006)
21. Hanaka, T., Kawai, K., Ono, H.: Computing $L(p, 1)$-labeling with combined parameters (2020). arXiv: 2009.10502
22. Hasunuma, T., Ishii, T., Ono, H., Uno, Y.: A linear time algorithm for $L(2, 1)$-labeling of trees. Algorithmica **66**(3), 654–681 (2013)
23. Hasunuma, T., Ishii, T., Ono, H., Uno, Y.: Algorithmic aspects of distance constrained labeling: a survey. Int. J. Netw. Comput. **4**(2), 251–259 (2014)
24. Jansen, B.M.P., Pieterse, A.: Optimal data reduction for graph coloring using low-degree polynomials. Algorithmica **81**(10), 3865–3889 (2019)
25. Knop, D., Masařík, T., Toufar, T.: Parameterized complexity of fair vertex evaluation problems. In: International Symposium on Mathematical Foundations of Computer Science (MFCS 2019), vol. 138, pp. 33:1–33:16 (2019)
26. Lenstra, H.W.: Integer programming with a fixed number of variables. Math. Oper. Res. **8**(4), 538–548 (1983)
27. Misra, N., Mittal, H.: Imbalance parameterized by twin cover revisited. In: Kim, D., Uma, R.N., Cai, Z., Lee, D.H. (eds.) COCOON 2020. LNCS, vol. 12273, pp. 162–173. Springer, Cham (2020). https://doi.org/10.1007/978-3-030-58150-3_13
28. Roberts, F.S.: T-colorings of graphs: recent results and open problems. Disc. Math. **93**(2), 229–245 (1991)

29. Todinca, I.: Coloring powers of graphs of bounded clique-width. In: Bodlaender, H.L. (ed.) WG 2003. LNCS, vol. 2880, pp. 370–382. Springer, Heidelberg (2003). https://doi.org/10.1007/978-3-540-39890-5_32
30. Zhou, X., Kanari, Y., Nishizeki, T.: Generalized vertex-colorings of partial k-trees. IEICE Trans. Fundamentals Electron. Commun. Comput. Sci. **E83-A**(4), 671–678 (2000)

On Compatible Matchings

Oswin Aichholzer[1], Alan Arroyo[2], Zuzana Masárová[2], Irene Parada[3],
Daniel Perz[1]([⊠]), Alexander Pilz[1], Josef Tkadlec[2],
and Birgit Vogtenhuber[1]

[1] Institute of Software Technology, Graz University of Technology, Graz, Austria
{oaich,daperz,apilz,bvogt}@ist.tugraz.at
[2] IST Austria, Klosterneuburg, Austria
{alanmarcelo.arroyoguevara,zuzana.masarova,josef.tkadlec}@ist.ac.at
[3] TU Eindhoven, Eindhoven, The Netherlands
i.m.de.parada.munoz@tue.nl

Abstract. A matching is compatible to two or more labeled point sets
of size n with labels $\{1,\ldots,n\}$ if its straight-line drawing on each of
these point sets is crossing-free. We study the maximum number of edges
in a matching compatible to two or more labeled point sets in general
position in the plane. We show that for any two labeled convex sets
of n points there exists a compatible matching with $\lfloor\sqrt{2n}\rfloor$ edges. More
generally, for any ℓ labeled point sets we construct compatible matchings
of size $\Omega(n^{1/\ell})$. As a corresponding upper bound, we use probabilistic
arguments to show that for any ℓ given sets of n points there exists
a labeling of each set such that the largest compatible matching has
$\mathcal{O}(n^{2/(\ell+1)})$ edges. Finally, we show that $\Theta(\log n)$ copies of any set of n
points are necessary and sufficient for the existence of a labeling such
that any compatible matching consists only of a single edge.

Keywords: Compatible graphs · Crossing-free matchings · Geometric
graphs

1 Introduction

For plane drawings of geometric graphs, the term *compatible* is used in two rather
different interpretations. In the first variant, *two* plane drawings of geometric
graphs are embedded on the *same* set P of points. They are called compatible
(to each other with respect to P) if their union is plane (see e.g. [5,19]). Note

A.A. funded by the Marie Skłodowska-Curie grant agreement No. 754411. Z.M. partially funded by Wittgenstein Prize, Austrian Science Fund (FWF), grant no. Z 342-
N31. I.P., D.P., and B.V. partially supported by FWF within the collaborative DACH
project *Arrangements and Drawings* as FWF project I 3340-N35. A.P. supported by
a Schrödinger fellowship of the FWF: J-3847-N35. J.T. partially supported by ERC
Start grant no. (279307: Graph Games), FWF grant no. P23499-N23 and S11407-N23
(RiSE).

© Springer Nature Switzerland AG 2021
R. Uehara et al. (Eds.): WALCOM 2021, LNCS 12635, pp. 221–233, 2021.
https://doi.org/10.1007/978-3-030-68211-8_18

that this is different to simultaneous planar graph embedding, as it is required not only that the two graphs are plane, but also that their union is crossing-free.

In the second setting, which is the one that we will consider in this work, *one* planar graph G is drawn straight-line on *two or more* labeled point sets (with the same label set). We say that G is compatible to the point sets if the drawing of G is plane for each of them (where each vertex of G is mapped to a unique label and thereby identified with a unique point of each point set). Note that the labelings of the point sets can be predefined or part of the solution. As an example, we mention the compatible triangulation conjecture [3]: *For any two sets P_1 and P_2 with the same number of points and the same number of extreme points, there is a labeling of the two sets such that there exists a triangulation which is compatible to both sets, P_1 and P_2.*

Motivation and Related Work. The study of the type of compatibility considered in this work (the second type from above) is motivated by applications in morphing [7,16,17], 2D shape animation [12], or cartography [22].

Compatible triangulations were first introduced by Saalfeld [22] for labeled point sets, who studied the construction of compatible triangulations using Steiner points (compatible triangulations do not always exist for pairs of labeled point sets). Aronov et al. [9] and Babikov et al. [11] showed that $O(n^2)$ Steiner points are always sufficient, while Pach et al. [21] showed that $\Omega(n^2)$ Steiner points are sometimes necessary. The compatible triangulation conjecture states that – in contrast to labeled point sets – two unlabeled point sets (in general position and with the same number of extreme points) can always be compatibly triangulated without using Steiner points. To date, the conjecture has only been proven for point sets with at most three interior points [3]. Krasser [20] showed that more than two point sets cannot always be compatibly triangulated.

Concerning compatible paths, Hui and Schaefer [18] showed that it is NP-hard to decide whether two labeled point sets admit a compatible spanning path. Arseneva et al. [10] presented efficient algorithms for finding a monotone compatible spanning path, or a compatible spanning path inside simple polygons (if they exist). Czyzowicz et al. [15] showed that any two labeled point sets admit a compatible path of length at least $\sqrt{2n}$ and also presented an $O(n^2 \log n)$ algorithm to find such a path for two convex point sets. In a similar direction, results from Czabarka and Wang [14] imply a lower bound of $(\sqrt{n-2}+2)/2$ on the length of the longest cycle compatible to two convex point sets.

In this paper we will focus on compatible matchings. To the best of our knowledge, previous results on (geometric) matchings study only compatibility of the first type, that is, where two matchings are embedded on the same point set. A well-studied question in this setting is whether any two perfect matchings can be transformed into each other by a sequence of steps such that at every step the intermediate graph is a perfect matching and the union of any two consecutive matchings is plane. Aichholzer et al. [5] proved that such a sequence of at most $O(\log n)$ steps always exists. Questions of whether any matching of a given point set can be transformed into any other and how many steps it takes (that is, the connectivity of and the distance in the so-called *reconfiguration*

Fig. 1. (a) There is no perfect matching compatible to the two labeled sets. (b) Any possible pair of matching edges crosses in exactly one of the three sets.

graph of matchings, as well as its other properties) have been investigated also for matchings on bicolored point sets and for edge-disjoint compatible matchings, see for example [4,8,19].

Our Results. We study the second type of compatibility for matchings on two or more point sets. This is a setting for which no previous comprehensive theory appears to exist. Let us mention that throughout this paper we denote unlabeled point sets with P and labeled point sets with $\mathcal{P}$. We start by considering convex point sets: given two unlabeled convex point sets P_1, P_2, both with n points, we study the largest guaranteed size $\mathrm{ccm}(n)$ of a compatible matching across all pairs of labelings of P_1 and P_2. More formally, $\mathrm{ccm}(n)$ is the minimum over all pairs of labelings of the maximum compatible matching size for the accordingly labeled n-point set pairs. The largest compatible matching for two labeled point sets is not necessarily perfect, see Fig. 1(a). In Sect. 2 we present upper and lower bounds on $\mathrm{ccm}(n)$. In particular, for any n that is a multiple of 10, we construct two labeled convex sets $\mathcal{P}_1, \mathcal{P}_2$ of n points each, for which the largest compatible matching has $2n/5$ edges. Using probabilistic arguments, we obtain an upper bound of $\mathrm{ccm}(n) = \mathcal{O}(n^{2/3})$. For the lower bound, we show that for any pair of labeled convex point sets $\mathcal{P}_1, \mathcal{P}_2$ there exists a compatible matching consisting of $\lfloor \sqrt{2n} \rfloor$ edges. This implies that $\mathrm{ccm}(n) = \Omega(\sqrt{n})$.

We further extend our study to consider ℓ point sets in general position instead of just two point sets in convex position. Given ℓ unlabeled sets $P_1, \ldots, P_\ell$ of n points in general position, we denote by $\mathrm{cm}(n; P_1, \ldots, P_\ell)$ the largest guaranteed size of a compatible matching across all ℓ-tuples of labelings of $P_1, \ldots, P_\ell$. We remark that the size n of the point sets is included in the notation only for the sake of clarity (since our bounds depend on n). In Sect. 3 we give bounds on $\mathrm{cm}(n; P_1, \ldots, P_\ell)$ for any sets $P_1, \ldots, P_\ell$ of n points in general position. Building on the ideas of the proofs for two convex sets, we show that $\mathrm{cm}(n; P_1, \ldots, P_\ell) = \mathcal{O}(n^{2/(\ell+1)})$ and that $\mathrm{cm}(n; P_1, \ldots, P_\ell) = \Omega(n^{1/\ell})$.

Finally, we investigate the question of how many labeled copies of a given unlabeled point set are needed so that the largest compatible matching consists of a single edge. Already for four points in convex position, three different sets are needed (and sufficient, see Fig. 1(b)). In Sect. 4 we prove that for any given set of $n \geq 5$ points in general position, $\Theta(\log n)$ copies of it are necessary and sufficient for the existence of labelings forcing that the largest compatible matching consists of a single edge.

For brevity, a plane matching that consists of k edges is called a *k-matching*. Due to lack of space, several proofs are only sketched. Full versions of those proofs can be found in the full version.

2   Two Convex Sets

Throughout this section we consider two labeled convex point sets $\mathcal{P}_1$, $\mathcal{P}_2$ consisting of n points each. Without loss of generality we assume that $\mathcal{P}_1$ is labeled $(1, 2, \ldots, n)$ in clockwise order and that $\mathcal{P}_2$ is labeled $(\pi(1), \pi(2), \ldots, \pi(n))$ in clockwise order for some permutation $\pi \colon [n] \to [n]$.

In the following, we present lower and upper bounds on the largest guaranteed size $\mathrm{ccm}(n)$ of a compatible matching of any two such sets. Starting with lower bounds, we present four pairwise incomparable results (Theorem 1), each of them giving rise to a polynomial-time algorithm for constructing a compatible k-matching with $k = \Omega(\sqrt{n})$ edges. The results are ordered by the size of the obtained compatible k-matching, where the last one gives the best lower bound for $\mathrm{ccm}(n)$, while the three other results yield compatible matchings of special structure. The second result can be generalized to any number of (not necessarily convex) sets (Theorem 3). We remark that [15] implies a lower bound of $\sqrt{2n}/2$, which is weaker than the fourth result.

Before stating the theorem, we introduce the notion of a shape of a matching on a convex point set which, informally stated, captures "how the matching looks". Consider a labeled point set $\mathcal{P}$ and a plane matching M on $\mathcal{P}$. Let $\mathcal{P}^M \subseteq \mathcal{P}$ be the points of $\mathcal{P}$ that are incident to an edge of M. The *shape* of M is the combinatorial embedding[1] of the union of M and the boundary of the convex hull of $\mathcal{P}^M$. Further, M is called *non-nested* if its shape is a cycle, that is, all edges of M lie on the boundary of the convex hull of $\mathcal{P}^M$. Note that the shape of M also determines the number of its edges (even though some or all of the edges might be "hidden" in the boundary of the convex hull of $\mathcal{P}^M$). We say that two matchings have the *same shape*, if their shapes are identical, possibly up to a reflection.

Theorem 1 (Lower bound for two convex sets). *For any two labeled convex sets $\mathcal{P}_1$, $\mathcal{P}_2$ of n points each, it holds that: (i) If $n \geq (2k - 2)^2 + 2$ then for any shape of a k-matching there exists a compatible k-matching having that shape in both $\mathcal{P}_1$ and $\mathcal{P}_2$. (ii) If $n \geq k^2 + 2k - 1$ then any maximal compatible matching consists of at least k edges. (iii) If $n \geq k^2 + k$ then there exists a compatible k-matching that is non-nested in both $\mathcal{P}_1$ and $\mathcal{P}_2$. (iv) If $n \geq \frac{1}{2}k^2 + k$ then there exists a compatible k-matching.*

---

[1] The combinatorial embedding fixes the cyclic order of incident edges for each vertex.

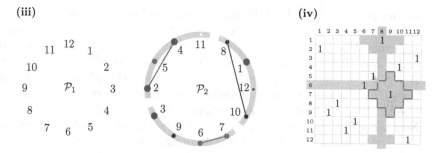

Fig. 2. Theorem 1, Claim (iii): Illustration with $n = 12$ points and $k = 3$ blocks (grey). After drawing an edge we switch the color of processed points (red to green to blue). Claim (iv): The permutation matrix Π and two 2-balls (yellow). A 2-ball centered at $[5, 7]$ would intersect a 2-ball at $[7, 9]$, so drawing the edge between points labeled 7, 9 forces us to discard at most 2 other points (6 and 8). (Color figure online)

Proof.

(i) By the circular Erdős-Szekeres Theorem [14], the permutation π contains a monotone subsequence σ having length $2k$. The sequence $S = \{x_i | i \in \sigma\}$ of points whose labels belong to σ has the same cyclic order in both sets $\mathcal{P}_1$, $\mathcal{P}_2$ (possibly once clockwise and once counter-clockwise), hence any plane matching on S in $\mathcal{P}_1$ is also plane in $\mathcal{P}_2$ and has the same shape.

(ii) Suppose we have already found a compatible matching M consisting of $m \le k-1$ edges. This leaves at least $n - 2m \ge k^2 + 1$ points yet unmatched. The unmatched points are split by the m matching edges into at most $m + 1 \le k$ subsets, both in $\mathcal{P}_1$ and in $\mathcal{P}_2$. Since there are at most k^2 different ways to choose one such subset from $\mathcal{P}_1$ and one from $\mathcal{P}_2$, there exist two yet unmatched points x, y that lie in the same subset in $\mathcal{P}_1$ and in the same subset in $\mathcal{P}_2$. Hence xy can be added to the matching M.

(iii) This claim is equivalent to Problem 5 given at IMO 2017.[2] For completeness we sketch a proof (see Fig. 2): split the perimeter of $\mathcal{P}_2$ into k contiguous blocks $B_1, \ldots, B_k$ consisting of $k + 1$ points each (that is, block B_1 consists of points labeled $\pi(1), \ldots, \pi(k+1)$ and so on). We aim to draw one matching edge per block. We process points x_i in order $i = 1, \ldots, n$ in which they appear in $\mathcal{P}_1$. Once some block, say $B^\star$, contains two processed points, say x_u and x_v, we draw edge $x_u x_v$, discard other already processed points and discard other points in $B^\star$. In this way, any time we draw an edge in some block, we discard at most one point from each other block. Since each block initially contains $k + 1$ points, we will eventually draw one edge in each block. The produced matching contains one edge per block, hence it is non-nested in $\mathcal{P}_2$. Since points x_i are processed in order $i = 1, \ldots, n$, the matching is also non-nested in $\mathcal{P}_1$.

(iv) The idea is to find two points x_i, x_j that are close to each other in the cyclic order in both $\mathcal{P}_1$ and $\mathcal{P}_2$. Then draw the edge $x_i x_j$, omit all points on the

---

[2] https://www.imo-official.org/problems/IMO2017SL.pdf, Problem C4.

shorter arcs of $x_i x_j$ in both $\mathcal{P}_1$ and $\mathcal{P}_2$, and proceed recursively.

Consider the permutation matrix Π given by π, that is, an $n \times n$ matrix such that $\Pi_{i,j} = 1$ if $\pi(i) = j$ and 0 otherwise. Given an integer $r > 0$ and a cell $\Pi_{i,j}$ containing a digit 1, the r-ball centered at $\Pi_{i,j}$ is a set $B(\Pi_{i,j}, r) = \{\Pi_{u,v} : |i - u| + |j - v| \le r\}$ of cells whose L_1-distance from $\Pi_{i,j}$ is at most r, where all indices are considered cyclically modulo n (see Fig. 2). Note that an r-ball contains $2r^2 + 2r + 1$ cells.

Now suppose n and r satisfy $n \le 2r^2 + 2r$ and consider r-balls centered at all n cells containing a digit 1. The balls in total cover $n \cdot (2r^2 + 2r + 1) > n^2$ cells, hence some two r-balls intersect and their centers $\Pi_{i,\pi(i)}, \Pi_{j,\pi(j)}$ have L_1-distance at most $2r$. This means that the shorter arcs between points labeled $\pi(i)$ and $\pi(j)$ contain, together in both point sets $\mathcal{P}_1$ and $\mathcal{P}_2$, at most $2r - 2$ other points. Drawing an edge $\pi(i)\pi(j)$ and removing these $2r - 2$ other points leaves convex sets in both $\mathcal{P}_1$ and $\mathcal{P}_2$ whose convex hulls do not intersect the matched edge $\pi(i)\pi(j)$.

The rest is induction. The claim holds for $k \in \{1, 2\}$. Suppose that $k = 2r$ is even and that $n = \frac{1}{2}k^2 + k = 2r^2 + 2r$. By the above argument, find a "short" edge $x_i x_j$ and remove up to $2r - 2$ other points. This leaves $n - 2r$ ($< 2r^2 + 2r$) points, so find another edge $x_u x_v$ and remove up to $2r - 2$ other points. This leaves $2r^2 - 2r = 2(r - 1)^2 + 2(r - 1)$ points and the induction applies. Last, note that the above shows that having $2r^2$ points implies a $(2r - 1)$-matching. Since $2r^2 = \lceil \frac{1}{2}(2r - 1)^2 + 2r - 1 \rceil$, the case of $k = 2r - 1$ odd and $n = \lceil \frac{1}{2}(2r - 1)^2 + 2r - 1 \rceil$ is also settled.   □

For the remainder of this section, we consider upper bounds on the size of compatible matchings for pairs of convex point sets.

We first describe an explicit construction of two labeled point sets $\mathcal{P}_{\text{id}}$ and $\mathcal{P}_\pi$, where n is a multiple of 10, the set $\mathcal{P}_{\text{id}}$ is labeled $(1, 2, \ldots, n)$ in clockwise order, and the set $\mathcal{P}_\pi$ is labeled $(\pi(1), \pi(2), \ldots, \pi(n))$ in clockwise order, by defining a specific permutation $\pi \colon [n] \to [n]$. We will show that any compatible matching of $\mathcal{P}_{\text{id}}$ and $\mathcal{P}_\pi$ misses at least $n/5$ of the points.

Our building block for π is the permutation $(2, 4, 1, 5, 3)$ of five elements. For labeling the $n = 5k$ points of $\mathcal{P}_\pi$ (with $k \ge 2$ even) we use the permutation $\pi = (2, 4, 1, 5, 3, 7, 9, 6, 10, 8, \ldots, 5(k-1)+2, 5(k-1)+4, 5(k-1)+1, 5(k-1)+5, 5(k-1)+3)$ that yields k blocks of 5 points each in both $\mathcal{P}_1$ and $\mathcal{P}_2$ (see Fig. 3).

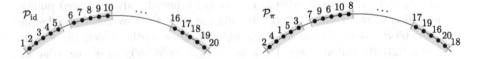

Fig. 3. The two labeled point sets $\mathcal{P}_{\text{id}}$ and $\mathcal{P}_\pi$ for the permutation π.

Proposition 1 (Constructive upper bound for two convex sets). *The largest compatible matching of the two labeled n-point sets $\mathcal{P}_{\text{id}}$ and $\mathcal{P}_\pi$ defined above contains $\frac{2}{5}n$ edges.*

Proof (sketch). Consider any compatible matching M of $\mathcal{P}_{\mathrm{id}}$ and $\mathcal{P}_\pi$. For the upper bound we show that M misses at least one point within each of the k blocks. To this end, we classify the edges of M into two types: those that connect two points in one block (we call them *short* edges) and those that connect two points from different blocks (we call them *long* edges). For each block B, we distinguish two cases:

Case 1: B contains at least one short edge. In this case we show by elementary casework that there is always at least one unmatched point in B.

Case 2: B contains no short edge. We argue that, under the assumption that all five points in B are matched (by a long edge), all those five edges in fact must go to the same block B', which we then show to be impossible.

To see that the bound is tight, note that within each block of $\mathcal{P}_\pi$ we can match the first two points and the next two points. This yields a compatible matching of $\mathcal{P}_{\mathrm{id}}$ and $\mathcal{P}_\pi$ with $2k = \frac{2}{5}n$ edges consisting only of short edges. $\qquad\square$

The above construction yields an upper bound of $\mathrm{ccm}(n) \leq \lceil \frac{2}{5}n \rceil$. However, this bound is not tight. We next show in a probabilistic way that there exists a permutation $\pi\colon [n] \to [n]$ for which the largest compatible matching consists of $k = \mathcal{O}(n^{2/3})$ edges. In Sect. 3, we will extend this approach to any number of point sets, not necessarily in convex position (Theorem 4).

Theorem 2 (Probabilistic upper bound for two convex sets). *Fix n and let $k = 4n^{2/3}$. Then two convex sets P_1, P_2 of n points each can be labeled such that the largest compatible matching consists of fewer than k edges.*

Proof. Let $\mathcal{P}_1$ be P_1 with labeling $(1, 2, \ldots, n)$ in clockwise order and let P_2 not yet be labeled. The idea for this proof is that for large n there are more ways to label P_2 than there are ways to draw a compatible k-matching.

For any $k \leq n$, let $f(k)$ be the number of plane k-matchings of P_i, $i \in \{1, 2\}$ (that is, matchings leaving $n - 2k$ points unmatched). As there are $\binom{n}{2k}$ ways to select the $2k$ points to be matched and the number of plane perfect matching on those points is $\frac{1}{k+1}\binom{2k}{k}$ (the k-th Catalan number), we obtain $f(k) = \binom{n}{2k} \cdot \frac{1}{k+1}\binom{2k}{k} \leq \frac{n!}{(n-2k)!\cdot k!\cdot k!}$.

Given two plane k-matchings, one of $\mathcal{P}_1$ and one of P_2, there are exactly $g(k) = (n-2k)! \cdot k! \cdot 2^k$ labelings of P_2 for which those two matchings constitute a compatible k-matching: there are $(n-2k)!$ ways to label the unmatched points of P_2, $k!$ ways to pair up the matching edges and 2^k ways to label their endpoints.

Therefore, $(f(k))^2 g(k)$ is an upper bound for the number of labelings π of P_2 such that there is a compatible k-matching for $\mathcal{P}_1$ and P_2 (P_2 with labeling π). On the other hand, there are $n!$ labelings of P_2 in total.

Our goal is to show that $(f(k))^2 \cdot g(k) < n!$. If we succeed, then there exists a labeling π of P_2 such that there is no compatible k-matching for $\mathcal{P}_1$ and P_2 (P_2 with labeling π). Canceling some of the factorials and using standard bounds $(n/e)^n < n! < n^n$ on the remaining ones (where e denotes Euler's number), we obtain

$$\frac{(f(k))^2 \cdot g(k)}{n!} \leq \frac{n! \cdot 2^k}{(n-2k)! \cdot (k!)^3} \leq \frac{n^{2k} \cdot 2^k}{(k/e)^{3k}} = \left(\frac{2e^3 n^2}{k^3}\right)^k.$$

For $k \geq 4n^{2/3}$, the above expression is less than one (we have $2e^3 < 4^3$), which completes the proof. □

3　Generalized and Multiple Sets

In this section we generalize our results in two ways, by considering point sets in general position and more than two sets. We again start with lower bounds. Theorem 3, which is a generalization of the second result of Theorem 1, implies that for any ℓ-tuple of point sets $P_1, \ldots, P_\ell$ we have $\mathrm{cm}(n; P_1, \ldots, P_\ell) = \Omega(n^{1/\ell})$.

Theorem 3 (Lower bound for multiple sets). *Let $\mathcal{P}_1, \mathcal{P}_2, \ldots, \mathcal{P}_\ell$ be labeled sets of n points each. If $n \geq k^\ell + 2k - 1$, then any maximal compatible matching consists of at least k edges.*

Proof. We extend the idea from the proof of Theorem 1, part (ii): suppose we have already found a compatible matching M consisting of $m \leq k - 1$ edges. This leaves at least $k^\ell + 2k - 1 - m \geq k^\ell + 2k - 1 - 2(k-1) = k^\ell + 1$ points yet unmatched. Imagine the ℓ point sets live in ℓ different planes. We process the m matching edges one by one. When an edge is processed, we extend it along its line in both directions until it hits another matching edge or an extension of a previously processed edge (in all ℓ planes). In this way, the m lines partition each plane into $m + 1 \leq k$ convex regions. By simple counting ($k^\ell + 1 > k^\ell$), there exist two yet unmatched points x, y that lie in the same region in each of the ℓ planes. Hence xy can be added to the matching M. □

Regarding upper bounds, the following theorem implies that for any fixed ℓ and any ℓ-tuple of point sets $P_1, \ldots, P_\ell$, we have $\mathrm{cm}(n; P_1, \ldots, P_\ell) = \mathcal{O}(n^{2/(\ell+1)})$.

Theorem 4 (Probabilistic upper bound for multiple sets). *Fix n and ℓ and let $k = 125 \cdot n^{2/(\ell+1)}$. Then any ℓ sets $P_1, \ldots, P_\ell$ of n points each, where each P_i is in general position, can be labeled such that the largest compatible matching consists of fewer than k edges.*

This theorem can be proven by extending the idea from the proof of Theorem 2 and combining results of Sharir, Sheffer and Welzl [24] and Sharir and Sheffer [23] on the number of triangulations and plane perfect matchings.

4　Forcing a Single-Edge Compatible Matching

In this section we consider the following question: Given an unlabeled point set P with n points, is there an integer ℓ such that there exist ℓ labelings of P for which every compatible matching has at most one edge? If ℓ exists, we denote as $\mathrm{force}(n; P)$ the minimum number ℓ of copies of P such that $\mathrm{cm}(n; P, \ldots, P) = 1$

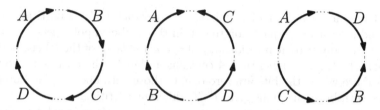

Fig. 4. Three labeled point sets obtained from different orders and orientations of four blocks A, B, C, and D.

(where P appears force($n; P$) times). Otherwise, we set force($n; P$) $= \infty$. In other words, we are asking for the existence (and minimal number) of labelings of the set P so that any pair of labeled edges crosses for at least one labeling. We remark that, again, the size n of the point sets is included in the notation only for the sake of clarity. Note that force($n; P$) $= \infty$ if and only if the straight-line drawing of K_n on P contains no crossing. Hence force($n; P$) is finite for any set P of $n \geq 5$ points.

We first focus on upper bounds and on the case when P is in convex position. We denote by cforce(n) the minimum number of copies of a convex set with n points that need to be labeled so that the largest compatible matching consists of only a single edge.

Let $b(n) = \lceil \log_2 n \rceil$, which is the number of bits that are needed to represent the labels 1 to n. We construct a family of $\frac{3}{2} b(n)^2$ labeled convex n-point sets such that all pairs of edges cross in at least one set. First consider three labeled convex point sets, which are obtained by partitioning the set of labels into four blocks A, B, C, D, and combining those blocks in different orders and orientations as depicted in Fig. 4. The order within a block is arbitrary, but identical for all three sets (up to reflection; those block orientations are indicated by arrows).

Lemma 1. *Consider three convex point sets* $\mathcal{P}_1$, $\mathcal{P}_2$, *and* $\mathcal{P}_3$ *that are labeled as in Fig. 4 for some partition* A, B, C, *and* D *of their label set. Then any pair of independent[3] edges, where none of them has both labels in one of the blocks* A, B, C, *and* D, *forms a crossing in at least one of* $\mathcal{P}_1$, $\mathcal{P}_2$, *and* $\mathcal{P}_3$.

The proof of this lemma, a case analysis depending on the number of blocks that contain at least one endpoint of the two edges, is deferred to the full version.

We next identify a small number of 4-partitions of the label set $\{1, 2, \ldots, n\}$ such that each edge pair fulfills the condition of Lemma 1 in at least one of the partitions (when the four subsets form blocks). This yields the following constructive upper bound for cforce(n).

Proposition 2 (Constructive upper bound on cforce(n)). *For any* $n \geq 4$ *and for* $b(n) = \lceil \log_2 n \rceil$, *we can define* $3\binom{b(n)}{2}$ *labeled convex sets of* n *points such that the largest matching compatible to all of them consists of a single edge.*

---

[3] Two edges are independent if they do not share an endpoint.

Proof. Given a convex set of n points, we construct $\binom{b(n)}{2}$ 4-partitions of the labels and use each such partition to obtain three labeled point sets as depicted in Fig. 4. For any two bit positions i, j, $0 \leq i \neq j < b(n)$, of the labels, partition the label set $\{1, 2, \ldots, n\}$ so that A contains all labels where those two bits are zero, B those where the bits are zero-one, C those with one-zero, and finally D the ones with both one. This gives $\binom{b(n)}{2}$ different partitions.

Now consider two arbitrary edges e and f. Then there is a bit position in which the two endpoints of e have different values, and the same is true for f. Let i and j, respectively, be those positions. If this would give $i = j$, then choose j arbitrarily but not equal to i. By Lemma 1, the edges e and f cross in one of the three labeled point sets for the partition generated for i and j.   □

The upper bound $\mathcal{O}(\log^2 n)$ of cforce(n) from Lemma 2 is constructive but it is not asymptotically tight. Next we present a probabilistic argument which shows that we actually have force($n; P$) $= \mathcal{O}(\log n)$ for any point set P of $n \geq 5$ points.

Lemma 2 (Probabilistic upper bound on force($n; P$)). *Given a set P of $n \geq 5$ points in general position, there exists a constant $c_P \geq 15/14$ such that* force($n; P$) $\leq \log_{c_P}(3\binom{n}{4}) = \mathcal{O}(\log n)$.

Proof. Fix P and let $\alpha_P \in (0, 1)$ be the proportion of 4-tuples of points in P that are in convex position. Note that since any 5-tuple of points contains at least one 4-tuple in convex position, we have $\alpha_P \geq 1/5$ (here we use $n \geq 5$).

There are $r = 3\binom{n}{4}$ pairs of non-incident edges. Fix one of them, say ac and bd. Note that when P is labeled uniformly at random, the edges ac, bd intersect with constant probability $\alpha_P/3$: indeed, the edges intersect if their 4 endpoints form a convex quadrilateral and the points a, b, c, d lie on its perimeter in two out of the six possible cyclic orders.

Now set $c_P = (1 - \alpha_P/3)^{-1} \geq 15/14$ and consider $\ell > \log_{c_P}(r)$ copies of P labeled independently and uniformly at random. We say that a pair of edges is *bad* if it is non-crossing in all ℓ point sets. Then any one fixed pair of edges is bad with probability $\rho = (1 - \alpha_P/3)^\ell < 1/r$. By linearity of expectation, the expected number of bad pairs of edges is $r \cdot \rho < 1$. Therefore there exists a labeling of the ℓ point sets for which no pair of edges is bad. In such a labeling, the largest compatible matching consists of a single edge.   □

We remark that for a fixed set P, one can often obtain a better bound on parameter α_P used in the proof and thus a better bound on c_P, which then gives a constant factor improvement on force($n; P$). Specifically, for sufficiently large n, finding the maximum constant α is a topic of high relevance in connection with the rectilinear crossing number of the complete graph, see [2] for a nice survey of this area. The currently best known bounds are $0.37997256 < \alpha < 0.38044919$ [1,6]. Moreover, when P is in convex position we have $\alpha_P = 1$ and thus the above proof implies cforce(n) $\leq \log_{3/2}(3\binom{n}{4})$. On the other hand, none of these observations leads to an asymptotic improvement on the upper bound of force($n; P$) or cforce(n). In the following we show that any such asymptotic improvement is in fact impossible.

Lemma 3 (Lower bound on force($n; P$)). *Fix $k \geq 1$ and let P be any set of $n = 2^k + 3$ points in general position. Then* force($n; P$) $\geq k + 2 = \Omega(\log n)$.

Proof. We use a similar argument as the one used in Theorem 3. Denote by $\mathcal{P}_1, \ldots, \mathcal{P}_{k+1}$ any $k + 1$ labeled copies of P. Take an arbitrary edge ab on the convex hull of $\mathcal{P}_{k+1}$. The line containing ab divides each of $\mathcal{P}_1, \ldots, \mathcal{P}_k$ into two parts (one possibly empty). Since there are $n-2 = 2^k +1 > 2^k$ unmatched points, there exist two points, say x, y, that lie in the same part, for each $i = 1, \ldots, k$. Thus the two edges xy and ab form a compatible 2-matching implying that force($n; P$) $\geq k + 2$. $\qquad \square$

Combining upper and lower bounds for force($n; P$) from Lemma 3 and 2, we obtain the following conclusion:

Theorem 5. *For every set P of $n \geq 5$ points in general position, it holds that*

$$\text{force}(n; P) = \Theta(\log n).$$

5 Future Research

Besides the presented results there are several related directions of research. Below we list some of them, together with open questions emerging from the previous sections.

1. A natural open problem is the computational complexity of finding compatible matchings of a certain size or even deciding their existence: How fast can we decide if two (or more) given (general or convex) labeled point sets have a perfect compatible matching (or a compatible matching of size k)?
2. Can we close the gaps between the lower and upper bounds, $\Omega(\sqrt{n})$ and $\mathcal{O}(n^{2/3})$ respectively, for the size of the largest compatible matchings for two labeled point sets (general, convex, or of any other particular order type)?
3. Can we improve the constructions to match their according probabilistic bounds?
4. Consider the following game version: two players alternately add an edge which must not intersect any previously added edge. The last player who is able to add such an edge wins. It is not hard to see that for a single set of points in convex position, this is the well-known game Dawson's Kayles, see e.g. [13]. This game can be perfectly solved using the nimber theory developed by Sprague-Grundy, see also [13] for a nice introduction to the area. An interesting generalization of Dawson's Kayles occurs when the players use two (or more) labeled (convex) point sets and add compatible edges.
5. What are tight bounds for the size of compatible paths and compatible cycles (see [14, 15])? In ongoing work we have obtained first results in this direction.

References

1. Ábrego, B.M., Fernández-Merchant, S., Leaños, J., Salazar, G.: A central approach to bound the number of crossings in a generalized configuration. In: LAGOS2007, pp. 273–278. ENDM (2007)
2. Ábrego, B.M., Fernández-Merchant, S., Salazar, G.: The rectilinear crossing number of K_n: closing in (or are we?). In: Pach, J. (ed.) Thirty Essays on Geometric Graph Theory, pp. 5–18. Springer (2013). https://doi.org/10.1007/978-1-4614-0110-0_2
3. Aichholzer, O., Aurenhammer, F., Hurtado, F., Krasser, H.: Towards compatible triangulations. TCS 296(1), 3–13 (2003)
4. Aichholzer, O., Barba, L., Hackl, T., Pilz, A., Vogtenhuber, B.: Linear transformation distance for bichromatic matchings. CGTA 68, 77–88 (2018)
5. Aichholzer, O., et al.: Compatible geometric matchings. CGTA 42(6–7), 617–626 (2009)
6. Aichholzer, O., Duque, F., Fabila-Monroy, R., Hidalgo-Toscano, C., García-Quintero, O.E.: An ongoing project to improve the rectilinear and the pseudolinear crossing constants. ArXiv e-Prints (2020)
7. Alamdari, S., et al.: How to morph planar graph drawings. SIAM J. Comput. 46(2), 824–852 (2017)
8. Aloupis, G., Barba, L., Langerman, S., Souvaine, D.: Bichromatic compatible matchings. CGTA 48(8), 622–633 (2015)
9. Aronov, B., Seidel, R., Souvaine, D.: On compatible triangulations of simple polygons. CGTA 3, 27–35 (1993)
10. Arseneva, E., et al.: Compatible paths on labelled point sets. In: CCCG 2018, pp. 54–60 (2020)
11. Babikov, M., Souvaine, D.L., Wenger, R.: Constructing piecewise linear homeomorphisms of polygons with holes. In: CCCG 1997 (1997)
12. Baxter III, W.V., Barla, P., Anjyo, K.: Compatible embedding for 2D shape animation. IEEE Trans. Vis. Comput. Graph. 15(5), 867–879 (2009)
13. Berlekamp, E.R., Conway, J.H., Guy, R.K.: Winning ways for your mathematical plays, vol. 1, 2 edn. A. K. Peters Ltd. (2001)
14. Czabarka, É., Wang, Z.: Erdő-Szekeres theorem for cyclic permutations. Involve J. Math. 12(2), 351–360 (2019)
15. Czyzowicz, J., Kranakis, E., Krizanc, D., Urrutia, J.: Maximal length common non-intersecting paths. In: CCCG 1996, pp. 185–189 (1996)
16. Erten, C., Kobourov, S.G., Pitta, C.: Intersection-free morphing of planar graphs. In: Liotta, G. (ed.) GD 2003. LNCS, vol. 2912, pp. 320–331. Springer, Heidelberg (2004). https://doi.org/10.1007/978-3-540-24595-7_30
17. Floater, M.S., Gotsman, C.: How to morph tilings injectively. JCAM 101, 117–129 (1999)
18. Hui, P., Schaefer, M.: Paired pointset traversal. In: Fleischer, R., Trippen, G. (eds.) ISAAC 2004. LNCS, vol. 3341, pp. 534–544. Springer, Heidelberg (2004). https://doi.org/10.1007/978-3-540-30551-4_47
19. Ishaque, M., Souvaine, D.L., Tóth, C.D.: Disjoint compatible geometric matchings. DCG 49, 89–131 (2013)
20. Krasser, H.: Kompatible Triangulierungen ebener Punktmengen. Master's thesis, Graz University of Technology (1999)
21. Pach, J., Shahrokhi, F., Szegedy, M.: Applications of the crossing number. Algorithmica 16(1), 111–117 (1996)

22. Saalfeld, A.: Joint triangulations and triangulation maps. In: SoCG 1987, pp. 195–204 (1987)
23. Sharir, M., Sheffer, A.: Counting triangulations of planar point sets. E-JC**18**(1) (2011)
24. Sharir, M., Sheffer, A., Welzl, E.: Counting plane graphs: perfect matchings, spanning cycles, and Kasteleyn's technique. JCTA **120**(4), 777–794 (2013)

Upward Point Set Embeddings
of Paths and Trees

Elena Arseneva[1] , Pilar Cano[2] , Linda Kleist[3(✉)] , Tamara Mchedlidze[4],
Saeed Mehrabi[5] , Irene Parada[6] , and Pavel Valtr[7]

[1] Saint Petersburg State University, Saint Petersburg, Russia
e.arseneva@spbu.ru
[2] Université Libre de Bruxelles, Brussels, Belgium
pilar.cano@ulb.ac.be
[3] Technische Universität Braunschweig, Brunswick, Germany
kleist@ibr.cs.tu-bs.de
[4] Utrecht University, Utrecht, Netherlands
mched@iti.uka.de
[5] Memorial University, St. John's, Canada
smehrabi@mun.ca
[6] TU Eindhoven, Eindhoven, The Netherlands
i.m.de.parada.munoz@tue.nl
[7] Faculty of Mathematics and Physics, Department of Applied Mathematics,
Charles University, Prague, Czech Republic

Abstract. We study upward planar straight-line embeddings (UPSE) of directed trees on given point sets. The given point set S has size at least the number of vertices in the tree. For the special case where the tree is a path P we show that: (a) If S is one-sided convex, the number of UPSEs equals the number of maximal monotone paths in P. (b) If S is in general position and P is composed by three maximal monotone paths, where the middle path is longer than the other two, then it always admits an UPSE on S. We show that the decision problem of whether there exists an UPSE of a directed tree with n vertices on a fixed point set S of n points is NP-complete, by relaxing the requirements of the previously known result which relied on the presence of cycles in the graph, but instead fixing position of a single vertex. Finally, by allowing extra points, we guarantee that each directed caterpillar on n vertices and with k switches in its backbone admits an UPSE on every set of $n2^{k-2}$ points.

Keywords: Upward planarity · Directed graph · Digraph · Tree · Caterpillar · Path · NP-completeness · Counting · Upper bound

1 Introduction

A classic result by Fary, Stein, Wagner [11,15,16], known as Fary's theorem, states that every planar graph has a crossing-free straight-line drawing. Given a

© Springer Nature Switzerland AG 2021
R. Uehara et al. (Eds.): WALCOM 2021, LNCS 12635, pp. 234–246, 2021.
https://doi.org/10.1007/978-3-030-68211-8_19

directed graph (called *digraph* for short), it is natural to represent the direction of the edges by an *upward* drawing, i.e., every directed edge is represented by a monotonically increasing curve. Clearly, it is necessary for the digraph to be acyclic in order to allow for an upward drawing.

In nice analogy to Fary's theorem, if a planar digraph has an upward planar drawing, then it also allows for an upward planar straight-line drawing [8]. In contrast, not every planar acyclic digraph has an upward planar drawing [8]. Nevertheless, some classes of digraphs always allow for such drawings. For instance, every directed tree has an upward planar straight-line drawing [7, p. 212].

In this work, we study upward planar straight-line drawings of directed trees on given point sets. An *upward planar straight-line embedding* (UPSE, for short) of a digraph $G = (V, E)$ on a point set S, where $|V| \leq |S|$, is an injection from V to S such that the induced straight-line drawing is planar (crossing-free) and upward, i.e., for the directed edge $uv \in E(G)$, the point representing u lies below the point representing v. As point-set embeddings of planar undirected graphs [3–6,14], UPSEs have been an active subject of research [1,2,12,13]. In the following we review the state of the art in relation to our problems.

Kaufmann et al. [12] showed that in case $|V| = |S|$ it is NP-complete to decide whether an upward planar digraph admits an UPSE on a given point set. We note that the upward planar digraph obtained in their reduction contains cycles in its underlying undirected graph. The same authors gave polynomial-time algorithm to decide if an upward planar digraph admits an UPSE on a convex point set.

A digraph whose underlying undirected structure is a simple path is called *oriented path*. For the class of oriented paths multiple partial results have been provided, by either limiting the class of the point set, the class of oriented paths, or by considering the case where $|S| > |V|$. In particular, by limiting the class of point sets, Binucci et al. showed that every oriented path admits an UPSE on every convex point set of the same size [2]. By limiting the class of oriented paths, it has been shown that the following subclasses of oriented paths always admit an UPSE on any general point set of the same size:

1. An oriented path with at most five switches[1] and at least two of its sections[2] having length[3] two [1].
2. An oriented path P with three switches [2].
3. An oriented path $P = (v_1, \ldots, v_n)$, so that if its vertex v_i is a sink, then its vertex v_{i+1} is a source [2].
4. An oriented path P such that its decomposition into maximal monotone paths $P_1, P_2 \ldots, P_r$ satisfies that $|P_i| \geq \sum_{j>i} |P_j|$ for every $i = 1, 2, \ldots, r - 1$ [6].

Given these partial results, it is an intriguing open problem whether every oriented path P admits an UPSE on every general point set S with $|P| = |S|$. In contrast to this question, there exists a directed tree T with n vertices and

---

[1] A vertex of a digraph which is either a source or a sink is called *switch*.

[2] A section of an oriented path is a subpath that connects two consecutive switches.

[3] The *length* of a path is the number of vertices in it.

a set S with n points in convex position such that T does not admit an UPSE on S [2]. While restricting the class of trees to directed caterpillars, Angelini et al. [1] have shown that an UPSE exists on any convex point set.

The variant of the problem where the point set is larger than the oriented path, was considered by Angelini et al. [1] and Mchedlidze [13]. They proved that every oriented path P with n vertices and k switches admits an UPSE on every general point set with $n2^{k-2}$ points [1] or on $(n-1)^2+1$ points Mchedlidze [13], respectively.

Our Contribution. In this paper we continue the study of UPSE of digraphs. We tackle the aforementioned open problem from multiple sides. Firstly, we show that the problem of deciding whether a digraph on n vertices admits an UPSE on a given set of n points remains NP-complete even for trees when one vertex lies on a predefined point. (Sect. 5). This strengthens the previously known NP-completeness result, where the underlying undirected structure contained cycles [12]. Thus, even if it is still possible that every oriented path admits an UPSE on every general point set, this new NP-completeness might foreshadow that a proof for this fact will not lead to a polynomial time construction algorithm. Secondly, we provide a new family of n-vertex oriented paths that admit an UPSE on any general set of n points (Section 3), extending the previous partial results [1,2]. Thirdly, by aiming to understand the degrees of freedom that one has while embedding an oriented path on a point set, we show that the number of different UPSEs of an n-vertex oriented path on a one-sided convex set of n points is equal to the number of sections the path contains (Sect. 2). Finally, as a side result, we study the case where the point set is larger than the graph and show that the upper bound $n2^{k-2}$ on the size of a general point set that hosts every oriented path [1] can be extended to caterpillars (Sect. 4), where k is the number of switches in the caterpillar. The proof is largely inspired by the corresponding proof for oriented paths. However, the result itself opens a new line of investigations of providing upper bounds on the size of general point sets that are sufficient for UPSE of families of directed trees.

Definitions. A set of points is called *general*, if no three of its points lie on the same line and no two have the same y-coordinate. The *convex hull* $H(S)$ of a point set S is the point set that can be obtained as a convex combination of the points of S. We say that a point set is in *convex position*, or is a *convex point set*, if none of its points lie in the convex hull of the others. Given a general point set S, we denote the lowest and the highest point of S by $b(S)$ and $t(S)$, respectively. A subset of points of a convex point set S is called *consecutive* if its points appear consecutively as we traverse the convex hull of S. A convex point set S is called *one-sided* if the points $b(S)$ and $t(S)$ are consecutive in it; refer to Fig. 1.

Let Γ be an UPSE of a digraph $G = (V, E)$ on a point set S. For every $v \in V$, $\Gamma(v)$ denotes the point of S where vertex v has been mapped to by Γ. A *directed tree* is a digraph, whose underlying graph is a tree. A digraph, whose underlying graph is a simple path is called *oriented path*. A *directed caterpillar* is a directed tree in which the removal of the vertices of degree 1 results in an oriented path: the *backbone*. For an oriented path $(v_1, v_2, \ldots, v_n)$, we call $v_i v_{i+1}$ a *forward* (resp., *backward*) edge if it is oriented from v_i to v_{i+1} (resp., from v_{i+1} to v_i). A vertex of a digraph with in-degree (resp., out-degree) equal to zero is called a *source* (resp., *sink*). A vertex of a digraph which is either a source or a sink is called *switch*. A subpath of an oriented path P connecting two of its consecutive switches is said to be *monotone* and called a *section* of P. A section is *forward* (resp., *backward*) if it consists of forward (resp., backward) edges.

2   Counting Embeddings of Paths on Convex Sets

In this section, we study the number of UPSEs that an n-vertex oriented path has on a one-sided convex set of n points. We show that this number is equal to the number of sections in the oriented path. We start with the following

Lemma 1. *Let P be an n-vertex oriented path with v_1 being one of its end-vertices and let S be a one-sided convex point set with $|S| = n$. For any two different UPSEs Γ_1 and Γ_2 of P on S, it holds that $\Gamma_1(v_1) \neq \Gamma_2(v_1)$.*

Proof. Let $\{p_1, \ldots, p_n\}$ be the points of S sorted by y-coordinate. For the sake of contradiction, assume that there exist two different UPSEs Γ_1 and Γ_2 of P on S with $\Gamma_1(v_1) = \Gamma_2(v_1)$. Additionally, assume that the considered counterexample is minimal, in the sense that for the vertex v_2 of P, adjacent to v_1, $\Gamma_1(v_2) \neq \Gamma_2(v_2)$. By [2, Lemma 3], vertices v_1 and v_2 lie on consecutive points of S. We assume that the edge $v_1 v_2$ is a forward edge; the case when $v_1 v_2$ is a backward edge is symmetric. Conditions $\Gamma_1(v_1) = \Gamma_2(v_1)$, $\Gamma_1(v_2) \neq \Gamma_2(v_2)$, and the fact that v_1, v_2 lie on consecutive points of S, imply that $\Gamma_1(v_1) = \Gamma_2(v_1) = p_1$, and $\Gamma_i(v_2) = p_n$, $\Gamma_j(v_2) = p_2$, with $i, j \in \{1, 2\}$. Embedding Γ_i is an UPSE of P when the edge $v_2 v_3$ is backward, while Γ_j is an UPSE of P when the edge $v_2 v_3$ is forward. We arrive to a contradiction. □

We are now ready to prove the result of this section.

Theorem 1. *An n-vertex oriented path P with k sections has exactly k UPSEs on a one-sided convex set of n points.*

Proof. We first show that P has at least k UPSEs. To do so, let $P = (v_1, \ldots, v_n)$ be an oriented path with k sections and let S be a one-sided convex point set with points $\{p_1, \ldots, p_n\}$ ordered by the increasing y-coordinate. Let v_l be the switch of P preceding v_n. Thus, the subpath of P between v_l and v_n is P's last section. Let denote the subpath of P between v_1 and v_l by P'. We prove the statement of the theorem by the following stronger induction hypothesis: there are k UPSEs of P on S, such that one of them maps v_n to p_n (resp., p_1) if the last section is forward (resp., backward). The base case of $k = 1$ is trivial. Assume that the k-th section of P is forward (resp., backward) and let $S' = \{p_1, \ldots, p_l\}$ (resp., $S' = \{p_{n-l+1}, \ldots, p_n\}$). By induction hypothesis, the path P' has $k - 1$ UPSEs on S', with one of them mapping v_l to p_1 (resp., p_n). Let Γ be one of them. Assume that $\Gamma(v_l) = p_i$, $p_i \in S'$. We shift every vertex that has been mapped to point p_j, $i < j \le l$ (resp., $n - l + 1 \le j < i$) by $n - l$ points up (resp., down); refer to Fig. 1(a). We map the k-th section to points $p_{i+1}, \dots, p_{i+n-l}$ (resp., $p_i - 1, \dots, p_{i-n+l}$). This gives us $k - 1$ UPSEs of P on S.

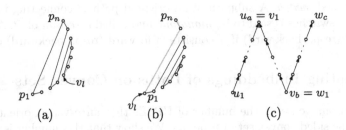

Fig. 1. (a–b) An illustration for the proof of Theorem 1, showing an extension of an UPSE of P' (black) to an UPSE of P (gray). In (a) v_l lies on a non-extreme point of S', in (b) v_l is an extreme point of S'. (c) Description of path P presented in Theorem 2.

Recall that, by induction hypothesis, P' has an UPSE Γ' on S' that maps v_l to p_1 (resp., p_n) on S', since its last section is backward (resp., forward). Thus, we can also extend Γ' by mapping the k-th section to points $p_{l+1}, \dots, p_n$ (resp., $p_1, \dots, p_{n-l}$); refer to Fig. 1(b). Hence, there exists a UPSE of P that maps v_n to p_n(resp., p_1) if the last section is forward (resp., backward). By Lemma 1, no two of the constructed embedding of P on S are the same. Thus, P has at least k UPSEs on S.

We now apply a counting argument to show that each oriented path with k sections has exactly k UPSEs on S. Note that the total number of possible UPSEs of different directed paths of size n on an n-point one-sided convex point set is $n \cdot 2^{n-2}$. To see this, note that an UPSE can be encoded by the start point and the position (clockwise or counterclockwise consecutive) of the next point. For the last choice the clockwise and the counterclockwise choices coincide. Moreover, the number of oriented paths with $n - 1$ edges and k sections is $\rho_k := 2\binom{n-2}{k-1}$ and the number of n-vertex oriented paths is $\sum_{k=1}^{n-1} \rho_k = 2^{n-1}$. Let η_k denote the number of UPSEs of all oriented paths on n vertices with k sections.

As shown above, an oriented path with k sections has at least k UPSEs on S. By the symmetry of the binomial coefficient, it holds that $\rho_k = \rho_{n-k}$. Therefore, the number of UPSEs of all oriented paths with k and $n - k$ sections evaluates to at least $k\rho_k + (n - k)\rho_{n-k} = \frac{n}{2}(\rho_k + \rho_{n-k})$. This implies

$$n2^{n-2} = \sum_{k=1}^{n-1} \eta_k \ge \sum_{k=1}^{n-1} k\rho_k = \frac{n}{2} \cdot \sum_{k=1}^{n-1} \rho_k = \frac{n}{2} \cdot 2^{n-1} = n2^{n-2}.$$

Consequently, each oriented path with k sections has exactly k UPSEs on S. □

3   Embedding of Special Directed Paths

In this section we present a family of oriented paths that always admit an UPSE on every general point set of the same size. For an illustration of paths in this family, consider Fig. 1(c).

Theorem 2. *Let P be an oriented path with three sections $P_1 = (u_1, \ldots, u_a)$, $P_2 = (v_1, \ldots, v_b)$ and $P_3 = (w_1, \ldots, w_c)$, where $u_a = v_1$, $v_b = w_1$ and $b \geq a, c$. Then path P admits an UPSE on any general set S of $n = a + b + c - 2$ points.*

Proof. We assume that P_1 is forward, otherwise, we rename the vertices of P by reading it off from vertex w_c. Let ℓ denote the line through $t(S)$ and $b(S)$, and let ℓ^- (resp., ℓ^+) be the halfplane on the left (resp., on the right) when walking from $b(S)$ to $t(S)$ along ℓ. We assume that both ℓ^- and ℓ^+ are closed, and hence $|S \cap \ell^+| + |S \cap \ell^-| = a + b + c$. We now consider two cases.

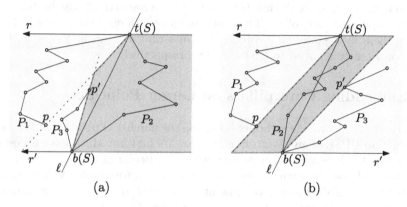

(a)   (b)

Fig. 2. An illustration of the proof of Theorem 2. (a) Embedding of P in Case 1. (b) Embedding of P in Case 2. Point set S_2 lies in the red convex polygon.

Case 1: $|S \cap \ell^+| \geq a + c$ or $|S \cap \ell^-| \geq a + c$. We consider the case $|S \cap \ell^-| \geq a + c$ (refer to Fig. 2(a), with $|S \cap \ell^+| \geq a + c$ being symmetric). We rotate a left heading horizontal ray r emanating from $t(S)$ in counter-clockwise direction until it hits the a-th point (including $t(S)$); we let p denote this point and let S_1 denote the points swept by r. Notice that $b(S) \notin S_1$ because $|S \cap \ell^-| > a$. We embed P_1 on the points in S_1, by sorting them by ascending y-coordinate.

From $|S \cap \ell^-| \geq a + c$ it follows that $|(S \setminus S_1) \cap \ell^-| \geq c$. Hence, we can embed P_3 on $S \cap \ell^-$. To do so, we rotate around $b(S)$ a left-heading horizontal ray r' in clockwise direction until it hits the c-th point of $(S \setminus S_1) \cap \ell^-$; denote these c points by S_3 and let p' be the last point hit by r'. We embed P_3 on the points of S_3 by sorting them by ascending y-coordinate.

Let $S_2 = (S \setminus (S_1 \cup S_3)) \cup \{b(S), t(S)\}$. Then, the polygonal region determined by the horizontal lines through $t(S)$ and $b(S)$, the line through $t(S)$ and p, the line through $b(S)$ and p', and the vertical line through the rightmost point of S_2 is a convex region that contains all points in S_2 and and has $t(s)$ (resp., $b(s)$) as its topmost (resp., bottommost) vertex. Therefore, we can embed P_2 onto S_2 by sorting them by descending y-coordinate. We observe that $u_a = v_1$ and $v_b = w_1$ have been consistently embedded on $b(S)$ and $t(S)$, respectively.

Case 2: $|S \cap \ell^+| \leq a + c$ and $|S \cap \ell^-| \leq a + c$. It follows that $|S \cap \ell^-| \geq b \geq a$ and $|S \cap \ell^+| \geq b \geq c$; refer to Fig. 2(b). Since $|S \cap \ell^-| \geq a$, we can construct the set $S_1 \subset S \cap \ell^-$ similarly to Case 1, and embed P_1 on its points. We then rotate a right-headed horizontal ray r' in counter-clockwise direction around $b(S)$ until it hits the c-th point p' of ℓ^+ and denote by S_3 the points swapped by r'. Since $|S \cap \ell^+| > c$, $t(S) \notin S_3$. We embed P_3 onto S_3 by sorting the points by ascending y-coordinate.

Finally, let $S_2 = (S \setminus (S_1 \cup S_3)) \cup \{b(S), t(S)\}$. We note that the polygonal region determined by the horizontal lines through $t(S)$ and $b(S)$, the line through $t(S)$ and p, and the line through $b(S)$ and p' is a convex polygon that contains all points of S_2. Also recall that $b(S)$ and $t(S)$ are respectively the bottommost and the topmost points of S_2. Thus, we can embed P_2 onto the points in S_2 by sorting them by descending y-coordinate. Note that $u_a = v_1$ and $v_b = w_1$ have been consistently mapped to $t(S)$ and $b(S)$, respectively. □

4   Embedding Caterpillars on Larger Point Sets

In this section, we provide an upper bound on the number of points that suffice to construct an UPSE of an n-vertex caterpillar. We first introduce some necessary notation. Let C be a directed caterpillar with n vertices, r of which, $v_1, v_2, \ldots, v_r$, form its backbone. For each vertex v_i $(i = 1, 2, \ldots, r)$, we denote by $A(v_i)$ (resp., $B(v_i)$) the set of degree-one vertices of C adjacent to v_i by outgoing (resp., incoming) edges. Moreover, we let $a_i = |A(v_i)|$ and $b_i = |B(v_i)|$.

Theorem 3. *Let C be a directed caterpillar with n vertices and k switches in its backbone. Then C admits an UPSE on any general point set S with $|S| \geq n2^{k-2}$.*

Proof. Assume that C contains r vertices $v_1, v_2, \ldots, v_r$ in its backbone. Let $C_{\alpha,\beta}$ denote the subgraph of C induced by the vertices $\bigcup_{i=\alpha}^{\beta} \{v_i\} \cup A(v_i) \cup B(v_i)$. We first make the following observation.

Observation 4 *If the backbone of C contains exactly one section, C has an UPSE on any general set S of n points.*

To see that, we sort the points in $S = \{p_1, \ldots, p_n\}$ by ascending y-coordinate and, assuming C is forward (the backward case is symmetric), we embed v_i on p_x, where $x = i + b_i + \sum_{j=1}^{i-1}(b_j + a_j)$. See to Fig. 3(a) for an illustration.

We now prove the theorem by induction on k. We assume as the induction hypothesis that a directed caterpillar C with i switches and n_i vertices has an UPSE on any general point set S of at least $n_i 2^{i-2}$ points. Let v denote the first vertex of the backbone of C. We additionally assume that if v is a source (resp.

sink), then v is mapped on the $(|B(v)|+1)$-th bottommost (resp. $(|A(v)|+1)$-th topmost) point of S. For $k = 2$, the backbone of C contains one section; hence, the induction hypothesis follows from Observation 4 and its proof.

We now consider a caterpillar C with $i+1$ switches and n_{i+1} vertices. Let v_1 and v_l denote the first and the second switches of the backbone of C, respectively. Let S be a set of at least $N = n_{i+1}2^{i-1}$ points. In the following, we only consider the case where the backbone of $C_{1,l}$ is forward; the backward case is symmetric.

We construct an UPSE of $C_{1,l-1}$ on the $c_1 = \sum_{i=1}^{l-1}(1 + b_i + a_i)$ lowest points of S by applying Observation 4. Let p denote the point where v_{l-1} is mapped. Let S' denote the unused points of S; thus, $|S'| = N - c_1$. We have $n_{i+1} = n_i + c_1$, where n_i is the number of vertices of $C_{l,r}$. Recall that $N = n_{i+1}2^{i-1}$. Therefore, $|S'| > n_i 2^{i-1}$. Let p' be the $(a_l + 1)$-th topmost point in S' and let ℓ denote the line through p and p'. Line ℓ partitions S' into two sets, so that for the largest, say S'', it holds that $|S''| \geq n_i 2^{i-2}$. Let S''' be the union of S'' with the set of points lying above v_l. Since $C_{l,r}$ contains i switches, by induction hypothesis, we can construct an UPSE of $C_{l,r}$ on S''' such that v_l is mapped on the $(a_l + 1)$-th topmost point of S''', which is the point p'. The only edge of the drawing of $C_{1,l}$ that interferes with the drawing of $C_{l,r}$ is (v_{l-1}, v_l); however, the drawing of $C_{l,r}$ (except for the edges incident to v_l) lies on one side of the edge (v_{l-1}, v_l).   □

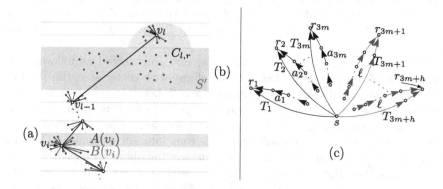

(a)   (b)   (c)

Fig. 3. Illustration of the proofs of (a) Observation 4; (b) Theorem 3; (c) An illustration of the tree T in the proof of Theorem 5.

5   NP-Completeness for Trees

In this section, we consider the following problem: Given a directed tree T with n vertices, a vertex v of T, a set S of n points in the plane, and a point p in S, does T have an UPSE on S which maps v to p? Our goal is to show the following.

Theorem 5. *UPSE of a directed tree with one fixed vertex is NP-complete.*

The 3-Partition problem is a strongly NP-complete problem by Garey and Johnson [9,10], which is formulated as follows: For a given multiset of $3m$ integers $\{a_1, \ldots, a_{3m}\}$, does there exist a partition into m triples $(a_{11}, a_{21}, a_{31}), \ldots,$ (a_{1m}, a_{2m}, a_{3m}), so that for each $j \in [m]$, $\sum_{i=1}^{3} a_{ij} = b$, where $b = (\sum_{i=1}^{3m} a_i)/m$. In this section, we present a reduction from 3-Partition. Without loss of generality, we may assume, possibly multiplying each a_i by 3, that each a_i is divisible by 3.

Given $3m$ integers $\{a_1, \ldots, a_{3m}\}$, where each a_i is divisible by 3, we construct an instance of our problem as follows. Let ℓ and h be two large numbers such that $\ell > mb$ and $h = m + 1$. We construct a tree T with $n := mb + h\ell + 1$ vertices, and a point set S with n points. As illustrated in Fig. 3(c), (the undirected version of) the tree T is a subdivision of a star, that is, T has a single vertex s of degree greater than two. Specifically, the degree of vertex s is $3m + h$, i.e., $3m + h$ paths meet in their initial vertex s; we call each such path a *branch* of T. Let $T_1, \ldots, T_{3m+h}$ denote the branches, respectively. For $i \in \{1, \ldots, 3m\}$, the branch T_i is a path with a_i edges and called *small*; for $i \in \{3m+1, \ldots, 3m+h\}$, the branch T_i is a path with ℓ edges and called *large*. Note that a_i and ℓ may also be interpreted as the number of vertices of a branch that are different from s.

For each branch T_i, we define its *root* r_i as follows. For a small branch T_i, r_i is the first vertex on T_i that is different from s; for a large branch, r_i is the second vertex different from s. For every branch, all edges are oriented so that its root is the unique sink of the branch. Thus the sinks of T are exactly the $3m + h$ roots of the branches, and the sources of T are s and the $3m + h$ leaves of T.

The point set S is depicted in Fig. 4(a) and constructed as follows. The lowest point of S is $p = (0, 0)$. Let E be an ellipse with center at p, (horizontal) semi-major axis 5 and (vertical) semi-minor axis 3. Let C be a convex x- and y-monotone curve above E and also above the point $(-3, 3)$, represented by the filled square mark in Fig. 4(a). Consider the cone defined by the upward rays from p with slopes -1, -2, and subdivide this cone into $2m + 1$ equal sectors by $2m$ upward rays from p. Let $s_0, \ldots, s_{2m}$ denote the obtained sectors ordered counter-clockwise. For each odd $i = 2k + 1$ with $0 < i \leq 2m$, consider the intersection between the sector s_i and the ellipse E (the orange arcs in Fig. 4(a)) and let B_k be the set of b equally spaced points on this intersection.

For each even $i = 2k$ with $0 \leq k \leq m$, consider the intersection c_k between sector s_i and curve C (the green arcs in Fig. 4(a)). Let the point set L_k be constructed as follows. Place the first point q'_k of L_k in the sector s_i slightly to the right of (and thus slightly below, to stay inside the sector) the topmost point of the arc c_k. Let, for $k > 0$, x_k be the point of intersection between c_k and the line through q'_k and the topmost point of the set B_k. For $k = 0$ let x_k be the highest point of c_k. Place the second point q_k of L_k on c_k slightly below x_k, but still above q'_k, see also Fig. 4(a). Place $\ell - 2$ points equally spaced on c_k below q_k. This concludes our description of T and S. In the remaining, the point sets B_k are called *small sets*, and the point sets L_k are called *large sets*.

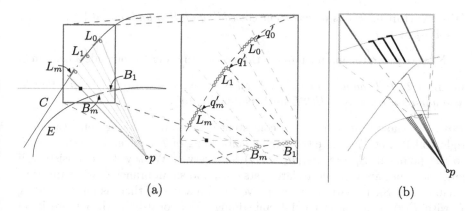

Fig. 4. An illustration for the proof of Theorem 5 with zoomed portions: (a) the point set S, and (b) a schematic illustration of a consistent embedding of T on S. (Color figure online)

To prove Theorem 5, we show that $\{a_1, \ldots a_{3m}\}$ admits a 3-partition if and only if T admits an UPSE Γ on S where $\Gamma(s) = p$. In particular, our proof is based on the fact that if such an UPSE exists, then it nearly fulfills the special property of being a *consistent embedding* of T on S (Lemma 2). An embedding Γ is consistent if $\Gamma(s) = p$ and each large branch of T is mapped to exactly one large point set; for a schematic illustration consider Fig. 4(b).

By construction of S, if the large branches are mapped to the large point sets, then the small branches of T must be subdivided into m groups each with a total of b vertices. Each such group then corresponds to a triple (a_{1i}, a_{2i}, a_{3i}) of the partition of $(a_1, \ldots, a_{3m})$ that sums up to b. Conversely, a 3-partition directly yields a consistent embedding. This proves the following fact:

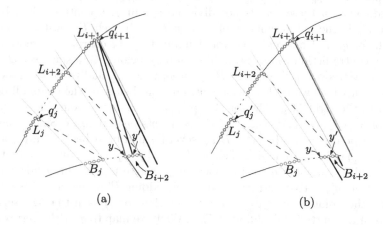

Fig. 5. (a) A partial embedding Γ_{i+1} of Γ that is not consistent, while Γ_i is. (b) A modification Γ' of Γ such that Γ'_{i+1} is consistent.

Observation 6. *T has a consistent embedding on S if and only if* $\{a_1, \ldots a_{3m}\}$ *has a 3-partition.*

Now we prove the main lemma of this section in order to conclude Theorem 5.

Lemma 2. *If T admits an UPSE* Γ *on S mapping s to p* $(\Gamma(s) = p)$, *then T also admits a consistent UPSE on S mapping s to p.*

Proof. To show that T admits a consistent embedding, we work our way from right to left. Let Γ_i denote the partial embedding of T induced by restricting Γ to the points in $S_i := \cup_{k \leq i}(L_k \cup B_k)$ for every i. We say Γ_i is consistent if large branches are mapped to large sets of S_i and small branches are mapped to small sets of S_i. In particular, we prove the following: If there is an embedding Γ with $\Gamma(s) = p$ whose partial embedding Γ_i is consistent, then there is an embedding Γ' with $\Gamma'(s) = p$ whose partial embedding Γ'_{i+1} is consistent.

For unifying notation, we define $B_0, B_{-1}, L_{-1} := \emptyset$. Suppose there is an embedding Γ such that Γ_i is consistent. Suppose that the partial embedding Γ_{i+1} of Γ is not consistent (otherwise let Γ' be Γ). Since q_{i+1} is the highest point of $S \setminus S_i$, $\Gamma(r_j) = q_{i+1}$ for some branch T_j. We distinguish two cases depending on whether T_j is a small or large branch. If T_j is a small branch, then B_{i+1} and q'_{i+1} is to the right of segment pq_{i+1}. Depending on whether T_j continues left or right, a set of small branches is mapped on $B_{i+1} \cup \{q'_{i+1}\}$ or $B_{i+1} \cup \{q_{i+1}, q'_{i+1}\}$, respectively. (In the latter case, T_j is one of the small branches mapped on $B_{i+1} \cup \{q_{i+1}, q'_{i+1}\}$.) However, the cardinality of these sets is not divisible by three; a contradiction. Therefore T_j is a large branch.

Then there exists a point $y \neq q'_{i+1}$ such that the segments py and yq_{i+1} represent the first two edges of T_j. If y belongs to a large set, then yq_{i+1} separates all points between y and q_{i+1} on C from p. Therefore, there are only three different options for the placement of y on a large set: q_{i+2}, q'_{i+2} or the left neighbour of q_{i+1} in L_{i+1}. If y is q_{i+2} or q'_{i+2}, then a number of small branches are mapped to $B_{i+1} \cup \{q'_{i+2}, q'_{i+1}\}$ or to $B_{i+1} \cup \{q'_{i+1}\}$, respectively. However, the cardinality of these sets is not divisible by three, a contradiction. In the case that y is the left neighbour of q_{i+1} in L_{i+1}, branch T_j must continue to the right of q_{i+1}. However, this would imply that T_j contains at most $b + 3$ points, a contradiction to $\ell > b + 3$. Therefore, y belongs to a small set B_j with $j > i + 1$. Moreover, y belongs to B_{i+2}, otherwise q'_{i+2} had to be the root of a large branch which would lead to a contradiction. Let A be the set of all points of B_{i+2} to the right of y. The set $A \cup \{q'_{i+1}\}$ is separated from the rest of the points of $S \setminus (S_i \cup \Gamma(T_j) \cup B_{i+1})$ by the segments py and yq_{i+1}. Since there are less than ℓ points in $A \cup \{q'_{i+1}\} \cup B_{i+1}$, several small branches of T are mapped to $A \cup \{q'_{i+1}\} \cup B_{i+1}$. In addition, since $|B_{i+1} \cup \{q'_{i+1}\}|$ is not divisible by three, for the small branch that maps its root to q'_{i+1}, the rest of this branch is mapped to the left of pq'_{i+1}. To construct a sought embedding Γ' we modify Γ as follows. Consider the point y' in B_{i+2} (y' lies to the right of y) such that the segment $y'q'_{i+1}$ is in the partial embedding of Γ_{i+1}. First, we map to q'_{i+1} the vertex of T_j previously mapped to y (i.e., $\Gamma^{-1}(y)$). Let $A' \subset A$ be all the points to the left of y' in A. For each vertex mapped to a point z in $A' \cup \{y'\}$, we map such vertex to the left neighbour of z in $A' \cup \{y\}$. Finally, the vertex that was mapped to q'_{i+1} is now mapped to y'. See Fig. 5.

The obtained embedding Γ' is such that Γ'_{i+1} is consistent. Indeed, in Γ' the branch T_j is embedded entirely on L_{i+1}; otherwise some point of L_{i+1} would be separated from p. And since $|B_{i+1}| = b < \ell$, a set of small branches is embedded on B_{i+1}. Together with the obvious fact that in any UPSE Γ of T on S, Γ_{-1} is consistent, the above observation implies the claim of the lemma.   □

6   Conclusion and Open Problems

In this paper, we continued the study of UPSE of directed graphs, specifically of paths, caterpillars, and trees. On the positive side, we showed that a certain family of n-vertex oriented paths admits an UPSE on any general n-point set and that any caterpillar can be embedded on a general point set if the set is large enough. Moreover, we provided the exact number of UPSEs for every path on a one-sided convex set. On the negative side, we proved that the problem of deciding whether a directed graph on n vertices admits an UPSE on a given set of n points remains NP-complete even for trees when one vertex lies on a pre-defined point. We conclude with a list of interesting open problems:

1. Given any oriented path P and any general point set S with $|P| = |S|$, does there exist an UPSE of P on S?
2. If the answer to the previous question turns out to be negative, what is the smallest constant c such that a set of c paths on a total of n vertices, has an UPSE on every general point set on n points? This problem could also be an interesting stepping stone towards a positive answer of the previous question.
3. Given a directed tree T on n vertices and a set S of n points, is it NP-hard to decide whether T has an UPSE on S?
4. Can the provided upper bound on the size of general point sets that host every caterpillar be improved to polynomial in the number of vertices, following the result in [13]? If yes, can this be extended to general directed trees?

Acknowledgement. This work was initiated at the 7th Annual Workshop on Geometry and Graphs, March 10-15, 2019, at the Bellairs Research Institute of McGill University, Barbados. We are grateful to the organizers and to the participants for a wonderful workshop. E. A. was partially supported by RFBR, project 20-01-00488. P. C. was supported by the F.R.S.-FNRS under Grant no. MISU F 6001 1. The work of P. V. was supported by grant no. 19-17314J of the Czech Science Foundation (GAČR).

References

1. Angelini, P., Frati, F., Geyer, M., Kaufmann, M., Mchedlidze, T., Symvonis, A.: Upward geometric graph embeddings into point sets. In: Brandes, U., Cornelsen, S. (eds.) GD 2010. LNCS, vol. 6502, pp. 25–37. Springer, Heidelberg (2011). https://doi.org/10.1007/978-3-642-18469-7_3

2. Binucci, C., et al.: Upward straight-line embeddings of directed graphs into point sets. Comput. Geom. **43**, 219–232 (2010)
3. Bose, P.: On embedding an outer-planar graph in a point set. Comput. Geom. **23**(3), 303–312 (2002)
4. Bose, P., McAllister, M., Snoeyink, J.: Optimal algorithms to embed trees in a point set. J. Graph Algorithms Appl. **1**(2), 1–15 (1997)
5. Cabello, S.: Planar embeddability of the vertices of a graph using a fixed point set is NP-hard. J. Graph Algorithms Appl. **10**(2), 353–363 (2006)
6. Cagirici, O., et al.: On upward straight-line embeddings of oriented paths. In: Abs. of the XVII Spanish Meeting on Computational Geometry, pp. 49–52 (2017)
7. Di Battista, G., Eades, P., Tamassia, R., Tollis, I.G.: Graph Drawing: Algorithms for the Visualization of Graphs. Prentice Hall, Upper Saddle River (1999)
8. Di Battista, G., Tamassia, R.: Algorithms for plane representations of acyclic digraphs. Theoret. Comput. Sci. **61**(2–3), 175–198 (1988)
9. Garey, M.R., Johnson, D.S.: Complexity results for multiprocessor scheduling under resource constraints. SIAM J. Comput. **4**(4), 397–411 (1975)
10. Garey, M.R., Johnson, D.S.: Computers and Intractability: A Guide to the Theory of NP-Completeness. W. H. Freeman & Co., London (1979)
11. István, F.: On straight-line representation of planar graphs. Acta Scientiarum Mathematicarum **11**(229–233), 2 (1948)
12. Kaufmann, M., Mchedlidze, T., Symvonis, A.: On upward point set embeddability. Comput. Geom. **46**(6), 774–804 (2013)
13. Mchedlidze, T.: Reprint of: upward planar embedding of an n-vertex oriented path on $O(n^2)$ points. Comput. Geom. **47**(3), 493–498 (2014)
14. Pach, J., Gritzmann, P., Mohar, B., Pollack, R.: Embedding a planar triangulation with vertices at specified points. Am. Math. Mon. **98**, 165–166 (1991)
15. Stein, S.K.: Convex maps. Proc. AMS **2**(3), 464–466 (1951)
16. Wagner, K.: Bemerkungen zum Vierfarbenproblem. Jahresbericht der Deutschen Mathematiker-Vereinigung **46**, 26–32 (1936)

2-Colored Point-Set Embeddings
of Partial 2-Trees

Emilio Di Giacomo[1(✉)], Jaroslav Hančl Jr.[2], and Giuseppe Liotta[1]

[1] Dipartimento di Ingegneria, Università degli Studi di Perugia, Perugia, Italy
{emilio.giacomo,giuseppe.liotta}@unipg.it
[2] Faculty of Mathematics and Physics, Charles University, Prague, Czech Republic
jarda.hancl@gmail.com

Abstract. Let G be a planar graph whose vertices are colored either red or blue and let S be a set of points having as many red (blue) points as the red (blue) vertices of G. A 2-colored point-set embedding of G on S is a planar drawing that maps each red (blue) vertex of G to a red (blue) point of S. We show that there exist properly 2-colored graphs (i.e., 2-colored graphs with no adjacent vertices having the same color) having treewidth two whose point-set embeddings may require linearly many bends on linearly many edges. For a contrast, we show that two bends per edge are sufficient for 2-colored point-set embedding of properly 2-colored outerplanar graphs. For separable point sets this bound reduces to one, which is worst-case optimal. If the 2-coloring of the outerplanar graph is not proper, three bends per edge are sufficient and one bend per edge (which is worst-case optimal) is sufficient for caterpillars.

Keywords: Point-set embedding · Curve complexity · Topological book embeddings · 2-page bipartite book embedding

1 Introduction

Let G be a planar graph with n vertices and let S be a set of n distinct points in the plane. A point-set embedding of G on S is a crossing free drawing of G such that every vertex of G is mapped to a distinct point of S. Point-set embeddings are a classical subject of investigation in graph drawing and computational geometry. The problem has been studied either assuming that the mapping of the vertices of G to the points of S is part of the input (see, e.g., [23]), or when the drawing algorithm can freely map any vertex of G to any distinct point of S (see, e.g., [21]), or when specific subsets of vertices of G can be freely mapped to specific subsets of points of S with the same cardinality (see, e.g., [11]). Recently, point-set embeddings in the "beyond planar" scenario, when a constant number of crossings per edge are allowed, have been also studied (see, e.g., [7,20]).

Work partially supported by: MIUR, grant 20174LF3T8 AHeAD: efficient Algorithms for HArnessing networked Data, and grant SVV–2020–260578.

In this paper we study 2-colored point-set embeddings: The input consists of a planar graph G with n vertices and of a set S of n distinct points in the plane. The graph G is 2-*colored*, that is its vertex set is partitioned into a subset V_R of red vertices and a subset V_B of blue vertices. If no two adjacent vertices have the same color, we say that G is *properly 2-colored*. Further, S is *compatible* with G, i.e. it has $|V_R|$ red points and $|V_B|$ blue points. We want to construct a planar drawing of G where every red vertex is mapped to a red point, every blue vertex is mapped to a blue point, and the number of bends per edge is small.

The literature about the 2-colored point-set embeddability problem is very rich and, for brevity, we mention here only some of the most relevant results. The case when no bends along the edges are allowed has been studied for properly 2-colored paths (see, e.g., [1,2,17]), 2-colored trees and forests (see, e.g., [6,10, 16,22,24]), and properly 2-colored cycles (see, e.g., [18]). When bends along the edges are allowed, Badent et al. prove that there are 2-colored graphs whose 2-colored point-set embeddings may require linearly many bends on linearly many edges [4]. On the positive side, outerplanar graphs always admit a 2-colored point set embedding with curve complexity at most five [9] which reduces to at most one when the input is a 2-colored path [13]. Variants of the 2-colored point-set embeddability problem, where the edges are orthogonal chains and only few bends per edge are allowed have also been studied (see, e.g., [12,19]). The main contributions of this paper are as follows.

- We show that for any $n \geq 14$ there exists a properly 2-colored bipartite graph G with $2n+4$ vertices and treewidth two for which a 2-colored point-set embedding of G may require at least $\frac{n-9}{4}$ bends on at least $\frac{n-5}{2}$ edges (Sect. 3). We remark that the known linear lower bound on the curve complexity of 2-colored point-set embeddings assumes the coloring not to be proper and the treewidth to be larger than two [4].
- We prove that for properly 2-colored outerplanar graphs two bends per edge are sufficient to compute a 2-colored point-set embedding. The curve complexity can be reduced to one (which is tight) if the point set is *linearly separable*, i.e., there exists one line with all the blue points on one side of the line and all the red points on the other side (Sect. 4).
- We extend the study to partial 2-trees whose coloring may not be proper. We show that curve complexity three is always sufficient for 2-colored point-set embeddings of outerplanar graphs and that one bend per edge is a tight bound for caterpillars (Sect. 5). This results improve over the known upper bounds of five and two, respectively, for these graph families [9,13].

We remark that the problem of computing a 2-colored point-set embedding of a graph G with few bends per edge is strictly related with the problem of computing a 2-*colored topological book embedding* with few spine crossings per edge. Let G be a 2-colored graph and let σ be a sequence of red and blue colors. A 2-colored topological book embedding of G is a planar drawing where all vertices are points along a horizontal line called *spine*, the edges are curves that can cross the spine, and the sequence of colors along the spine coincides with σ. By establishing bounds on the number of spine crossings per edge in a 2-colored topological book embedding of G, we shall establish bounds on the curve complexity of the 2-colored point-set embeddings of G.

2 Book Embeddings and Point-Set Embeddings

Let G be a planar graph. A *2-page book embedding* of G is a planar drawing of G where all vertices lie on a horizontal line ℓ called the *spine* and each edge is drawn in one of the two half-planes defined by ℓ. A *2-page topological book embedding* of G is a planar drawing of G such that all vertices still lie on the spine ℓ but the edges can cross the spine. Each crossing between an edge an the spine is called a *spine crossing*.

Let G be a 2-colored planar graph. A *2-colored sequence* is a sequence of colors such that each element is either red or blue. A 2-colored sequence σ is *compatible* with G if the number of red (blue) elements in σ is equal to the number of red (blue) vertices in G. A (topological) 2-page book embedding of a 2-colored graph G is *consistent* with a given 2-colored sequence σ if the sequence of the vertex colors along the spine coincides with σ. Let S be a 2-colored point set. Set S is *compatible* with G if its number of red (blue) points is equal to the number of red (blue) vertices of G. Assume that the points of S have different x-coordinates (if not we can rotate the plane so to achieve this condition) and project the points of S on the x-axis. We denote by $seq(S)$ the left-to-right sequence of colors of the projected points.

In the rest of the paper we will use of the following theorem.

Theorem 1 [13]. *Let G be a 2-colored planar graph and let σ be a compatible 2-colored sequence. If G admits a 2-page topological book embedding with at most one spine crossing per edge consistent with σ, then G admits a 2-colored point-set embedding on any compatible 2-colored point set S for which $seq(S) = \sigma$ such that every edge that crosses the spine in the topological book embedding has two bends, while every edge that does not cross the spine has one bend.*

Note: If this paper is printed black and white, grey points stand for red points.

3 Properly 2-Colored Partial 2-Trees

In this section we prove that there exist partial 2-trees whose 2-colored point-set embeddings may have linear curve complexity on linearly many edges.

Theorem 2. *For every $n \geq 14$, there exists a properly 2-colored partial 2-tree G with $2n + 4$ vertices and a 2-colored point set compatible with G such that every 2-colored point-set embedding of G on S has at least $\frac{n-5}{2}$ edges each having at least $\frac{n-9}{4}$ bends.*

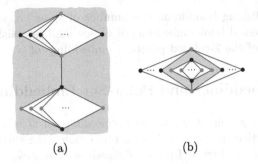

(a)   (b)

Fig. 1. (a) A properly partial 2-tree G for the proof of Theorem 2. (b) A different planar embedding of G. In both figures the face colored gray is the only face containing both red and blue vertices of degree 2.

Proof. Let G be the properly 2-colored partial 2-tree of Fig. 1(a), which consists of two copies of $K_{2,n}$ connected by an edge; in the first copy of $K_{2,n}$ there are n blue vertices and 2 red vertices, while in the second copy there are n red vertices and 2 blue vertices. The additional edge connects a high-degree vertex of the first copy to a high-degree vertex of the second copy. Let S be a set of points along a horizontal line ℓ whose colors are alternately red and blue. Let Γ be any point-set embedding of G on S. If we connect the vertices of Γ with additional edges in the order they appear along ℓ and we also connect the last vertex along ℓ to the first one, we obtain a Hamiltonian cycle that alternates red and blue vertices and possibly crosses some edges of G. If an edge of G is crossed b times by this cycle, then it has at least $b+1$ bends. We now show that for any cycle C that passes through all vertices of G alternating between red and blue vertices there are at least $\frac{n-5}{2}$ edges that are crossed at least $\frac{n-9}{4} - 1 = \frac{n-13}{4}$ times. Assume first that n is even. Whatever is the planar embedding of G there exists only one face f that contains both red and blue vertices of degree 2 (see Fig. 1 for two different planar embeddings of G). We assign a level to the degree-2 blue vertices as follows. The degree-2 blue vertices of face f are vertices of level 0; the degree-2 blue vertices that are on face f after the removal of vertices of level i are vertices of level $i+1$. The number of degree-2 blue vertices is n, which is even, and there are two such vertices at each level; hence the number of levels is $n/2$. The blue vertices of level i ($0 \leq i \leq n/2 - 1$) together with their adjacent vertices (i.e. the two red vertices of high degree) form a 4-cycle C_i that separates the blue vertices of level $j > i$ from the degree-2 red vertices. The number of blue vertices of level $j > i$ are $n - 2(i+1)$; each of them has two incident edges in C. At most 4 of these edges can be incident to the two high-degree red vertices and therefore at least $2(n - 2(i+1)) - 4$ of them have to cross the edges of all the 4-cycles C_l with $l \leq i$. Thus we have at least i cycles crossed at least $2(n - 2(i+1)) - 4$ times, which implies that there are at least i edges (one per cycle) crossed at least $\frac{2(n-2(i+1))-4}{4}$ times. For $i = \frac{n}{4} - 1$, we obtain that there are at least $\frac{n}{4} - 1$ edges each crossed at least $\frac{n-12}{4}$ times. By applying the same argument as above to the red vertices of degree 2, we obtain that there are at least $\frac{n}{4} - 1$ edges incident to the degree-2 red vertices each crossed at least $\frac{n-12}{4}$ times. Thus in total we have at least $\frac{n}{2} - 2 = \frac{n-4}{2}$ edges each crossed at least $\frac{n-12}{4}$ times. If n is odd we remove one blue and one red vertex of degree 2 and by the same argument as above, there are at least $\frac{n-5}{2}$ edges each crossed at least $\frac{n-13}{4}$ times.   □

It may be worth recalling that a linear lower bound on the curve complexity of 2-colored planar graphs was already known [4], but this lower bound holds for graphs with treewidth larger than two and with a 2-coloring that is not proper.

4   Properly 2-Colored Outerplanar Graphs

Theorem 2 motivates the study of meaningful families of properly 2-colored graphs having treewidth two and such that a bound on the number of bends per edge is independent from the number of vertices. In this section we show that for the properly 2-colored outerplanar graphs two bends per edge are always sufficient and that one bend per edge is sufficient for special configurations of the points. We start with two lemmas about 2-colored topological book embeddings.

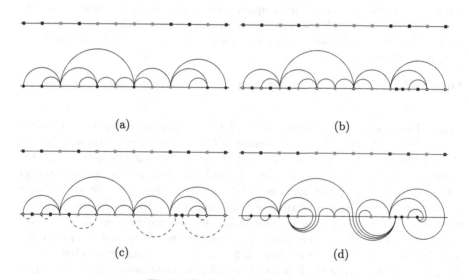

(a)

(b)

(c)

(d)

Fig. 2. Illustration of Lemma 1.

Lemma 1. *Let G be a properly 2-colored outerplanar graph and let σ be a compatible 2-colored sequence. G admits a 2-page topological book embedding consistent with σ and at most one spine crossing per edge.*

Sketch of Proof. Compute a 1-page book embedding Γ of G without taking into account the colors of the vertices. The sequence of the vertex colors along the spine of Γ is in general different from the given sequence σ (see Fig. 2(a)). The idea now is to move some vertices and allow spine crossings in order to make the sequence of colors along the spine coincide with σ. To this aim it is sufficient to remap only the vertices of one color, say the blue color. Add to the spine of Γ new blue positions in such a way that the original red positions and these new blue positions define a sequence of colors that coincides with σ (see the small blue squares in Fig. 2(b)). Using a matching bracket algorithm we add matching edges in the second page between the blue vertices and these new positions (see Fig. 2(c)). These edges do not cross each other and do not cross the original edges. We now replace each original blue vertex v with as many spine crossings as the number of edges incident to them. Each edge is assigned to one of these spine crossings in such a way to avoid crossings between these edges and each spine crossing is connected to the new location matched to v (see Fig. 2(d)). In this way we obtain a topological book embedding of G consistent with σ. The edges cross the spine at their blue end-vertices. Since G is properly 2-colored, each edge has only one blue end-vertex and thus crosses the spine only once.

When the 2-colored sequence is *separable*, i.e. all elements of a color precede those of the other color, we can avoid spine crossings. A 2-page book embedding where all vertices of a partition set appear consecutive along the spine is called a *2-page bipartite book embedding*. It may be worth remarking that 2-page bipartite book embeddings find applications, for example, in the study of hybrid planarity testing problems [3,8].

Lemma 2. *An outerplanar bipartite graph has a 2-page bipartite book embedding.*

Proof. The proof is a consequence of [5,14,15]. Let G be an outerplanar bipartite graph. We construct a $(2,2)$-track layout of G, i.e., a drawing of G such that: (i) the vertices are placed on two different horizontal lines called *tracks*, (ii) no two vertices on a same track are adjacent, (iii) and the edges are partitioned into two sets such that the edges in the same set do not cross each other. Starting from a red vertex we number the vertices of G according to a BFS traversal where the vertices adjacent to a vertex are visited in the order induced by an outerplanar embedding of G. The red vertices are placed on one of the two tracks ordered from left to right according to their BFS number; analogously the blue vertices are placed on the other track ordered from left to right according to their BFS number. The edges can be partitioned into two color classes: one containing the edges whose blue vertex has a lower BFS number than the red vertex, and the other one containing all the other edges (i.e. those whose red vertex has a lower BFS number than the blue one). By construction no two edges in the same set cross. Starting from the constructed $(2,2)$-track layout a 2-page bipartite book embedding is computed by applying Lemma 13 of [14]. □

We are now ready to give the main result of this section.

Theorem 3. *Let G be a properly 2-colored outerplanar graph and let S be a compatible 2-colored point set. G admits a 2-colored point-set embedding on S with at most two bends per edge. If S is separable, G admits a 2-colored point-set embedding on S with at most one bend per edge, which is worst-case optimal.*

Proof. By Lemma 1, G admits a 2-page topological book embedding consistent with σ and at most one spine crossing per edge. By Theorem 1 G has a 2-colored point-set embedding on S with at most two bends per edge. If S is separable, by Lemma 2 and Theorem 1 follows that the number of bends per edge reduces to one. This is worst case optimal since a straight-line 2-colored point-set embedding on a separable 2-colored point-set may not exist even when G is a properly 2-colored cycle [18].   □

5   Non Proper 2-Colorings

The results in Sect. 4 consider outerplanar graphs with a proper 2-coloring. In the following we extend the results of the previous section to 2-colored outerplanar graphs whose coloring may not be proper. A variant of the construction in the proof of Theorem 3 and an ad-hoc geometric strategy to draw the edges give rise to the following theorem. See Fig. 3 for an illustration of this variant of Theorem 3 and the appendix for a complete proof.

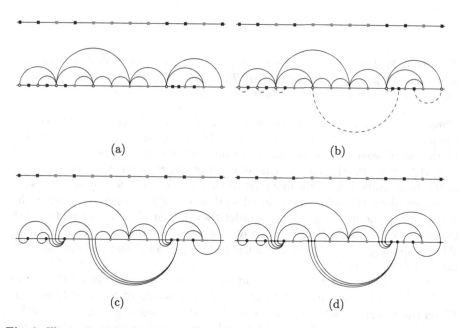

(a)   (b)

(c)   (d)

Fig. 3. Illustration for Theorem 4: Edges whose endvertices are both blue make two spine crossings. These edges are subdivided with a dummy vertex.

Theorem 4. *Let G be a 2-colored outerplanar graph and let S be a 2-colored point set compatible with G. G admits a 2-colored point-set embedding on S with at most three bends per edge.*

In the special case of caterpillars, we show a construction that achieves one bend per edge, which is worst-case optimal since a straight-line 2-colored point-set embedding may not exist even for properly 2-colored paths [2].

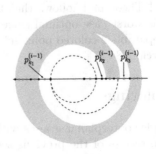

Fig. 4. Illustration for the proof of Theorem 5. The grey region represents the already computed book embedding; the free points are small squares while the used points are disks. The distinguished vertices are highlighted. Examples of free points that are top- and/or bottom-visible are connected with dashed curves. See the appendix for a larger figure.

Theorem 5. *Let T be a (not necessarily proper) 2-colored caterpillar and let S be a 2-colored point set compatible with T. T has a 2-colored point-set embedding on S with at most one bend per edge, which is worst-case optimal.*

Proof. Based on Theorem 1, we prove that T has a 2-page book embedding compatible with $seq(S)$. Let $v_1, v_2, \ldots, v_s$ be the spine of T, let V_i^R and V_i^B be the set of leaves connected to v_i of color red and blue, respectively, and let $n_i^c = |V_i^c|$ ($c \in \{R, B\}$). For the ease of description, we present the construction of the book embedding as a mapping of the vertices of T to any set of points S' placed along the spine and colored so that $seq(S') = seq(S)$. We denote by $p_1, p_2, \ldots, p_n$ the points in S' in the order they appear along the spine from left to right. We denote by $c(x)$ the color of x, where x is either a vertex or a point. We call *spine edge* an edge that belongs to the spine of T and *leaf edge* an edge incident to a leaf in T.

The algorithm works in s steps; at Step i the vertex v_i and all the leaves in $V_i = V_i^R \cup V_i^B$ are mapped to $n_i^R + n_i^B + 1$ points of S' suitably chosen. When the algorithm adds a vertex v to the book embedding constructed so far mapping it to a point p, it also adds an edge connecting v to some other vertex already in the book embedding. We say that *vertex v is mapped to p with a top (resp. bottom) edge* to mean that v is mapped to p and the edge connecting v to the existing book embedding is drawn in the top (resp. bottom) page. Denote by Γ_i the book embedding constructed at the end of Step i. A point is *used* in Γ_i if a vertex has been mapped to it, *free* otherwise. Let p_j and p_k be two points of S', we say that p_j and p_k are *top-connectible* (resp. *bottom-connectible*) if it is possible to connect p_j and p_k by a curve in the top (resp. bottom) page without intersecting any other edge of Γ_i. Denote by $p_{k_1}^{(i)}$ the first free point and by $p_{k_3}^{(i)}$ the last free point at the end of Step i. Also, denote by $p_{k_2}^{(i)}$ the last free point such that $c(p_{k_2}^{(i)}) \neq c(p_{k_3}^{(i)})$. Points $p_{k_1}^{(i)}$, $p_{k_2}^{(i)}$, and $p_{k_3}^{(i)}$ are the *distinguished points* at the end of Step i. In our notation of the three distinguished points, the apex i indicates the Step i, while k_1, k_2 and k_3 are the indices of the points in the sequence $p_1, p_2, \ldots, p_n$. Note that all points before $p_{k_1}^{(i)}$ and all points after $p_{k_3}^{(i)}$ are used; moreover, all points between $p_{k_2}^{(i)}$ and $p_{k_3}^{(i)}$ with the same color as $p_{k_2}^{(i)}$ are used. The algorithm maintains the following invariants at the end of Step i (see Fig. 4 for a schematic illustration):

I1 Let p_j and p_h be two points connected by a spine edge e such that p_j is to the left of p_h. If p_h is red, edge e is in the bottom page; if p_h is blue, edge e is in the top page.

I2 Let p_j and p_h be two points connected by a leaf edge e such that the leaf vertex of e is mapped to p_h. If p_h is red, edge e is in the bottom page; if p_h is blue, edge e is in the top page.

I3 Let p_j be the point where v_i is mapped and let p_h be a free point. If $p_j \neq p_{k_3}^{(i-1)}$ then p_j and p_h are top-connectible if $c(p_h) = B$ and bottom-connectible if $c(p_h) = R$. If $p_j = p_{k_3}^{(i-1)}$ then p_j and p_h are top-connectible if $c(p_h) = c(p_j) = B$ and bottom-connectible if $c(p_h) = c(p_j) = R$.

I4 Let p_j and p_h be two free points. If $k_1 \leq j, h \leq k_2$, p_j and p_h are both top- and bottom-connectible. If $k_2 < j \leq k_3$, p_j and p_h are top-connectible if $c(p_j) = B$ and bottom-connectible if $c(p_j) = R$.

I5 If p_j is an used point such that $k_2 < j < k_3$ and $c(p_j) = p_{k_3}^{(i)}$, then the vertex mapped to p_j is a leaf.

I6 If p_j is an used point such that $k_2 < j < k_3$ and $c(p_j) \neq p_{k_3}^{(i)}$, then no spine edge is incident ot p_j from bottom (resp. top) if $c(p_j) = B$ (resp. $c(p_j) = R$).

Assume that $c(p_1) = c(v_1)$; if this is not the case, we can add a dummy vertex at the beginning of the spine (with no leaves) and a dummy point at the beginning of S' giving them the same color. At Step 1 we first map v_1 to p_1; then we map all the leaves in V_1^R to the n_1^R rightmost red points with bottom edges. Finally, we map all vertices of V_1^B to the n_1^B rightmost blue points with top edges. At the end of Step 1, Invariants I1, I2, I5 and I6 trivially hold. Invariants I3 and I4 are maintained because the leaves adjacent to v_1 are mapped to the rightmost red and blue points.

Now, assume that at the end of Step $i - 1$ the invariants hold; we describe how to execute Step i. Vertex v_i is always mapped to one of the distinguished points $p_{k_1}^{(i-1)}$, $p_{k_2}^{(i-1)}$ and $p_{k_3}^{(i-1)}$. In order to preserve Invariant I1 we will draw the edge (v_{i-1}, v_i) in the top or in the bottom page depending on the color of its rightmost end-vertex. We distinguish various cases depending on which of the three distinguished points has the same color as v_i (since $p_{k_2}^{(i-1)}$ and $p_{k_3}^{(i-1)}$ have different colors, at least one of the three points as the same color as v_i); in particular, we use points $p_{k_2}^{(i-1)}$ and $p_{k_3}^{(i-1)}$ only when it is not possible to use $p_{k_1}^{(i-1)}$, i.e., when $c(v_i) \neq (p_{k_1}^{(i-1)})$. We assume that the color of v_i is red (the case when it is blue is analogous exchanging red with blue and top with bottom). For space reasons, we shall just describe where the algorithm maps v_i and its adjacent leaves. See the appendix for a proof that Γ_i is a 2-page book embedding that maintains Invariants I1–I6.

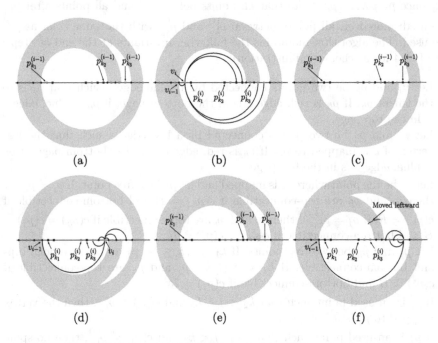

Fig. 5. Proof of Theorem 5: (a)–(b) Case 1. (c)–(d) Case 2. (e)–(f) Case 3.

Case 1: $c(v_i) = c(p_{k_1}^{(i-1)})$. Refer to Figs. 5(a) and 5(b). If $p_{k_1}^{(i-1)}$ is to the right of the point p_j representing v_{i-1} we map v_i to $p_{k_1}^{(i-1)}$ with a bottom edge. If $p_{k_1}^{(i-1)}$ is to the left of p_j we map v_i to $p_{k_1}^{(i-1)}$ with a top edge. We map all the leaves in V_i^R to the n_i^R rightmost free red points with bottom edges. Finally, we map all vertices of V_i^B to the n_i^B rightmost free blue points with top edges.

Case 2: $c(v_i) \neq c(p_{k_1}^{(i-1)})$ **and** $c(v_i) = c(p_{k_2}^{(i-1)})$. Refer to Figs. 5(c) and 5(d). We map v_i to $p_{k_2}^{(i-1)}$ with a bottom edge. We then map all the leaves in V_i^R to the rightmost n_i^R free red points with bottom edges. Finally, we map the leaves in V_i^B to the n_i^B leftmost free points to the right of $p_{k_2}^{(i-1)}$ with top edges; if the free points to the right of $p_{k_2}^{(i-1)}$ are not enough, we just use the n_i^B rightmost free points. In other words, if the free points to the right of $p_{k_2}^{(i-1)}$ are not enough, we use all of them plus those to the left of $p_{k_2}^{(i-1)}$ that are the closest to $p_{k_2}^{(i-1)}$.

Case 3: $c(v_i) \neq c(p_{k_1}^{(i-1)})$ **and** $c(v_i) \neq c(p_{k_2}^{(i-1)})$. Refer to Figs. 5(e) and 5(f). In this case $c(p_{k_3}^{(i-1)}) = c(v_i)$ because $c(p_{k_3}^{(i-1)}) \neq c(p_{k_2}^{(i-1)})$. We map v_i to $p_{k_3}^{(i-1)}$ with a bottom edge. In order to maintain Invariant I2 we need to connect v_i to its blue leaves by using edges in the top page. However, this may not be possible because there could be edges in the top page incident to blue vertices mapped between $p_{k_2}^{(i-1)}$ and $p_{k_3}^{(i-1)}$. Let $Q' = \{p_1', \ldots, p_{n_i^B}'\}$ be the rightmost blue points (possibly used) that are to the left of $p_{k_3}^{(i-1)}$. Informally speaking, the idea is to "move leftward" the book embedding until all points of Q' become top-connectible to $p_{k_3}^{(i-1)}$. As it is proven in the appendix Invariant I5 and I6 allow us to do so without introducing edge crossings. Once the points in Q' are moved, we map v_i to $p_{k_3}^{(i-1)}$ and connect v_i to all the points in Q' with top edges. Finally, we map the leaves in V_i^R to the n_i^R rightmost free red points with bottom edges. $\square$

Figure 6(c) shows a 2-page book embedding constructed according to Theorem 5.

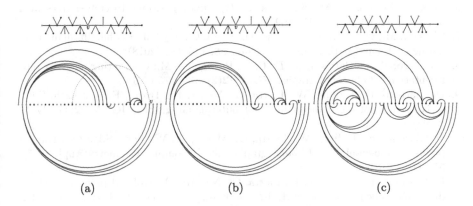

(a)   (b)   (c)

Fig. 6. Illustration for Theorem 5. See the appendix for a larger figure.

6   Open Problems

We conclude by listing three open problems that we find interesting.

OP.1: Are the upper bounds on the curve complexity stated in Theorem 3 and in Theorem 4 tight?

OP.2: Theorem 5 shows that 2-colored caterpillars admit 2-page book embeddings for any compatible 2-colored sequence. Can this be extended to 2-colored trees?

OP.3: Would it be possible to trade curve complexity with edge crossings? For example, does every properly 2-colored 2-tree with n vertices admit a 2-colored point set embedding with $o(n)$ curve complexity and $o(n)$ crossings per edge? A good starting point is the work of Claverol et al. [7].

References

1. Abellanas, M., Garcia-Lopez, J., Hernández-Peñver, G., Noy, M., Ramos, P.A.: Bipartite embeddings of trees in the plane. DAM **93**(2–3), 141–148 (1999)
2. Akiyama, J., Urrutia, J.: Simple alternating path problem. Discrete Math. **84**, 101–103 (1990)
3. Angelini, P., Da Lozzo, G., Di Battista, G., Frati, F., Patrignani, M., Rutter, I.: Intersection-link representations of graphs. JGAA **21**(4), 731–755 (2017)
4. Badent, M., Di Giacomo, E., Liotta, G.: Drawing colored graphs on colored points. Theor. Comput. Sci. **408**(2–3), 129–142 (2008)
5. Bannister, M.J., Devanny, W.E., Dujmović, V., Eppstein, D., Wood, D.R.: Track layouts, layered path decompositions, and leveled planarity. Algorithmica **81**(4), 1561–1583 (2019)
6. Bose, P., McAllister, M., Snoeyink, J.: Optimal algorithms to embed trees in a point set. JGAA **2**(1), 1–15 (1997)
7. Claverol, M., Olaverri, A.G., Garijo, D., Seara, C., Tejel, J.: On hamiltonian alternating cycles and paths. Comput. Geom. **68**, 146–166 (2018)
8. Da Lozzo, G., Di Battista, G., Frati, F., Patrignani, M.: Computing nodetrix representations of clustered graphs. JGAA **22**(2), 139–176 (2018)
9. Di Giacomo, E., Didimo, W., Liotta, G., Meijer, H., Trotta, F., Wismath, S.K.: $k-$colored point-set embeddability of outerplanar graphs. JGAA **12**(1), 29–49 (2008)
10. Di Giacomo, E., Didimo, W., Liotta, G., Meijer, H., Wismath, S.K.: Constrained point-set embeddability of planar graphs. Int. J. Comput. Geometry Appl. **20**(5), 577–600 (2010)
11. Di Giacomo, E., Gąsieniec, L., Liotta, G., Navarra, A.: Colored point-set embeddings of acyclic graphs. In: Frati, F., Ma, K.-L. (eds.) GD 2017. LNCS, vol. 10692, pp. 413–425. Springer, Cham (2018). https://doi.org/10.1007/978-3-319-73915-1_32
12. Di Giacomo, E., Grilli, L., Krug, M., Liotta, G., Rutter, I.: Hamiltonian ortho-geodesic alternating paths. J. Discrete Algorithms **16**, 34–52 (2012)
13. Di Giacomo, E., Liotta, G., Trotta, F.: On embedding a graph on two sets of points. Int. J. Found. Comput. Sci. **17**(5), 1071–1094 (2006)
14. Dujmović, V., Pór, A., Wood, D.R.: Track layouts of graphs. Discret. Math. Theor. Comput. Sci. **6**(2), 497–522 (2004)
15. Felsner, S., Liotta, G., Wismath, S.: Straight-line drawings on restricted integer grids in two and three dimensions. JGAA **7**(4), 363–398 (2003)
16. Ikebe, Y., Perles, M.A., Tamura, A., Tokunaga, S.: The rooted tree embedding problem into points in the plane. Discrete Comput. Geometry **11**(1), 51–63 (1994). https://doi.org/10.1007/BF02573994

17. Kaneko, A., Kano, M., Suzuki, K.: Path coverings of two sets of points in the plane. In: Pach, J. (ed.) Towards a Theory of Geometric Graphs, Volume 342 of Contemporary Mathematics. American Mathematical Society (2004)

18. Kaneko, A., Kano, M., Yoshimoto, K.: Alternating hamilton cycles with minimum number of crossing in the plane. IJCGA **10**, 73–78 (2000)

19. Kano, M.: Discrete geometry on red and blue points on the plane lattice. In: Proceedings of JCCGG 2009, pp. 30–33 (2009)

20. Kaufmann, M.: On point set embeddings for k-planar graphs with few bends per edge. In: Catania, B., Královič, R., Nawrocki, J., Pighizzini, G. (eds.) SOFSEM 2019. LNCS, vol. 11376, pp. 260–271. Springer, Cham (2019). https://doi.org/10.1007/978-3-030-10801-4_21

21. Kaufmann, M., Wiese, R.: Embedding vertices at points: few bends suffice for planar graphs. J. Graph Algorithms Appl. **6**(1), 115–129 (2002)

22. Pach, J., Törőcsik, J.: Layout of rooted tree. In: Trotter, W.T. (ed.) Planar Graphs (DIMACS Series in Discrete Mathematics and Theoretical Computer Science), vol. 9, pp. 131–137. American Mathematical Society (1993)

23. Pach, J., Wenger, R.: Embedding planar graphs at fixed vertex locations. Graph. Combin. **17**(4), 717–728 (2001)

24. Tokunaga, S.: On a straight-line embedding problem graphs. Discrete Math. **150**, 371–378 (1996)

Better Approximation Algorithms for Maximum Weight Internal Spanning Trees in Cubic Graphs and Claw-Free Graphs

Ahmad Biniaz[(✉)]

School of Computer Science, University of Windsor, Windsor, Canada
ahmad.biniaz@gmail.com

Abstract. Given a connected vertex-weighted graph G, the maximum weight internal spanning tree (MaxwIST) problem asks for a spanning tree of G that maximizes the total weight of internal nodes. This problem is NP-hard and APX-hard, with the currently best known approximation factor 1/2 (Chen et al., Algorithmica 2019). For the case of claw-free graphs, Chen et al. present an involved approximation algorithm with approximation factor 7/12. They asked whether it is possible to improve these ratios, in particular for claw-free graphs and cubic graphs.

For cubic graphs we present an algorithm that computes a spanning tree whose total weight of internal vertices is at least $\frac{3}{4} - \frac{3}{n}$ times the total weight of all vertices, where n is the number of vertices of G. This ratio is almost tight for large values of n. For claw-free graphs of degree at least three, we present an algorithm that computes a spanning tree whose total internal weight is at least $\frac{3}{5} - \frac{1}{n}$ times the total vertex weight. The degree constraint is necessary as this ratio may not be achievable if we allow vertices of degree less than three.

With the above ratios, we immediately obtain better approximation algorithms with factors $\frac{3}{4} - \epsilon$ and $\frac{3}{5} - \epsilon$ for the MaxwIST problem in cubic graphs and claw-free graphs having no degree-2 vertices, for any $\epsilon > 0$. The new algorithms are short (compared to that of Chen et al.) and fairly simple as they employ a variant of the depth-first search algorithm. Moreover, they take linear time while previous algorithms for similar problem instances are super-linear.

Keywords: Approximation algorithm · Vertex-weighted tree · Internal spanning tree · Cubic graph · Claw-Free graph · Depth-first search

1 Introduction

The problems of computing spanning trees with enforced properties have been well studied in the fields of algorithms and graph theory. In the last decades,

Supported by NSERC. See [3] for a full version of the paper.

R. Uehara et al. (Eds.): WALCOM 2021, LNCS 12635, pp. 260–271, 2021.
https://doi.org/10.1007/978-3-030-68211-8_21

a number of works have been devoted to the problem of finding spanning trees with few leaves. Besides its own theoretical importance as a generalization of the Hamiltonian path problem, this problem has applications in the design of cost-efficient communication networks [19] and pressure-consistent water supply [4].

The problem of finding a spanning tree of a given graph having a minimum number of leaves (MinLST) is NP-hard—by a simple reduction from the Hamiltonian path problem—and cannot be approximated within a constant factor unless P = NP [16]. From an optimization point of view, the MinLST problem is equivalent to the problem of finding a spanning tree with a maximum number of internal nodes (the MaxIST problem). The MaxIST is NP-hard—again by a reduction form the Hamiltonian path problem—and APX-hard as it does not admit a polynomial-time approximation scheme (PTAS) [15]. Although the MinLST is hard to approximate, there are constant-factor approximations algorithms for the MaxIST problem with successively improved approximation ratios 1/2 [17,19], 4/7 [18] (for graphs of degree at least two), 3/5 [12], 2/3 [13], 3/4 [15], and 13/17 [5]. The parameterized version of the MaxIST problem, with the number of internal vertices as the parameter, has also been extensively studied, see e.g. [4,8,9,14,17].

This paper addresses the weighted version of the MaxIST problem. Let G be a connected vertex-weighted graph where each vertex v of G has a non-negative weight $w(v)$. The *maximum weight internal spanning tree* (MaxwIST) problem asks for a spanning tree T of G such that the total weight of internal vertices of T is maximized.

The MaxwIST problem has received considerable attention in recent years. Let n and m denote the number of vertices and edges of G, respectively, and let Δ be the maximum vertex-degree in G. Salamon [18] designed the first approximation algorithm for this problem. Salamon's algorithm runs in $O(n^4)$ time, has approximation ratio $1/(2\Delta - 3)$, and relies on the technique of locally improving an arbitrary spanning tree of G. By extending the local neighborhood in Salamon's algorithm, Knauer and Spoerhase [12] obtained the first constant-factor approximation algorithm for the problem with ratio $1/(3 + \epsilon)$, for any constant $\epsilon > 0$.

Very recently, Chen et al. [6] present an elegant approximation algorithm with ratio 1/2 for the MaxwIST problem. Their algorithm, which is fairly short and simple, works as follows. Given a vertex-weighted graph G, they first assign to each edge (u, v) of G the weight $w(u) + w(v)$. The main ingredient of the algorithm is the observation that the total internal weight of an optimal solution for the MaxwIST problem is at most the weight of a maximum-weight matching in G. Based on this, they obtain a maximum-weight matching M and then augment it to a spanning tree T in such a way that the heavier end-vertex of every edge of M is an internal node in T. This immediately gives a 1/2 approximate solution for the MaxwIST problem in G. The running time of the algorithm is dominated by the time of computing M which is $O(n \min\{m \log n, n^2\})$.

The focus of the present paper is the MaxwIST problem in cubic graphs (3-regular graphs) and claw-free graphs (graphs not containing $K_{1,3}$ as an induced subgraph). We first review some related works for these graph classes, and then we provide a summary of our contributions.

1.1 Related Works on Cubic Graphs and Claw-Free Graphs

The famous Hamiltonian path problem is a special case of the MaxIST problem that seeks a spanning tree with $n-2$ internal nodes, where n is the total number of vertices. The Hamiltonian path problem is NP-hard in cubic graphs [10] and in line graphs (which are claw-free) [2]. Therefore, the MaxIST problem (and consequently the MaxwIST problem) is NP-hard in both cubic graphs and claw-free graphs.

For the unweighted version, Salamon and Wiener [19] present approximation algorithms with ratios 2/3 and 5/6 for the MaxIST problem in claw-free graphs and cubic graphs, respectively. Although the first ratio has been improved to 13/17 (even for arbitrary graphs) [5], the ratio 5/6 is still the best known for cubic graphs. Binkele-Raible et al. [4] studied the parameterized version of the MaxIST problem in cubic graphs. They design an FPT-algorithm that decides in $O^*(2.1364^k)$ time whether a cubic graph has a spanning tree with at least k internal vertices. The Hamiltonian path problem (which is a special case with $k = n-2$) arises in computer graphics in the context of stripification of triangulated surface models [1,11]. Eppstein studied the problem of counting Hamiltonian cycles and the traveling salesman problem in cubic graphs [7].

For the weighted version, Salamon [18] presented an approximation algorithm with ratio 1/2 for the MaxwIST problem in claw-free graphs (this is obtained by adding more local improvement rules to their general $1/(2\Delta - 3)$-approximation algorithm). In particular, they show that if a claw-free graph has degree at least two, then one can obtain a spanning tree whose total internal weight is at least 1/2 times the total vertex weight. Chen et al. [6] improved this approximation ratio to 7/12; they obtain this ratio by extending their own simple 1/2-approximation algorithm for general graphs. In contrast to their first algorithm which is simple, this new algorithm (restricted to claw-free graphs) is highly involved and relies on twenty-five pages of detailed case analysis.

For the MaxwIST problem in cubic graphs, no ratio better than 1/2 (which holds for general graphs) is known. Chen et al. [6] asked explicitly whether it is possible to obtain better approximation algorithms for the MaxwIST problem in cubic graphs and claw-free graphs.

1.2 Our Contributions

We study the MaxwIST problem in cubic graphs and claw-free graphs. We obtain approximation algorithms with better factors for both graph classes. For cubic graphs we present an algorithm, namely A1, that achieves a tree whose total internal weight is at least $\frac{3}{4} - \frac{3}{n}$ times the total vertex weight. This ratio (with respect to the total vertex weight) is almost tight if n is sufficiently large.

For claw-free graphs of degree at least three we present an algorithm, namely A2, that achieves a tree whose total internal weight is at least $\frac{3}{5} - \frac{1}{n}$ times the total vertex weight. This ratio (with respect to the total vertex weight) may not be achievable if we drop the degree constraint.

With the above ratios, immediately we obtain better approximation algorithms with factors $\frac{3}{4} - \epsilon$ and $\frac{3}{5} - \epsilon$ for the MaxwIST problem in cubic graphs and claw free graphs (without degree-2 vertices). For cubic graphs if $n \leqslant \frac{3}{\epsilon}$ then we find an optimal solution by checking all possible spanning trees, otherwise we run our algorithm A1. This establishes the approximation factor $\frac{3}{4} - \epsilon$. The factor $\frac{3}{5} - \epsilon$ for claw-free graphs is obtained analogously by running A2 instead.

Simplicity. In addition to improving the approximation ratios, the new algorithms (A1 and A2) are relatively short compared to that of Chen et al. [6]. The new algorithms are not complicated either. They involve an appropriate use of the depth-first search (DFS) algorithm that selects a relatively-large-weight vertex in every branching step. The new algorithms take linear time, while previous algorithms for similar problem instances are super-linear.

1.3   Preliminaries

Let G be a connected undirected graph. A *DFS-tree* in G is the rooted spanning tree that is obtained by running the depth-first search (DFS) algorithm on G from an arbitrary vertex called the *root*. It is well-known that the DFS algorithm classifies the edges of an undirected graph into *tree edges* and *backward edges*. Backward edges are the non-tree edges of G. These edges have the following property that we state in an observation.

Observation 1. *The two end-vertices of every non-tree edge of G belong to the same path in the DFS-tree that starts from the root. In other words, one end-vertex is an ancestor of the other.*

Let T be a DFS-tree in G. For every vertex v we denote by $d_T(v)$ the degree of v in T. For every two vertices u and v we denote by $\delta_T(u, v)$ the unique path between u and v in T. For every edge e in G, we refer to the end-vertex of e that is closer to the root of T by the *higher end-vertex*, and refer to other end-vertex of e by the *lower end-vertex*.

If G is a vertex-weighted graph and S is a subset of vertices of G, then we denote the total weight of vertices in S by $w(S)$.

2   The MaxwIST Problem in Cubic Graphs

Let G be a connected vertex-weighted cubic graph with vertex set V such that each vertex $v \in V$ has a non-negative weight $w(v)$. For each vertex v let $N(v)$ be the set containing v and its three neighbors. Let r be a vertex of G with minimum $w(N(r))$. Observe that $w(N(r)) \leqslant 4w(V)/n$, where $n = |V|$. We employ a greedy version of the DFS algorithm that selects—for the next node of the traversal—a node x that maximizes the ratio $\frac{w(x)}{u(x)}$ where $u(x)$ is the number of unvisited neighbors of x. If $u(x) = 0$ then the ratio is $+\infty$. Let T be the tree obtained

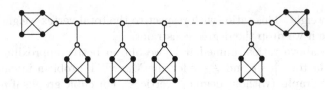

Fig. 1. A cubic graph in which black and white vertices have weights 1 and 0, respectively.

by running this greedy DFS algorithm on G starting from r. Notice that T is rooted at r. Also notice that, during the DFS algorithm, for every vertex x (with $x \neq r$) we have $u(x) \in \{0, 1, 2\}$.

In the rest of this section we show that T is a desired tree, that is, the total weight of internal vertices of T is at least $3/4 - 3/n$ times the total weight of all vertices. This ratio is almost tight for large n. For example consider the cubic graph in Fig. 1 where every black vertex has weight 1 and every white vertex has weight 0. In any spanning tree of this graph at most three-quarter of black vertices can be internal. The following lemma plays an important role in our analysis. In the rest of this section the expression "while processing x" refers to the moment directly after visiting vertex x and before following its children in the DFS algorithm.

Lemma 1. *Let $x_0 x_1'$ be a backward edge with x_0 lower than x_1'. Let*

$$\delta(x_0, x_1') = (x_1' x_1, x_1 x_2', x_2' x_2, \ldots, x_{k-1} x_k', x_k' x_k) \quad \text{with } k \geqslant 1$$

be a path in G such that each $x_i' x_i$ is a tree-edge where x_i is the child of x_i' on the path $\delta_T(x_i', x_0)$, each $x_i x_{i+1}'$ is a backward edge with x_i lower than x_{i+1}', and $u(x_k) = 2$ while processing x_k'. Let $u(x_0)$ be the number of unvisited neighbors of x_0 while processing x_1'. Then, it holds that $u(x_0) \in \{1, 2\}$, and

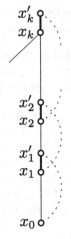

$$w(x_1) + w(x_2) + \cdots + w(x_k) \geqslant \frac{2w(x_0)}{u(x_0)}.$$

Moreover, $d_T(x_i) = d_T(x_i') = 2$ for all $i \in \{1, \ldots, k-1\}$ if $k > 1$, and either x_k' is the root or it has degree 2 in T.

Proof. In the figure to the right, dotted lines represent backward edges, bold-solid lines represent tree-edges, every normal-solid line represents a path in T, and the dashed line represents either a tree edge or a backward edge connecting x_k to a lower vertex.

First we determine potential values of $u(x_0)$. At every step of the DFS algorithm the number of unvisited neighbors of the next node—to be traversed—is at most 2, because of the 3-regularity. Thus, $u(x_0) \leqslant 2$. Since $x_0 x_1'$ is a backward edge, there is a node, say x_0', on the path $\delta_T(x_1', x_0)$ such that $x_0' x_0$ is a tree edge (it might be the case that $x_0' = x_1$). Thus at the moment x_1' was processed, x_0' was an unvisited neighbor of x_0, and thus $u(x_0) \geqslant 1$. Therefore, $u(x_0) \in \{1, 2\}$.

Now we prove the inequality. For every $i \in \{1, \ldots, k\}$ it holds that $\frac{w(x_i)}{u(x_i)} \geqslant \frac{w(x_{i-1})}{u(x_{i-1})}$ while processing x_i', because otherwise the greedy DFS would have selected x_{i-1} instead of x_i. If $k = 1$, then by the statement of the lemma we have $u(x_1) = 2$ (as x_2' is undefined), and thus $\frac{w(x_1)}{2} \geqslant \frac{w(x_0)}{u(x_0)}$ and we are done. Assume that $k \geqslant 2$. For every $i \in \{1, \ldots, k-1\}$ it holds that $u(x_i) = 1$ while processing x_i', because x_i has a visited neighbor x_{i+1}' and an unvisited neighbor which is x_i's child on the path $\delta_T(x_i, x_0)$. For every $i \in \{2, \ldots, k\}$ it holds that $u(x_{i-1}) = 2$ while processing x_i', because x_{i-1} has two unvisited neighbors which are x_{i-1}' and x_{i-1}'s child on the path $\delta_T(x_{i-1}, x_0)$. By the statement of the lemma $u(x_k) = 2$ while processing x_k'. Therefore,

$$w(x_k) \geqslant w(x_{k-1}),$$

$$w(x_1) \geqslant \frac{w(x_0)}{u(x_0)}, \text{ and}$$

$$w(x_i) \geqslant \frac{w(x_{i-1})}{2} \text{ for } i \in \{2, \ldots, k-1\}.$$

The above inequalities imply that

$$w(x_k) \geqslant \frac{1}{2^{k-2}} \cdot \frac{w(x_0)}{u(x_0)}, \text{ and}$$

$$w(x_i) \geqslant \frac{1}{2^{i-1}} \cdot \frac{w(x_0)}{u(x_0)} \text{ for } i \in \{1, \ldots, k-1\}.$$

Therefore,

$$w(x_1) + w(x_2) + \cdots + w(x_k) \geqslant \left(\frac{1}{2^0} + \frac{1}{2^1} + \cdots + \frac{1}{2^{k-3}} + \frac{1}{2^{k-2}} + \frac{1}{2^{k-2}} \right) \cdot \frac{w(x_0)}{u(x_0)}$$

$$= 2 \cdot \frac{w(x_0)}{u(x_0)}.$$

To verify the degree constraint notice that each vertex $x \in \{x_1, \ldots, x_{k-1}, x_1', \ldots, x_{k-1}'\}$ has a child and a parent in T, and also it is incident to a backward edge. Therefore $d_T(x) = 2$. The vertex x_k' has a child in T, and also it is incident to a backward edge. If x_k' has a parent in T then $d_T(x_k') = 2$ otherwise it is the root. □

Let L be the set of nodes of T that do not have any children (the leaves); L does not contain the root. Consider any leaf a in L. Let b_1' and c_1' be the higher end-vertices of the two backward edges that are incident to a. It is implied from Observation 1 that both b_1' and c_1' lie on $\delta_T(r, a)$. Thus we can assume, without loss of generality, that c_1' is an ancestor of b_1'.

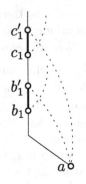

Start from a, follow the backward edge ab_1', then follow the tree edge $b_1'b_1$ where b_1 is the child of b_1' on $\delta_T(b_1', a)$, and then follow non-tree and tree edges alternately and find the path $\delta(a, b_1') = (b_1'b_1, b_1b_2', \ldots, b_k'b_k)$ with $k \geqslant 1$, that satisfies the conditions of the path $\delta(x_0, x_1')$ in Lemma 1 where a plays the role of x_0, b_is play the roles of x_is, and b_i's play the roles of x_i's. (If $u(b_i) < 2$ while processing b_i', then b_i must be the lower endpoint of some backward edge b_ib_{i+1}'.) Observe that such a path exists and it is uniquely defined by the pair (a, b_1') because $u(b_k) = 2$ while processing b_k' and $d_T(b_i) = d_T(b_i') = 2$ for all $i \in \{1, \ldots, k-1\}$ if $k > 1$. While processing b_1' we have $u(a) = 1$. Therefore, by Lemma 1 we get

$$w(b_1) + \cdots + w(b_k) \geqslant 2w(a).$$

Analogously, find the path $\delta(a, c_1') = (c_1'c_1, c_1c_2' \ldots, c_l'c_l)$ with $l \geqslant 1$, by following the backward edge ac_1'. Since $u(a) = 2$ while processing c_1', Lemma 1 implies that

$$w(c_1) + \cdots + w(c_l) \geqslant w(a).$$

Adding these two inequalities, we get

$$w(b_1) + \cdots + w(b_k) + w(c_1) + \cdots + w(c_l) \geqslant 3w(a). \qquad (1)$$

Consider the sets $\{b_1, \ldots, b_k\}$ and $\{c_1, \ldots, c_l\}$ for all leaves in L; notice that there are $2|L|$ sets. All elements of these sets are internal vertices of T as they have a parent and a child. Each such parent, possibly except the root of T, is incident to exactly one backward edge. Thus if the root is incident to at most one backward edge, then these sets do not share any vertex. If the root is incident to two backward edges then it has only one child in the tree which we denote it by r_c. In this case the sets can share (only) r_c. Moreover only two sets can share r_c (because only two backward edges are incident to the root).

Let I be a set that contains all internal nodes of T except the root. Then $V = I \cup L \cup \{r\}$. Based on the above discussion and Inequality (1) we have

$$w(I) \geqslant 3w(L) - w(r_c) = 3\left(w(V) - w(I) - w(r)\right) - w(r_c).$$

By rearranging the terms and using the fact that $w(N(r)) \leqslant 4w(V)/n$ we have

$$4w(I) \geqslant 3w(V) - 3w(r) - w(r_c) \geqslant 3w(V) - 3w(N(r)) \geqslant 3w(V) - 12w(V)/n$$

Dividing both sides by $4w(V)$ gives the desired ratio

$$\frac{w(I)}{w(V)} \geqslant \frac{3}{4} - \frac{3}{n}.$$

Therefore, T is a desired tree. As discussed in Sect. 1.2 we obtain a $\left(\frac{3}{4} - \epsilon\right)$-approximation algorithm for the MaxwIST problem in cubic graphs. Because of the 3-regularity, the number of edges of every n-vertex cubic graph is $O(n)$. Therefore, the greedy DFS algorithm takes $O(n)$ time.

Theorem 1. *There exists a linear-time $\left(\frac{3}{4} - \epsilon\right)$-approximation algorithm for the maximum weight internal spanning tree problem in cubic graphs, for any $\epsilon > 0$.*

A Comparison. The 5/6 approximation algorithm of Salamon and Wiener [19] for unweighted cubic graphs is also based on a greedy DFS. However, there are major differences between their algorithm and ours: (i) The DFS algorithm of [19] selects a vertex with minimum number of unvisited neighbors in every branching step. This criteria does not guarantee a good approximation ratio for the weighted version. Our branching criteria depends on the number of unvisited neighbors and the weight of a node. (ii) The ratio 5/6 is obtained by a counting argument that charges every leaf of the DFS tree to five internal nodes. The counting argument does not work for the weighted version. The weight of a leaf could propagate over many internal nodes, and thus to bound the approximation ratio more powerful ingredients and analysis are required, such as our Lemma 1.

3   The MaxwIST Problem in Claw-Free Graphs

Let G be a connected vertex-weighted claw-free graph with vertex set V such that each vertex $v \in V$ is of degree at least 3 and it has a non-negative weight $w(v)$.

Our algorithm for claw-free graphs is more involved than the simple greedy DFS algorithm for cubic graphs. For cubic graphs we used the DFS-tree directly because we were able to charge the weight of every internal vertex (except the root) to exactly one leaf as every internal vertex is incident to at most one backward edge. However, this is not the case for claw-free graphs—every internal vertex of a DFS-tree can be incident to many backward edges. To overcome this issue, the idea is to first compute a DFS-tree using a different greedy criteria and then modify the tree.

Here we employ a greedy version of the DFS algorithm that selects a maximum-weight vertex in every branching step. Let T be the tree obtained by running this greedy DFS algorithm on G starting from a minimum-weight vertex r. Notice that T is rooted at r, and $w(r) \leqslant w(V)/n$, where $n = |V|$. In the rest of this section we modify T to obtain another spanning tree T' whose total internal weight at least $3/5 - 1/n$ times its total vertex weight. This ratio may not be achievable if we allow vertices of degree less than 3. For example consider the claw-free graph in Fig. 2 where every black vertex has weight 1 and every white vertex has weight 0. In any spanning tree of this graph at most half of black vertices can be internal.

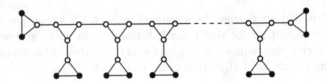

Fig. 2. A claw-free graph in which black and white vertices have weights 1 and 0, respectively.

3.1 Preliminaries: Some Properties of T

The following lemma, though very simple, plays an important role in the design of our algorithm.

Lemma 2. *The tree T is a binary tree, i.e., every node of T has at most two children.*

Proof. If a node $v \in T$ has more than two children, say $v_1, v_2, v_3, \ldots$, then by Observation 1 there are no edges between v_1, v_2, and v_3 in G. Therefore, the subgraph of G that is induced by $\{v, v_1, v_2, v_3\}$ is a $K_{1,3}$. This contradicts the fact that G is claw-free. □

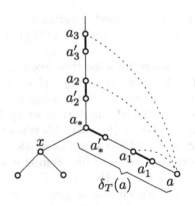

In the following description, "a leaf of T" refers to a node of T that does not have any children, so the root is not a leaf even if it has degree 1. For every leaf a of T, we denote by $a_1, a_2, \ldots, a_k$ the higher end-vertices of the backward edges incident to a while walking up T from a to the root. Since each vertex of G has degree at least three, every leaf a is incident to at least two backward edges, and thus $k \geqslant 2$. For each $i \in \{1, \ldots, k\}$ we denote by a_i' the child of a_i on the path $\delta_T(a_i, a)$; by Observation 1 such a path exists (it might be the case that $a_i = a_{i+1}'$ for some i). Our greedy choice in the DFS algorithm implies that for each i

$$w(a_i') \geqslant w(a).$$

In the figure to the right, dotted lines represent backward edges, bold-solid lines represent tree-edges, and every normal-solid line represents a path in T. By Lemma 2, T is a binary tree and thus its vertices have degrees 1, 2, and 3. For every leaf a of T, we denote by a_* the degree-3 vertex of T that is closest to a, and by a_*' the child of a_* on the path $\delta_T(a_*, a)$; it might be the case that $a_*' = a$. If such a degree-3 vertex does not exist (T is a path), then we set a_* to be the root r. We refer to the path $\delta_T(a_*, a)$ as the *leaf-branch* of a, and denote it by

$\delta_T(a)$ (because this path is uniquely defined by a). A leaf-branch is *short* if it contains only two vertices a and a_*, and it is *long* otherwise. A *deep branching-vertex* is a degree-3 vertex of T that is incident to two leaf-branches. Thus the subtree of every deep branching-vertex has exactly two leaves. In the figure to the right x is a deep branching-vertex, but a_* is not. The following lemma comes in handy for the design of our algorithm.

Lemma 3. *Every internal node (of T) is adjacent to at most two leaves (of T) in G.*

Proof. For the sake of contradiction assume that an internal node v is adjacent to more than two leaves, say $l_1, l_2, l_3, \ldots$ in G. By Observation 1 there are no edges between l_1, l_2, l_3 in G. Therefore, the subgraph of G induced by $\{v, l_1, l_2, l_3\}$ is a $K_{1,3}$. This contradicts the fact that G is claw-free. $\qquad\square$

It is implied by Lemma 3 that every internal node of T is an endpoint of at most two backward edges that are incident to leaves of T.

3.2   Obtaining a Desired Tree from T

We assign to every vertex $v \in V$ a charge equal to the weight of v. Thus every vertex v holds the charge $w(v)$. In a general picture, our algorithm (described in the full version of the paper [3]) distributes the charges of internal nodes between the leaves and at the same time modifies the tree T to obtain another tree T' in which every leaf a (excluding r) has at least $2.5w(a)$ charge. The algorithm does not touch the charge of r. In the end, T' would be our desired tree. Therefore, if L' is the set of leaves of T' (not including r) and I' is the set of internal nodes of T' (again not including r) then we have

$$w(I') \geqslant 2.5w(L') - w(L') = 1.5w(L') = 1.5(w(V) - w(I') - w(r)).$$

By rearranging the terms and using the fact that $w(r) \leqslant w(V)/n$ we get

$$2.5w(I') \geqslant 1.5w(V) - 1.5w(V)/n.$$

Dividing both sides by $2.5w(V)$ gives the desired ratio

$$\frac{w(I')}{w(V)} \geqslant \frac{3}{5} - \frac{3}{5n} > \frac{3}{5} - \frac{1}{n}.$$

The algorithm is described in detail in the full version of the paper [3]; in particular it shows how to obtain T' from T. The running time analysis and inclusion of degree-1 vertices are also described in [3]. The following theorem summarizes our result.

Theorem 2. *There exists a linear-time $\left(\frac{3}{5} - \epsilon\right)$-approximation algorithm for the maximum weight internal spanning tree problem in claw-free graphs having no degree-2 vertices, for any $\epsilon > 0$.*

4 Conclusions

Although the ratio $3/4 - 3/n$ for cubic graphs is almost tight and cannot be improved beyond $3/4$ (with respect to total vertex weight) and the ratio $3/5 - 1/n$ (with respect to total vertex weight) may not be achievable for claw-free graphs of degree less than 3, approximation factors better than $3/4 - \epsilon$ and $3/5 - \epsilon$ might be achievable. A natural open problem is to improve the approximation factors further. It would be interesting to drop the "exclusion of degree-2 vertices" from the $(3/5 - \epsilon)$-approximation algorithm for claw-free graphs. Also, it would be interesting to use our greedy DFS technique to obtain better approximation algorithms for the MaxwIST problem in other graph classes.

References

1. Arkin, E.M., Held, M., Mitchell, J.S.B., Skiena, S.S.: Hamiltonian triangulations for fast rendering. Vis. Comput. **12**(9), 429–444 (1996). https://doi.org/10.1007/BF01782475. Also in ESA 1994
2. Bertossi, A.A.: The edge Hamiltonian path problem is NP-complete. Inf. Process. Lett. **13**(4/5), 157–159 (1981)
3. Biniaz, A.: Better approximation algorithms for maximum weight internal spanning trees in cubic graphs and claw-free graphs. arXiv:2006.12561 (2020)
4. Binkele-Raible, D., Fernau, H., Gaspers, S., Liedloff, M.: Exact and parameterized algorithms for MAX INTERNAL SPANNING TREE. Algorithmica **65**(1), 95–128 (2013). https://doi.org/10.1007/s00453-011-9575-5. Also in WG 2009
5. Chen, Z.-Z., Harada, Y., Guo, F., Wang, L.: An approximation algorithm for maximum internal spanning tree. J. Comb. Optim. **35**(3), 955–979 (2018). https://doi.org/10.1007/s10878-017-0245-7
6. Chen, Z.Z., Lin, G., Wang, L., Chen, Y., Wang, D.: Approximation algorithms for the maximum weight internal spanning tree problem. Algorithmica **81**(11–12), 4167–4199 (2019). https://doi.org/10.1007/s00453-018-00533-w. Also in COCOON 2017
7. Eppstein, D.: The traveling salesman problem for cubic graphs. J. Graph Algorithms Appl. **11**(1), 61–81 (2007). Also in WADS 2003
8. Fomin, F.V., Gaspers, S., Saurabh, S., Thomassé, S.: A linear vertex kernel for maximum internal spanning tree. J. Comput. Syst. Sci. **79**(1), 1–6 (2013). Also in ISAAC 2009
9. Fomin, F.V., Grandoni, F., Lokshtanov, D., Saurabh, S.: Sharp separation and applications to exact and parameterized algorithms. Algorithmica **63**(3), 692–706 (2012). https://doi.org/10.1007/s00453-011-9555-9. Also in LATIN 2010
10. Garey, M.R., Johnson, D.S., Tarjan, R.E.: The planar Hamiltonian circuit problem is NP-complete. SIAM J. Comput. **5**(4), 704–714 (1976)
11. Gopi, M., Eppstein, D.: Single-strip triangulation of manifolds with arbitrary topology. Comput. Graph. Forum **23**(3), 371–380 (2004). Also in SoCG 2004
12. Knauer, M., Spoerhase, J.: Better approximation algorithms for the maximum internal spanning tree problem. Algorithmica **71**(4), 797–811 (2015). https://doi.org/10.1007/s00453-013-9827-7. Also in WADS 2009
13. Li, W., Cao, Y., Chen, J., Wang, J.: Deeper local search for parameterized and approximation algorithms for maximum internal spanning tree. Inf. Comput. **252**, 187–200 (2017). Also in ESA 2014

14. Li, W., Wang, J., Chen, J., Cao, Y.: A $2k$-vertex kernel for maximum internal spanning tree. In: Dehne, F., Sack, J.-R., Stege, U. (eds.) WADS 2015. LNCS, vol. 9214, pp. 495–505. Springer, Cham (2015). https://doi.org/10.1007/978-3-319-21840-3_41
15. Li, X., Zhu, D.: Approximating the maximum internal spanning tree problem via a maximum path-cycle cover. In: Ahn, H.-K., Shin, C.-S. (eds.) ISAAC 2014. LNCS, vol. 8889, pp. 467–478. Springer, Cham (2014). https://doi.org/10.1007/978-3-319-13075-0_37. Also in arXiv:1409.3700 under the title: A 4=3-approximation algorithm for finding a spanning tree to maximize its internal vertices
16. Lu, H.-I., Ravi, R.: The power of local optimization: approximation algorithms for maximum-leaf spanning tree. In: Proceedings of the 13th Annual Allerton Conference on Communication, Control and Computing, pp. 533–542 (1996)
17. Prieto, E., Sloper, C.: Either/or: using VERTEX COVER structure in designing FPT-algorithms—The case of k-INTERNAL SPANNING TREE. In: Dehne, F., Sack, J.-R., Smid, M. (eds.) WADS 2003. LNCS, vol. 2748, pp. 474–483. Springer, Heidelberg (2003). https://doi.org/10.1007/978-3-540-45078-8_41
18. Salamon, G.: Approximating the maximum internal spanning tree problem. Theoret. Comput. Sci. **410**(50), 5273–5284 (2009). Also in MFCS 2007
19. Salamon, G., Wiener, G.: On finding spanning trees with few leaves. Inf. Process. Lett. **105**(5), 164–169 (2008)

APX-Hardness and Approximation for the k-Burning Number Problem

Debajyoti Mondal[1], N. Parthiban[2], V. Kavitha[2], and Indra Rajasingh[3]([✉])

[1] Department of Computer Science, University of Saskatchewan, Saskatoon, Canada
dmondal@cs.usask.ca
[2] Department of Computer Science and Engineering,
SRM Institute of Science and Technology, Chennai, India
parthiban24589@gmail.com, kavitha.psk@gmail.com
[3] School of Advanced Sciences, Vellore Institute of Technology,
Chennai, India
indrarajasingh@yahoo.com

Abstract. Consider an information diffusion process on a graph G that starts with $k > 0$ burnt vertices, and at each subsequent step, burns the neighbors of the currently burnt vertices, as well as k other unburnt vertices. The k-*burning number* of G is the minimum number of steps $b_k(G)$ such that all the vertices can be burned within $b_k(G)$ steps. Note that the last step may have smaller than k unburnt vertices available, where all of them are burned. The 1-burning number coincides with the well-known burning number problem, which was proposed to model the spread of social contagion. The generalization to k-burning number allows us to examine different worst-case contagion scenarios by varying the spread factor k.

In this paper we prove that computing k-burning number is APX-hard, for any fixed constant k. We then give an $O((n + m) \log n)$-time 3-approximation algorithm for computing k-burning number, for any $k \geq 1$, where n and m are the number of vertices and edges, respectively. Finally, we show that even if the burning sources are given as an input, computing a burning sequence itself is an NP-hard problem.

Keywords: Network analysis · Burning number · APX-hard · Approximation · k-Burning

1 Introduction

We consider an information diffusion process that models a social contagion over time from a theoretical point of view. At each step, the contagion propagates from the infected people to their neighbors, as well as a few other people in the network become infected. The burning process, proposed by Bonato et al. [6,7],

The work of D. Mondal is partially supported by NSERC, and by two CFREF grants coordinated by GIFS and GIWS.

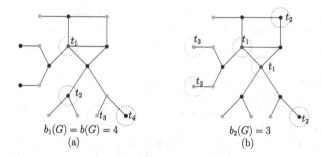

$b_1(G) = b(G) = 4$
(a)

$b_2(G) = 3$
(b)

Fig. 1. The process of burning a graph G. The unburnt vertices that have been chosen to burn at round i (except for the neighbors of the previously burned vertices) are labelled with t_i. (a) A 1-burning with 4 rounds, which is also the minimum possible number of rounds to burn all the vertices with 1-burning, i.e., $b(G) = 4$. (b) A 2-burning with 3 rounds, which is the minimum possible, i.e., $b_2(G) = 3$.

provides a simple model for such a social contagion process. Specifically, the *burning number* $b(G)$ of a graph G is the minimum number of discrete time steps or rounds required to burn all the vertices in the graph based on the following rule. One vertex is burned in the first round. In each subsequent round t, the neighbors of the existing burnt vertices and a new unburnt vertex (if available) are burned. If a vertex is burned, then it remains burnt in all the subsequent rounds. Figure 1(a) illustrates an example of the burning process. The vertices that are chosen to burn directly at each step, form the *burning sequence*.

In this paper we examine *k-burning number* for a graph, which generalizes the burning number by allowing to directly burn k unburnt vertices at each round; see Fig. 1(b). Throughout the paper, we use the notation $b_k(G)$ to denote the k-burning number of a graph G. Note that in the case when $k = 1$, the 1-burning number $b_1(G)$ coincides with the original burning number $b(G)$. The burning process can be used to model a variety of applications, e.g, the selection of the vertices in social networks (e.g., LinkedIn or Facebook) to fast spread information to the target audience with a pipeline of steady new recruits. It may also be used in predictive models to examine the worst-case spread of disease among humans or crop fields. The generalization of a burning process to k-burning allows us to use k as a model parameter, i.e., one can choose a cost-effective value for k to increase the probability of reaching the target audience.

Related Work: The problem of computing the burning number of a graph is NP-complete, even for simple graph classes such as trees with maximum degree three, and for forests of paths [4]. A rich body of research examines upper and lower bounds for the burning number for various classes of graphs. Bonato et al. [4,6] showed that for every connected graph G, $b(G) \leq 2\sqrt{n} - 1$, where n is the number of vertices, and conjectured that the upper bound can be improved to $\lceil \sqrt{n} \rceil$. While the conjecture is still open, Land and Lu [17] improved this bound to $\frac{\sqrt{6n}}{2}$. However, the $\lceil\sqrt{n}\rceil$ upper bound holds for spider graphs [9] and for p-caterpillars with at least $2\lceil\sqrt{n}\rceil - 1$ vertices of degree 1 [13].

The burning number problem has received considerable attention in recent years and nearly tight upper and lower bounds have been established for various well-known graph classes including generalized Petersen graphs [22], p-caterpillars [13], graph products [20], dense and tree-like graphs [15] and theta graphs [18]. The NP-hardness of the burning number problem motivated researchers to study the parameterized complexity and approximation algorithms. Kare and Reddy [16] gave a fixed-parameter tractable algorithm to compute burning number parameterized by neighborhood diversity, and showed that for cographs and split graphs the burning number can be computed in polynomial time. Bonato and Kamali [8] showed that the burning number of a graph is approximable within a factor of 3 for general graphs and 2 for trees. They gave a polynomial-time approximation scheme (PTAS) forests of paths, and a polynomial-time algorithm when the number of paths is fixed. They also mentioned that 'it might be possible that a PTAS exists for general graphs'.

A closely related model that relates to the burning process is the firefighter model [12]. In a firefighter problem, a fire breaks out at a vertex, and at each subsequent step, the fire propagates to the undefended neighbors and the firefighter can defend a vertex from burning. The burnt and defended vertices remain so in the next steps. The problems seek to maximize the number of defended vertices. This problem does not have a constant factor approximation [2], which indicates that it is very different than the burning number problem. A variant of firefighter problem where $b \geq 2$ vertices can be defended at each step has been shown not to be approximable within a constant factor [3]. There are many information diffusion models and broadcast scheduling methods in the literature [5,19,23], but the k-burning process seems to differ in the situation that at each step it allows k new sources to appear anywhere in the graph, i.e., some new burn locations may not be in close proximity of the currently burnt vertices.

Our Contribution. In this paper, we generalize the concept of burning number of a graph to k-burning number. We first prove that computing burning number is APX-hard, settling the complexity question posed by Bonato and Kamali [8]. We then show that the hardness result holds for k-burning number, for any fixed k. We prove that k-burning number is 3-approximable in polynomial time, for any $k \geq 1$, where a 3-approximation algorithm was known previously for the case when $k = 1$ [8]. Finally, we show that even if the burning sources are given as an input, computing a burning sequence itself is an NP-hard problem.

2   Preliminaries

In this section we introduce some notation and terminology. Given a graph G, the k-burning process on G is a discrete-time process defined as follows: Initially, at time $t = 0$, all the vertices are unburnt. At each time step $t \geq 1$, the neighbors of the previously burnt vertices are burnt. In addition, k new unburnt vertices are burned directly, which are called the *burning sources* appearing at the tth step. If the number of available unburnt vertices is less than k, then all of them are burned. The burnt vertices remain in that state in the subsequent steps. The process ends when all vertices of G are burned. The k-burning number of a graph G, denoted by $b_k(G)$, is the minimum number of rounds needed for the process to end. For $k = 1$, we omit the subscript k and use the notation $b(G)$.

The burning sources are chosen at every successive round form an ordered list, which is referred to as a *k-burning sequence*. A burning sequence corresponding to the minimum number of steps $b_k(G)$ is referred to as a *minimum burning sequence*. We use the notation $L(G, k)$ to denote the length of a minimum k-burning sequence.

Let $G = (V, E)$ be a graph with n vertices and m edges. A vertex cover is a set $S \subseteq V$ such that at least one end-vertex of each edge belongs to S. A *dominating set* of G is a set $D \subseteq V$ such that every vertex in G is either in D or adjacent to a vertex in D. An *independent set* of G is a set of vertices such that no two vertices are adjacent in G. A *minimum vertex cover* (resp., minimum independent and dominating set) is a vertex cover (resp., independent and dominating set) with the minimum cardinality. An independent set Q is called *maximal* if one cannot obtain a larger independent set by adding more vertices to Q, i.e., every vertex in $V \setminus Q$ is adjacent to a vertex in Q.

3 APX-Hardness

In this section we show that computing burning number is an APX-hard problem, which settles the complexity question posed by Bonato and Kamali [8]. We then show that the k-burning number problem is APX-hard for any $k \in O(1)$.

3.1 APX-Hardness for Burning Number

A graph is called *cubic* if all its vertices are of degree three. We will reduce the minimum vertex cover problem in cubic graphs, which is known to be APX-hard [1]. Given an instance $G = (V, E)$ of the minimum vertex cover, we construct a graph G' of the burning number problem. We then show that a polynomial-time approximation scheme (PTAS) for the burning number in $G' = (V', E')$ implies a PTAS for the minimum vertex cover problem, which contradicts that the minimum vertex cover problem is APX-hard.

Construction of G'. The graph $G' = (V', E')$ will contain vertices that correspond to the vertices and edges of G. Figures 2(a)–(c) illustrate an example for the construction of G' from G. To keep the illustration simple, we used a maximum degree three graph instead of a cubic graph.

To construct V', we first make a set S by taking a copy of the vertex set V. We refer to S as the *v-vertices* of G'. For every edge $(u, v) \in E$, we include two vertices uv and vu in V', which we refer to as the *e-vertices* of G'. In addition, we add $(2n + 3)$ isolated vertices in V', where $n = |V|$. For every edge $(u, v) \in E$, we add three edges in E': $(u, uv), (v, vu)$ and (uv, vu). Figure 2(b) illustrates the resulting graph. We then divide the edge (uv, vu) with $2n$ division

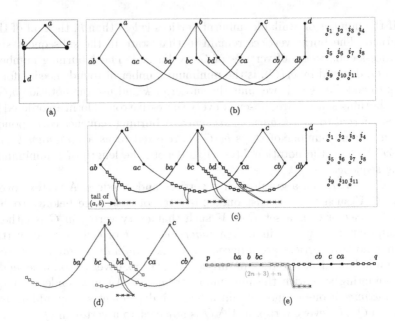

Fig. 2. Illustration for the construction of G' from G. To keep the illustration simple, here we use a maximum degree three graph instead of a cubic graph. (a) G, (b) construction idea, and (c) G', where v- and e-vertices are shown in black disks, d-vertices are shown in squares, tail vertices are shown in cross, and isolated vertices are shown in unfilled circles. (d) H_{bc}. (e) Burning H_{bc} with burning sources p and q, which are outside of H_{bc}.

vertices. We refer to these division vertices as the *d-vertices* of G'. We also add an n-vertex path and connect one end with the median division vertices of the path $u, uv, \ldots, vu, v$. We refer to this n-vertex path as the tail of edge (u, v). Figure 2(c) illustrates the resulting graph.

This completes the construction of G'. Note that the number of vertices and edges in G' is $O(n)$, and it is straightforward to compute G' in $O(n)$ time.

Reduction. In the following, we show how to compute a burning sequence in G' from a vertex cover in G, and vice versa.

Lemma 1. *If G has a vertex cover of size at most q, then G' has a burning sequence of length at most $(q + 2n + 3)$, and vice versa.*

Proof. We will use the idea of a neighborhood of a vertex. By a *r-hop neighborhood* of a vertex u in G', we denote the vertices that are connected to u by a path of at most r edges.

Vertex Cover to Burning Sequence: Let $\mathcal{C}$ be a vertex cover of G of size at most q. In G', we create a burning sequence S by choosing the v-vertices of $\mathcal{C}$ as the burning sources (in any order), followed by the burning of the $(2n + 3)$

isolated vertices. Note that we need at most q rounds to burn the v-vertices in G' that correspond to the nodes in $\mathcal{C}$, and in the subsequent $(2n+3)$ rounds, we can burn the isolated vertices.

We now show that all the vertices are burnt within $(q+2n+3)$ rounds. First observe that after q rounds, all the v-vertices corresponding to $\mathcal{C}$ are burnt. Since $\mathcal{C}$ is a vertex cover, all the v-vertices that do not belong to $\mathcal{C}$ are within $(2n+3)$-hop neighborhood from some vertex in $\mathcal{C}$. Therefore, all v-vertices will be burnt within the next $(2n+3)$ rounds. Similarly, all the e-vertices, d-vertices and tail vertices are within $(2n+3)$-hop neighborhood from some vertex in $\mathcal{C}$, and thus they will be burnt within the next $(2n+3)$ rounds. Since the isolated vertices are chosen as the burning sources for the last $(2n+3)$ rounds, all the vertices of G' will be burnt within $(q+2n+3)$ rounds.

Burning Sequence to Vertex Cover: We now show how to transform a given burning sequence S of length $(q+2n+3)$ into a vertex cover $\mathcal{C}$ of G such that $|\mathcal{C}| \leq q$. Let S be the burning sources of the given burning sequence for G'.

For every edge $(b,c) \in E$, we define H_{bc} to be a subgraph of G' induced by the $(n+1)$-hop neighborhood of b and c, as well as the vertices on the path $b, bc, \ldots, cb, c$, and the vertices of the tail associated to (b,c), e.g., see Fig. 2(d). For every H_{bc} and for each burning source w in it, we check whether w is closer to b than c. If b (resp., c) has a smaller shortest path distance to w, then we include b (resp., c) into $\mathcal{C}$. We break ties arbitrarily.

We now prove that $\mathcal{C}$ is a vertex cover of G. Suppose for a contradiction that there exists an edge $(b,c) \in E$, where neither b nor c belongs to $\mathcal{C}$. Then every burning source s in G' is closer to some v-vertex other than b and c. In other words, H_{bc} is empty of any burning source. Since H_{bc} contains an induced path of $(n+1) + 1 + (2n+2) + 1 + (n+1) = (4n+6)$ vertices and a tail of n vertices, burning all the vertices by placing burning sources outside H_{bc} would take at least $(\frac{4n+6}{2} + n + 1)$ steps, which is strictly larger than $(q+2n+3)$, e.g., see Fig. 2(d). Therefore, by construction of $\mathcal{C}$, at least one of b and c must belong to $\mathcal{C}$.

It now suffices to show that the size of $\mathcal{C}$ is at most q. Since there are $(2n+3)$ isolated vertices in G', they must correspond to $(2n+3)$ burning sources in the burning sequence. The remaining q burning sources are distributed among the graphs H_{uv}. Therefore, $\mathcal{C}$ can have at most q vertices.   □

We now have the following theorem.

Theorem 1. *The burning number problem is APX-hard.*

Proof. Let G be an instance of the vertex cover problem in a cubic graph, and let G' be the corresponding instance of the burning number problem. By Lemma 1, if G has a vertex cover of size at most q, then G' has a burning sequence of length at most $(q+2n+3)$, and vice versa. Let C^* be a minimum vertex cover in G. Then $b(G') \leq |C^*| + 2n + 3$.

Let $\mathcal{A}$ be a $(1+\varepsilon)$-approximation algorithm for computing the burning number, where $\varepsilon > 0$. Then the burning number computed using $\mathcal{A}$ is at most

$(1 + \varepsilon)b(G')$. By Lemma 1, we can use the solution obtained from $\mathcal{A}$ to compute a vertex cover $\mathcal{C}$ of size at most $(1 + \varepsilon)b(G') - 2n - 3$ in G. Therefore, $\frac{|C|}{|C^*|} = \frac{(1+\varepsilon)b(G')-2n-3}{|C^*|} = \frac{b(G')+\varepsilon b(G')-2n-3}{|C^*|} \leq \frac{(|C^*|+2n+3)+\varepsilon b(G')-2n-3}{|C^*|} = 1 + \frac{\varepsilon b(G')}{|C^*|}$.

Note that G' has n v-vertices, $(2n+3)$ isolated vertices, $2|E|$ e-vertices, $n|E|$ tail vertices and $2n|E|$ d-vertices. Since $|E| \leq 3n/2$, the total number of vertices in G' without the isolated vertices is upper bounded by $n + 3n + n^2 + 3n^2 \leq 4n^2 + 4n \leq 5n^2$, for any $n > 4$. Since the burning number of a connected graph with r vertices is bounded by $2\sqrt{r}$ [4], the burning number of G' is upper bounded by $(2n + 3) + 2\sqrt{5n^2} < 8n$, where the term $(2n + 3)$ corresponds to the isolated vertices in G'. In other words, we can always burn the connected component first, and then the isolated vertices. Furthermore, by Brooks' theorem [10], $|C^*| > n/3$.

We thus have $\frac{|C|}{|C^*|} \leq 1 + \frac{\varepsilon b(G')}{|C^*|} \leq 1 + \frac{8n\varepsilon}{|C^*|} \leq 1 + \frac{8n\varepsilon}{n/3} = 1 + 24\varepsilon$, which implies a polynomial-time approximation scheme for the minimum vertex cover problem. Hence the APX-hardness of burning number problem follows from the APX-hardness of minimum vertex cover. □

Hardness for Connected Graphs: Note that in our reduction, G' was disconnected. However, we can prove the hardness even for connected graphs as follows. Let G be the input cubic graph, and let v be a vertex in G. We create another graph H by adding two vertices w and z in a path v, w, z. It is straightforward to see that the size of a minimum vertex cover of H is exactly one plus the minimum vertex cover of G. We now carry out the transformation into a burning number instance G' using H, but instead of using $(2n + 3)$ isolated vertices, we connect them in a path $P = (w, Q, Q', i_1, Q', i_2, Q', \dots, i_{2n+3}, Q')$, where Q is a sequence of $(q + 2n + 2)$ vertices, Q' is a sequence of $(2n + 2)$ vertices, and $i_1, \dots, i_{2n+3}$ are the vertices corresponding to the (previously) isolated vertices. Note that $P \setminus \{u, Q\}$ has $(2n + 2)(2n + 3) + (2n + 3) = (2n + 3)^2$ vertices. Since the burning number of a path of r vertices is $\lceil \sqrt{r} \rceil$ [4], any burning sequence will require $(2n + 3)$ burning sources for $P \setminus \{u, Q\}$.

Note given a vertex cover $\mathcal{C}$ in H of length q, if w is not in $\mathcal{C}$, then $\mathcal{C}$ must contain z. Hence we can replace z by w. Therefore, we can burn all the vertices within $(q+2n+3)$ rounds by burning w first and then the other vertices of $\mathcal{C}$, and then the vertices of $P \setminus \{u, Q\}$ using the known algorithm for burning path [4]. On the other hand, if a burning sequence of length $(q + 2n + 3)$ is provided, then $(2n + 3)$ sources must be used to burn $P \setminus \{u, Q\}$. Since they are at least $(q+2n+3)$ distance apart from the vertices of H, at most q burning sources are distributed in H, implying a vertex cover of size q. We thus have the following corollary.

Corollary 1. *The burning number problem is APX-hard, even for connected graphs.*

The generalization of the APX-hardness proof for k-burning number is included in the full version [21].

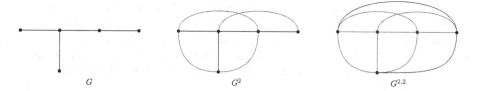

Fig. 3. Illustration for the construction of G^2 and $G^{2,2}$ from G.

4 Approximation Algorithms

Bonato and Kamali [8] gave an $O((n+m)\log n)$-time 3-approximation for burning number We leverage Hochbaum and Shmoys's [14] framework for designing the approximation algorithm and give a generalized algorithm for computing k-burning number. For convenience, we first describe the 3-approximation algorithm in terms of Hochbaum and Shmoys's [14] framework (Fig. 3).

4.1 Approximating Burning Number

Here we show that for connected graphs, the burning number can be approximated within a factor of 3 in $O((n + m)\log n)$ time. Let G^i be the ith power of G, i.e., the graph obtained by taking a copy of G and then connecting every pair of vertices with distance at most i with an edge. We now have the following lemma.

Lemma 2. *Let G be a connected graph and assume that $b(G) = t$. Then G^t must have a dominating set of size at most t.*

Proof. Since $b(G) = t$, all the vertices are burnt within t rounds. Therefore, every vertex in G must have a burning source within its t-hop neighborhood. Consequently, each vertex in G^t, which does not correspond to a burning source in G, must be adjacent to at least one burning source. One can now choose the set of burning sources as the dominating set in G^t.   □

For convenience, we define another notation $G^{i,j}$, which is the jth power of G^i. Although $G^{i,j}$ coincides with G^{i+j}, we explicitly write i,j. Let $M_{i,2}$ be a maximal independent set of $G^{i,2}$. We now have the following lemma, which follows from the observation in [14] that the size of a minimum dominating set in G is at least the size of a maximal independent set in G^2. However, we give a proof for completeness.

Lemma 3. *The size of a minimum dominating set in G^i is at least $|M_{i,2}|$.*

Proof. Let Q be a minimum dominating set in G^i. It suffices to prove that for each vertex v in $(M^{i,2} \setminus Q)$, there is a distinct vertex in $(Q \setminus M_{i,2})$ dominating v (i.e., in this case, adjacent to v).

Let $\{p, q\} \subset (M_{i,2} \setminus Q)$ be two vertices in G^i, which are dominated by a vertex $w \in Q$ in G^i. Since w is adjacent to p, q in $G^{i,2}$ and $M_{i,2}$ is an independent set, we must have $w \in (Q \setminus M_{i,2})$. Since w is adjacent to two both p, q in G^i, p, q will be adjacent in $G^{i,2}$, which contradicts that they belong to the independent set $M_{i,2}$. Therefore, each vertex in $(M^{i,2} \setminus Q)$, must be dominated by a distinct vertex in $(Q \setminus M_{i,2})$.   □

Assume that $b(G) = t$. By Lemma 3, G^t must have a dominating set of size at least $|M_{t,2}|$. By Lemma 2, the size of a minimum dominating set Q in G^t is upper bounded by t. We thus have the condition $|M_{t,2}| \leq |Q| \leq t$.

Corollary 2. *Let G be a graph with burning number t and let $M_{t,2}$ be a maximal independent set in $G^{t,2}$. Then $|M_{t,2}| \leq t$.*

Note that for any other positive integer $k < t$, the condition $|M_{k,2}| \leq k$ is not guaranteed. We use this idea to approximate the burning number. We find the smallest index j, where $1 \leq j \leq n$, that satisfies $|M_{j,2}| \leq j$ and prove that the burning number cannot be less than j.

Lemma 4. *Let j' be a positive integer such that $j' < j$. Then $b(G) \neq j'$.*

Proof. Since j is the smallest index satisfying $|M_{j,2}| \leq j$, for every other $M_{j',2}$, with $j' < j$ we have $|M_{j',2}| \geq j' + 1$. Suppose for a contradiction that $b(G) = j'$, then by Lemma 2, $G^{j'}$ will have a dominating set of size at most j'. But by Lemma 3, $G^{j'}$ has a minimum dominating set of size at least $|M_{j',2}| \geq j' + 1$. □

The following theorem shows how to compute a burning sequence in G of length $3j$. Since j is a lower bound on $b(G)$, this gives us a 3-approximation algorithm for the burning number problem.

Theorem 2. *Given a connected graph G with n vertices and m edges, one can compute a burning sequence of length at most $3b(G)$ in $O((n + m) \log n)$ time.*

Proof. Note that Lemma 4 gives a lower bound for the burning number. We now compute an upper bound. We burn all the vertices of $M_{j,2}$ in any order. Since every maximal independent set is a dominating set, $M_{j,2}$ is a dominating set in $G^{j,2}$. Therefore, after the jth round of burning, every vertex of G can be reached from some burning source by a path of at most $2j$ edges. Thus all the vertices will be burnt in $|M_{j,2}| + 2j \leq 3j$ steps. Since j is a lower bound on $b(G)$, we have $|M_{j,2}| + 2j \leq 3j \leq 3b(G)$.

It now suffices to show that the required j can be computed in $O((n + m) \log n)$ time. Recall that j is the smallest index satisfying $|M_{j,2}| \leq j$. For any $j' > j$, we have $|M_{j',2}| \leq |M_{j,2}| \leq j < j'$. Therefore, we can perform a binary search to find j in $O(\log n)$ steps. At each step of the binary search, we need to compute a maximal independent set $M_{r,2}$ in a graph $G^{r,2} = G^{2r}$, where $1 \leq r \leq n$. To compute $M_{r,2}$, we repeatedly insert an arbitrary vertex w of G into $M_{r,2}$ and then delete w along with its r-hop neighborhood in G following a breadth-first order. Figure 4 illustrates such a process. Since every edge is considered at most once, and the process takes $O(m + n)$ time. Hence the total time is $O((n + m) \log n)$.   □

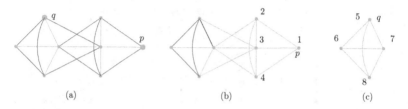

Fig. 4. Illustration for computing $M_{r,2}$, when $r = 1$. (a) $G^{1,2} = G^2$, where the edges of G is shown in black, and a maximal independent set $M_{1,2} = \{p, q\}$. (b)–(c) Computation of $M_{1,2}$, where the numbers represent the order of vertex deletion.

4.2   Approximating k-Burning Number

It is straightforward to generalize Lemma 2 for k-burning number, i.e., if $b_k(G) = t$, then the size of a minimum dominating set Q in G^t is at most kt. By Lemma 6, G^t must have a dominating set of size at least $|M_{t,2}|$. Therefore, we have $|M_{t,2}| \leq |Q| \leq kt$.

Let j be the smallest index such that $|M_{j,2}| \leq kj$. Then for any $j' < j$, we have $|M_{j',2}| > kj'$, i.e., every minimum dominating set in $G^{j'}$ must be of size larger than kj'. We thus have $b_k(G) \neq j'$. Therefore, j is a lower bound on $b_k(G)$.

To compute the upper bound, we first burn the vertices of $M_{j,2}$. Since $|M_{j,2}| \leq kj$, this requires at most j steps. Therefore, after j steps, every vertex has a burning source within its $2j$-hop neighborhood. Hence all the vertices can be burnt within $3j \leq 3b_k(G)$ steps.

Theorem 3. *The k-burning number of a graph can be approximated within a factor of 3 in polynomial time.*

5   Burning Scheduling Is NP-Hard

It is tempting to design heuristic algorithms that start with an arbitrary set of burning sources and then iteratively improve the solution based on some local modification of the set. However, we show that even when a set of k burning sources are given as an input, computing a burning sequence (i.e., burning scheduling) using those sources to burn all the vertices in k rounds is NP-hard.

We reduce the NP-hard problem 3-SAT [11]. Given an instance I of 3-SAT with m clauses and n variables, we design a graph G with $O(n^2 + m)$ vertices and edges, and a set of $2n$ burning sources. We prove that an ordering of the burning sources to burn all the vertices within $2n$ rounds can be used to compute an affirmative solution for the 3-SAT instance I, and vice versa (e.g., see Fig. 5). Due to space constraints, we include the details in the full version [21].

6   Directions for Future Research

A natural open problem is to find an improved approximation algorithm for k-burning number. One can also investigate whether existing approaches to compute burning number for various graph classes can be extended to obtain nearly

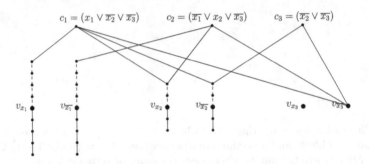

Fig. 5. Illustration for the construction of G, where the given sources are shown in large disks.

tight bounds for their k-burning number. For example, the burning number of an n-vertex path is $\lceil \sqrt{n} \rceil$ [6], which can be generalized to $\lceil \sqrt{n}/k \rceil$ for k-burning, as shown in the full version [21].

It would also be interesting to examine the edge burning number, where a new edge is burned at each step, as well as the neighboring unburnt edges of the currently burnt edges are burned. The goal is to burn all the edges instead of all the vertices. Edge burning number can be different than the burning number, e.g., one can burn the vertices of every wheel graph in two rounds, but the edge burning number can be three. Given a graph, can we efficiently determine whether the burning number is equal to its edge burning number?

Acknowledgement. We thank Payam Khanteimouri, Mohammad Reza Kazemi, and Zahra Rezai Farokh for an insightful discussion that resulted into the addition of tail vertices while constructing G' in the proof of Lemma 1.

References

1. Alimonti, P., Kann, V.: Some APX-completeness results for cubic graphs. Theoret. Comput. Sci. **237**(1), 123–134 (2000)
2. Anshelevich, E., Chakrabarty, D., Hate, A., Swamy, C.: Approximability of the firefighter problem - computing cuts over time. Algorithmica **62**(1–2), 520–536 (2012)
3. Bazgan, C., Chopin, M., Ries, B.: The firefighter problem with more than one firefighter on trees. Discret. Appli. Math. **161**(7), 899–908 (2013)
4. Bessy, S., Bonato, A., Janssen, J.C.M., Rautenbach, D., Roshanbin, E.: Burning a graph is hard. Discret. Appl. Math. **232**, 73–87 (2017)
5. Bonato, A., Gunderson, K., Shaw, A.: Burning the plane: densities of the infinite cartesian grid. Graphs Comb. (2020, to appear)
6. Bonato, A., Janssen, J., Roshanbin, E.: How to burn a graph. Internet Math. **12**(1–2), 85–100 (2016)
7. Bonato, A., Janssen, J., Roshanbin, E.: Burning a graph as a model of social contagion. In: Bonato, A., Graham, F.C., Pralat, P. (eds.) WAW 2014. LNCS, vol. 8882, pp. 13–22. Springer, Cham (2014). https://doi.org/10.1007/978-3-319-13123-8_2

8. Bonato, A., Kamali, S.: Approximation algorithms for graph burning. In: Gopal, T.V., Watada, J. (eds.) TAMC 2019. LNCS, vol. 11436, pp. 74–92. Springer, Cham (2019). https://doi.org/10.1007/978-3-030-14812-6_6

9. Bonato, A., Lidbetter, T.: Bounds on the burning numbers of spiders and path-forests. Theor. Comput. Sci. **794**, 12–19 (2019)

10. Brooks, R.L.: On colouring the nodes of a network. In: Mathematical Proceedings of the Cambridge Philosophical Society, vol. 37, no. 2, pp. 194–197 (1941)

11. Garey, M.R., Johnson, D.S.: Computers and Intractability: A Guide to the Theory of NP-Completeness. W. H. Freeman & Co., London (1979)

12. Hartnell, B.: Firefighter! an application of domination. In: Proceedings of the 20th Conference on Numerical Mathematics and Computing (1995)

13. Hiller, M., Triesch, E., Koster, A.M.C.A.: On the burning number of p-caterpillars. CoRR abs/1912.10897 (2019)

14. Hochbaum, D.S., Shmoys, D.B.: A unified approach to approximation algorithms for bottleneck problems. J. ACM **33**(3), 533–550 (1986)

15. Kamali, S., Miller, A., Zhang, K.: Burning two worlds: algorithms for burning dense and tree-like graphs. CoRR abs/1909.00530 (2019)

16. Kare, A.S., Vinod Reddy, I.: Parameterized algorithms for graph burning problem. In: Colbourn, C.J., Grossi, R., Pisanti, N. (eds.) IWOCA 2019. LNCS, vol. 11638, pp. 304–314. Springer, Cham (2019). https://doi.org/10.1007/978-3-030-25005-8_25

17. Land, M.R., Lu, L.: An upper bound on the burning number of graphs. In: Bonato, A., Graham, F.C., Pralat, P. (eds.) WAW 2016. LNCS, vol. 10088, pp. 1–8. Springer, Cham (2016). https://doi.org/10.1007/978-3-319-49787-7_1

18. Liu, H., Zhang, R., Hu, X.: Burning number of theta graphs. Appl. Math. Comput. **361**, 246–257 (2019)

19. Middendorf, M.: Minimum broadcast time is NP-complete for 3-regular planar graphs and deadline 2. Inf. Process. Lett. **46**(6), 281–287 (1993)

20. Mitsche, D., Pralat, P., Roshanbin, E.: Burning number of graph products. Theor. Comput. Sci. **746**, 124–135 (2018)

21. Mondal, D., Parthiban, N., Kavitha, V., Rajasingh, I.: APX-hardness and approximation for the k-burning number problem. CoRR abs/2006.14733 (2020). https://arxiv.org/abs/2006.14733

22. Sim, K., Tan, T.S., Wong, K.: On the burning number of generalized Petersen graphs. Bull. Malays. Math. Sci. Soc. **41**, 1657–1670 (2017). https://doi.org/10.1007/s40840-017-0585-6

23. Singh, S.S., Singh, K., Kumar, A., Shakya, H.K., Biswas, B.: A survey on information diffusion models in social networks. In: Luhach, A.K., Singh, D., Hsiung, P.-A., Hawari, K.B.G., Lingras, P., Singh, P.K. (eds.) ICAICR 2018. CCIS, vol. 956, pp. 426–439. Springer, Singapore (2019). https://doi.org/10.1007/978-981-13-3143-5_35

Efficient Enumeration of Non-isomorphic Distance-Hereditary Graphs and Ptolemaic Graphs

Kazuaki Yamazaki[✉], Mengze Qian, and Ryuhei Uehara

School of Information Science, Japan Advanced Institute
of Science and Technology (JAIST), Nomi, Japan
{torus711,qianmengze,uehara}@jaist.ac.jp

Abstract. Recently, a general framework for enumerating every non-isomorphic element in a graph class was given. Applying this framework, some graph classes have been enumerated using supercomputers, and their catalogs are provided on the web. Such graph classes include the classes of interval graphs, permutation graphs, and proper interval graphs. Last year, the enumeration algorithm for the class of Ptolemaic graphs that consists of graphs that satisfy Ptolemy inequality for the distance was investigated. They provided a polynomial time delay algorithm, but it is far from implementation. From the viewpoint of graph classes, the class is an intersection of the class of chordal graphs and the class of distance-hereditary graphs. In this paper, using the recent framework for enumerating every non-isomorphic element in a graph class, we give enumeration algorithms for the classes of distance-hereditary graphs and Ptolemaic graphs. For distance-hereditary graphs, its delay per graph is a bit slower than a previously known theoretical enumeration algorithm, however, ours is easy for implementation. In fact, although the previously known theoretical enumeration algorithm has never been implemented, we implemented our algorithm and obtained a catalog of distance-hereditary graphs of vertex numbers up to 14. We then modified the algorithm for distance-hereditary graphs to one for Ptolemaic graphs. Its delay can be the same as one for distance-hereditary graphs, which is much efficient than one proposed last year. We succeeded to enumerate Ptolemaic graphs of vertex numbers up to 15.

Keywords: Distance-hereditary graph · Enumeration · Graph isomorphism · Ptolemaic graph

1 Introduction

Nowadays, we have to process huge amounts of data in the area of data mining, bioinformatics, etc. Finding a common structure from a huge amount of data is an important problem when we deal with such data. On the other hand, since 1980s, there has been many research on the structures in graphs in the area of

© Springer Nature Switzerland AG 2021
R. Uehara et al. (Eds.): WALCOM 2021, LNCS 12635, pp. 284–295, 2021.
https://doi.org/10.1007/978-3-030-68211-8_23

graph theory and graph algorithms [4]. From the viewpoints of both applications and graph theory, it is natural to consider that two graphs are different if and only if they are non-isomorphic. However, in general, it is unknown whether graph isomorphism can be solved efficiently, even on quite restricted graph classes (see, e.g., [15]). On the other hand, even if we restrict ourselves to the graph classes that allow us to solve graph isomorphism efficiently, such graph classes still have certain structures with many applications.

From these backgrounds, efficient enumeration of all non-isomorphic graphs that belong to some specific graph classes have been investigated in literature. There are two early results [12,13]: In these papers, the authors gave efficient enumeration algorithms for proper interval graphs and bipartite permutation graphs. Later, these algorithms had been implemented [6,8], and the lists of these graphs are published on a website [16]. For example, proper interval graphs are enumerated up to $n = 23$ so far, where n is the number of vertices. Since these graph classes have simple structures, the file sizes are the bottleneck to publish on the website. They are efficient since they are designed for specific graph classes based on their own geometric structures.

On the other hand, a general framework that gives us to enumerate all non-isomorphic graphs in some graph classes is investigated and proposed in [18]. In order to avoid duplicates, the graph isomorphism can be solved efficiently for the class. In this context, enumeration algorithms for interval graphs and permutation graphs have been developed, and the data sets of these graph classes are on public on the above website. The class of permutation graphs is enumerated up to $n = 8$, and the class of interval graphs is enumerated up to $n = 12$ on the website. Recently, the result has been updated to $n = 15$ for interval graphs [11]. For these graph classes, since their structures are more complicated than proper interval graphs and bipartite permutation graphs, the computation time is the bottleneck to enumerate.

In this context, an enumeration algorithm for Ptolemaic graphs was investigated last year [14]. To solve graph isomorphism efficiently, they used a tree structure called CL-tree proposed in [17]. The CL-tree represents laminar structure on cliques of a Ptolemaic graph. They succeeded to propose an enumeration algorithm for non-isomorphic Ptolemaic graphs, however, they only give a polynomial bound for each delay between two Ptolemaic graphs since their algorithm is quite complicated, and hence the enumeration algorithm has not yet implemented. From the viewpoint of graph theory, the class of Ptolemaic graphs is the intersection of classes of chordal graphs and distance-hereditary graphs. We note that the class of chordal graph is well investigated (see, e.g., [3,4]), however, the graph isomorphism problem is intractable unless the graph isomorphism problem can be solved efficiently in general (see, e.g., [15] for further details). On the other hand, the graph isomorphism can be solved efficiently for distance-hereditary graphs [9].

In this paper, we first focus on enumeration of distance-hereditary graphs. In [9], the authors proposed a tree structure DH-tree to solve graph isomorphism efficiently. Using the DH-tree, they also claimed that they can enumerate the class of distance-hereditary graphs in polynomial time delay; more precisely, they claimed that they can enumerate distance-hereditary graphs in $O(n)$ time delay up to graphs of vertices n in [9, Theorem 32]. However, it is not accurate. They actually enumerate all DH-trees for the class of distance-hereditary graphs with $O(n)$ delay for each tree. From a DH-tree, we should use $O(n + m)$ time to generate each corresponding distance-hereditary graphs, where n and m are the numbers of vertices and edges, respectively. Therefore, when their algorithm is used to enumerate all distance-hereditary graphs themselves, its delay should be estimated as $O(n^2)$. In fact, as far as the authors know, this enumeration algorithm has never been implemented. We first put this DH-tree structure to the general framework proposed in [18]. We then obtain a simple enumeration algorithm for distance-hereditary graphs up to n vertices which runs with $O(n^3)$ delay. Although the running time is not as good as one in [9], we can implement it and obtain a catalog, which is available on the web [16], of non-isomorphic distance-hereditary graphs up to $n = 15$. (On the web, we put the catalog up to $n = 14$ since the data file for $n = 15$ is too huge.)

Next we turn to the enumeration of Ptolemaic graphs. The class of Ptolemaic graphs is the intersection of the classes of chordal graphs and distance-hereditary graphs. Using the properties of chordal graphs, we can obtain a characterization of Ptolemaic graphs similar to DH-tree mentioned in [2]. Precisely, when we introduce additional constraint into generation rules for DH-tree, we can obtain a canonical tree structure for Ptolemaic graphs. Therefore, modifying the enumeration algorithm for distance-hereditary graphs, we can obtain an enumeration algorithm for Ptolemaic graphs. If we modify the enumeration algorithm for distance-hereditary graphs to one for Ptolemaic graphs straightforwardly, we need $O(n)$ extra cost (or extra delay) for checking the constraint. However, using the technique for amortized complexity, the extra cost can be reduced. Thus we obtain an enumeration algorithm for Ptolemaic graphs and its delay is $O(n^3)$, which is (theoretically) the same as distance-hereditary graphs. We have modified the enumeration algorithm for distance-hereditary graphs to enumerate Ptolemaic graphs, and obtained a catalog of non-isomorphic Ptolemaic graphs up to $n = 15$, which is also available on the web [16].

2   Preliminaries

We consider only simple graphs $G = (V, E)$ with no self-loop and multi edges. We assume $V = \{v_0, v_1, \ldots, v_{n-1}\}$ for some n and $|E| = m$. Let K_n denote the complete graph of n vertices and P_n denote the path of n vertices of length $n - 1$. We define a *graph isomorphism* between two graphs $G_0 = (V_0, E_0)$ and $G_1 = (V_1, E_1)$ as follows. The graph G_0 is isomorphic to G_1 if and only if there is a one-to-one mapping $\phi : V_0 \to V_1$ such that for any pair of vertices $u, v \in V_0$, $\{u, v\} \in E_0$ if and only if $\{\phi(u), \phi(v)\} \in E_1$. We denote by $G_0 \sim G_1$ for two isomorphic graphs G_0 and G_1.

The *neighborhood* of a vertex v in a graph $G = (V, E)$ is the set $N_G(v) = \{u \in V \mid \{u, v\} \in E\}$. The *degree* of a vertex v is $|N_G(v)|$ which is denoted by $\deg_G(v)$.

For a subset U of V, we denote by $N_G(U)$ the set $\{v \in V \mid v \in N(u)$ for some $u \in U\}$. If no confusion can arise, we will omit the index G. We denote the closed neighborhood $N(v) \cup \{v\}$ by $N[v]$. Given a graph $G = (V, E)$ and a subset U of V, the *induced subgraph* by U, denoted by $G[U]$, is the graph (U, E'), where $E' = \{\{u, v\} \mid u, v \in U$ and $\{u, v\} \in E\}$. Given a graph $G = (V, E)$, its *complement* is defined by $\bar{E} = \{\{u, v\} \mid \{u, v\} \notin E\}$, which is denoted by $\bar{G} = (V, \bar{E})$. A vertex set I is an *independent set* if $G[I]$ contains no edges, and then the graph $\bar{G}[I]$ is said to be a *clique*. Two vertices u and v are said to be a pair of *twins* if $N(u) \setminus \{v\} = N(v) \setminus \{u\}$. For a pair of twins u and v, we say that they are *strong twins* if $\{u, v\} \in E$, and *weak twins* if $\{u, v\} \notin E$.

Given a graph $G = (V, E)$, a sequence of the distinct vertices $v_0, v_1, \ldots, v_l$ is a *path*, denoted by $(v_0, v_1, \ldots, v_l)$, if $\{v_j, v_{j+1}\} \in E$ for each $0 \le j < l$. The *length* of a path is the number l of edges on the path. For two vertices u and v, the *distance* of the vertices, denoted by $d(u, v)$, is the minimum length of the paths joining u and v. A *cycle* is a path beginning and ending at the same vertex. An edge which joins two vertices of a cycle but it is not an edge of the cycle is a *chord* of the cycle. A graph is *chordal* if each cycle of length at least 4 has a chord. Given a graph $G = (V, E)$, a vertex $v \in V$ is *simplicial* in G if $G[N(v)]$ is a clique in G.

Given a graph $G = (V, E)$ and a subset U of V, an induced connected subgraph $G[U]$ is *isometric* if the distances of pairs of vertices in $G[U]$ are the same as in G. A graph G is *distance-hereditary* if G is connected and every induced path in G is isometric. In other words, a connected graph G is distance-hereditary if and only if all induced paths are shortest paths. A connected graph G is *Ptolemaic* if for any four vertices u, v, w, x of G, $d(u, v)d(w, x) \le d(u, w)d(v, x) + d(u, x)d(v, w)$.

The following characterization of Ptolemaic graphs is due to Howorka [7]:

Theorem 1. *The following conditions are equivalent: (1) G is Ptolemaic; (2) G is distance-hereditary and chordal; (3) for all distinct non-disjoint maximal cliques P, Q of G, $P \cap Q$ separates $P \setminus Q$ and $Q \setminus P$.*

2.1 Generation Rules and Tree Structures

In this section, we give generation rules for distance-hereditary graphs and Ptolemaic graphs, and tree structures based on them. We first introduce two basic operations for generation of a graph. For a vertex v in a graph $G = (V, E)$, we *add pendant* when we add a new vertex u into V with an edge $\{u, v\}$ into E. (The vertex u of degree 1 is called *pendant*, and the vertex v is called *neck* of the pendant.) For a vertex v in a graph $G = (V, E)$, we *split v into weak (and strong) twins* when we add a new vertex u into V so that $N(u) = N(v)$ (and $N[u] = N[v]$, respectively). For the classes of distance-hereditary graphs and Ptolemaic graphs, the following characterizations are known:

Theorem 2 ([2], **Theorem 1**). *A vertex is a distance-hereditary graph. Let $G = (V, E)$ be a distance-hereditary graph, and v be any vertex in G. Then a* *graph obtained by either (1) adding a pendant $u \notin V$ to v, (2) splitting v into weak twins, or (3) splitting v into strong twins is a distance-hereditary graph.*

We here note that in [14, Corollary 1], the authors used (1) and (3) in Theorem 2 as a characterization of the class of Ptolemaic graphs. However, it not correct; a simple counter example is shown in [2, Figure 11], which is a Ptolemaic graph that cannot be generated by applying only (1) and (3). The correct characterization is given as follows:

Theorem 3 ([2],**Corollary 6**). *A vertex is a Ptolemaic graph. Let $G = (V, E)$ be a Ptolemaic graph, and v be any vertex in G. Then a graph obtained by either (1) adding a pendant $u \notin V$ to v, (2') splitting v into weak twins if v is simplicial in G, or (3) splitting v into strong twins is a Ptolemaic graph.*

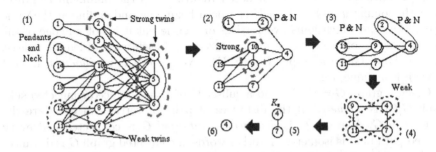

Fig. 1. A distance-hereditary graph G and its contracting/pruning process.

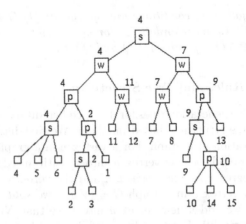

Fig. 2. The DH-tree $T(G)$ derived from the distance-hereditary graph G in Fig. 1.

It is easy to see that K_1 and K_2 are Ptolemaic graphs (and hence distance-hereditary graphs). Hereafter, to simplify the notations and arguments, we assume that every graph has at least three vertices. For a given distance-hereditary graphs $G = (V, E)$, we define three families of vertex sets as follows:

$$\mathcal{S} = \{S \mid x, y \in S \text{ if } N[x] = N[y] \text{ and } |S| \geq 2\},$$
$$\mathcal{W} = \{W \mid x, y \in W \text{ if } N(x) = N(y), |S| \geq 2, \text{ and } |N(x) = N(y)| > 1|\},$$
$$\mathcal{P} = \{P \mid x, y \in W \text{ if } x \text{ is a pendant and } y \text{ is its neck}\}.$$

We note that a neck can have two or more pendants. In [9, Lemma 6], they show that these families are disjoint, and we can obtain the reverse process of the generation process of a distance-hereditary graph (Fig. 1) based on the family and Theorem 2. Based on this property, they introduced the DH-tree $\mathcal{T}(G)$ of a distance-hereditary graph $G = (V, E)$ (Fig. 2): (0) the set of leaves of $\mathcal{T}(G)$ has one-to-one mapping to V, (1) each inner node[1] has its label from s, w, and p, corresponding to "strong twins", "weak twins", and "pendants & neck". Children of an inner node with label s (and w) are generated as strong (and weak, resp.) twins, while children of an inner node with label p are generated as pendants such that the leftmost child corresponds to the neck and the other children are pendants. We here note that the nodes with labels s or w are unordered, however, the nodes with label p should be partially ordered since the leftmost child indicates the neck of the other children which correspond to pendants. Some other properties of a DH-tree $\mathcal{T}(G)$ of a distance-hereditary graph G can be found in [9, Lemma 10]. Based on the properties, it is proved in [9] that (1) a graph G is a distance-hereditary graph if and only if its normalized DH-tree $\mathcal{T}(G)$ is uniquely constructed, and hence (2) two distance-hereditary graphs G_1 and G_2 are isomorphic if and only if $\mathcal{T}(G_1)$ is isomorphic to $\mathcal{T}(G_2)$ (as two labeled trees with restrictions above). That is, a distance-hereditary graph $G = (V, E)$ can be represented by the corresponding DH-tree $\mathcal{T}(G)$ with $2|V| - 1$ vertices uniquely up to isomorphism.

Lemma 1 ([9], Theorem 32). *DH-trees of distance-hereditary graphs with at most n vertices can be enumerated in $O(n)$ time for each, with $O(n^2)$ space.*

As mentioned in Introduction, the algorithm in [9, Theorem 32] enumerates DH-trees of distance-hereditary graphs with at most n vertices. For a DH-tree, it is not difficult to construct the corresponding distance-hereditary graph $G = (V, E)$ in $O(n + m)$ time, where $n = |V|$ and $m = |E|$. Thus we can obtain the following immediately by Lemma 1:

Theorem 4. *Non-isomorphic distance-hereditary graphs with at most n vertices can be enumerated in $O(n^2)$ time for each.*

As far as the authors know, the enumeration algorithm in Theorem 4 has never been implemented.

For Ptolemaic graphs, Uehara and Uno gave a tree structure named CL-tree, which is based on the laminar structure of intersections of maximal cliques

---

[1] In this paper, the original graph G has its "vertices", while corresponding tree structure $\mathcal{T}(G)$ has its "nodes" to distinguish them.

in a Ptolemaic graph [17]. The CL-tree for a Ptolemaic graph has the same property of DH-tree for a distance-hereditary graph: (1) a graph G is a Ptolemaic graph if and only if its normalized CL-tree is uniquely constructed in linear time, and hence (2) two Ptolemaic graphs G_1 and G_2 are isomorphic if and only if their corresponding CL-trees are isomorphic (as two labeled trees with some restrictions). Based on this characterization, Tran and Uehara showed the following theorem for Ptolemaic graphs [14]:

Theorem 5. *Non-isomorphic Ptolemaic graphs with at most n vertices can be enumerated in polynomial time for each.*

However, their algorithm consists of two phases; it first enumerates the CL-tree structure, and then assigns vertices of the corresponding Ptolemaic graphs. The reason is that each node in the CL-tree corresponds to a set of strong twins in the Ptolemaic graph $G = (V, E)$, and hence there are many partitions of the vertex set V into these subsets. Therefore, although it is polynomial, the exact upper bound of the polynomial in Theorem 5 is not clear in their algorithm, and it is quite difficult to implement it from the practical point of view.

3   Enumeration Algorithms

3.1   Enumeration Algorithm for Distance-Hereditary Graphs

We first give an enumeration algorithm for distance-hereditary graphs:

Lemma 2. *Distance-hereditary graphs with at most n vertices can be enumerated in $O(n^3)$ time for each.*

Comparing to Theorem 4, this algorithm is a bit less efficient from the theoretical point of view. However, the algorithm in Theorem 4 had never been implemented as far as the authors know, while our algorithm in Lemma 2 has been implemented and enumerated up to $n = 15$. In fact, for such a "small" n, we consider the factor n is not an essential matter.

Our algorithm uses a general framework proposed in [18], which is based on *reverse search* invented by Avis and Fukuda [1]. In the context of enumeration of distance-hereditary graphs, we first define *family tree* $\hat{T} = (\hat{V}, \hat{E})$ as follows. Each element in $\hat{V}$ is a DH-tree, and each arc (T_1, T_2) in $\hat{E}$ joins T_1 and T_2 if T_1 is the *parent* of T_2. We define the root of the family tree $\hat{T}$ is K_2, and K_1 and K_2 are output first as a special graph. (Recall that we assume that our distance-hereditary graphs have at least three vertices.) The definition of parent-child relationship between two DH-trees is essential. Once the parent-child relationship is defined, our reverse search algorithm traverses the family tree $\hat{T}$ in the breadth-first search manner. As we will show later, each children DH-tree has one more vertex than the parent DH-tree. Therefore, we can enumerate all non-isomorphic DH-trees corresponding to non-isomorphic distance-hereditary graphs of up to n vertices by traversing all elements of depth $n - 2$ in the family tree from the root K_2 of depth 0. (Here, the *depth* of a family tree is defined in a natural way;

the root has depth 0, and the other node has depth $d + 1$ where d is the depth of its parent node.)

Now we turn to the parent-child relationship of two DH-trees T_1 and T_2. The basic idea is the same as the enumeration algorithm for trees [10]: We first introduce the order for layout of the DH-tree; a tree is drawn in the *right-heavy* manner so that, for each node p, its children are in the right-heavy order. (We note that the leftmost child of the inner node with label p is a special fixed node to represent the neck of the pendants which correspond to the siblings in the DH-tree.) Then, two distance-hereditary graphs G_1 and G_2 are isomorphic if and only if the corresponding DH-trees are not only isomorphic, but also their layouts are the same. Then, for two distance-hereditary graphs $G_1 = (V_1, E_1)$ and $G_2 = (V_2, E_2)$ with $|V_1| = |V_2| - 1$, G_1 is the (unique) parent of G_2 if and only if $T(G_1)$ is obtained from $T(G_2)$ by removing the rightmost leaf of the first node having only leaves found by the DFS on the DH-tree $T(G_2)$.

As shown in [9], each normalized DH-tree of a distance-hereditary graph of n vertices can be represented in the standardized string representation of length $O(n)$. Based on the parent-child relationship and the standard string representation, our enumeration algorithm in Lemma 2 is described as follows (in each line, the last comment indicates the time complexity of the line):

0. Put the DH-tree[2] of the root K_2 into a queue Q; /* $O(1)$ time. */
1. Pick up one DH-tree T from Q; /* $O(n)$ time. */
2. Construct the distance-hereditary graph G represented by T and output G; /* $O(n^2)$ time since G has $O(n^2)$ edges */
3. If G contains less than n vertices, construct the set C, represented by trie, of candidates of children of T through steps 3.1 to 3.4; /* $|C| = O(n)$. */
   3.1. Generate each graph G' obtained by adding pendant u to each vertex v in V; /* $O(n)$ ways with $O(1)$ time. */
   3.2. Generate each graph G' obtained by splitting each vertex v in V to weak twins; /* $O(n)$ ways with $O(n)$ time for each to copy the edges. */
   3.3. Generate each graph G' obtained by splitting each vertex v in V to strong twins; /* $O(n)$ ways with $O(n)$ time for each to copy the edges. */
   3.4. Each graph G' generated in steps 3.1–3.3 is encoded into a canonical string that represents corresponding to the DH-tree $T(G')$ and put into a trie to remove duplicates (up to isomorphism) as follows;
      3.4.1. Generate the DH-tree $T(G')$; /* $O(n^2)$ time as step 2. */
      3.4.2. Normalize $T(G')$ in the right-heavy manner to make it canonical up to isomorphism; /* $O(n)$ time. */
      3.4.3. The canonical string representation of $T(G')$ is put into a trie to remove duplicates; /* $O(n)$ time. */

---

[2] The DH-tree for K_2 is the only exception that it is not well-defined. It can be seen as two strong twins, or one of them is a pendant of the other. From the practical point of view, it is better to initialize the queue Q by two nodes for K_3 and a path (u, v, w) of three vertices.

4. For each candidate string representation of $T(G')$ in C, check if it is a child of T as follows, and put it into Q if the parent of $T(G')$ is T;
   /* The number of iterations is $O(n)$ since $|C| = O(n)$. */
   4.0. Construct the DH-tree $T(G')$ from its string representation;
   /* $O(n)$ time. */
   4.1. Construct the unique parent T' from $T(G')$; /* $O(n)$ time. */
   4.2. If T' is isomorphic to T, put $T(G')$ into the queue since the parent T' of $T(G')$ is T. /* $O(n)$ time. */

Proof. (of Lemma 2) The correctness of the framework follows the arguments in [18]. Therefore, we discuss time complexity. Let $G = (V, E)$ be any distance-hereditary graph with $|V| = n$. A key observation is that we have one-to-one corresponding between the set V of vertices of G and the set of leaves of the DH-tree $T(G)$. Since $T(G)$ has $n - 1$ inner nodes, $T(G)$ can be represented in $O(n)$ space. By arguments in [9], the canonical string representation of length $O(n)$ of $T(G)$ can be constructed in $O(n)$ time and vice versa. A *trie* is a data structure (also known as *prefix tree*) to represent a set. (The details can be found in [9] and omitted here.) Using a trie, we can obtain a set (without duplicates) of sequences of integers from a multiset (with duplicates) of sequences of integers in linear time of the number of total elements in the multiset. Thus, by using trie to represent the set C in step 3, step 3.4.3 (registration of a candidate) can be done in linear time of the size of a candidate, and we can remove the duplicates in C if they exist. Using a queue Q to keep the DH-trees, we can traverse the DH-trees in the breadth-first manner. Thus, in total, distance-hereditary graphs with at most n vertices can be enumerated in $O(n^3)$ time for each.   □

3.2   Enumeration Algorithm for Ptolemaic Graphs

In this section, we give an enumeration algorithm for Ptolemaic graphs:

Theorem 6. *Ptolemaic graphs with at most n vertices can be enumerated in $O(n^3)$ time for each.*

We modify the algorithm shown in the proof of Lemma 2 based on Theorem 3. Naively, we can enumerate Ptolemaic graphs by replacing the step 3.2 by the following step 3.2':

3.2'. Generate each graph G' obtained by splitting each vertex v in V to weak twins if v is simplicial in V;
   /* $O(n)$ ways with $O(n)$ time for each to copy the edges. */

The original step 3.2 has n candidates v in V, and then it adds $\deg(v)$ edges into E for each candidate v. Thus this step contributes $O(n^2)$ time in total running time $O(n^3)$. If we straightforwardly check whether $N(v)$ is a clique in G or not, it takes $O(n^2)$ time. Then the modified algorithm runs in $O(n^4)$ time for each.

In order to reduce the time to check whether a given vertex v in V is simplicial or not, we introduce a function $s : V \to \{true, false\}$ that indicates $s(v) = true$ if v is simplicial and $s(v) = false$ otherwise.

Lemma 3. *For each generation rule (1), (2'), and (3) in Theorem 3, the update of the function $s(v)$ can be done in $O(n)$ time, where $n = |V|$.*

Proof. For notational simplicity, we start from K_2 that consists of two vertices u and v. We initialize $s(u) = s(v) = true$ since they are simplicial. Let $G = (V, E)$ be a Ptolemaic graph with $|V| \geq 2$. We have three cases to consider.

(1) Adding a pendant $u \notin V$ to $v \in V$: In this case, it is easy to see that we have $s(u) = true$ and $s(v) = false$. For the other vertices w in G, $s(w)$ is not changed. Thus this step requires $O(1)$ time to update.

(3) Splitting $v \in V$ into strong twins u and v with $u \notin V$: In this case, we add u into V and make $N[u] = N[v]$, especially, we add an edge $\{u, v\}$ into E. Then it is easy to see that $s(u)$ is equal to $s(v)$. For the other vertex w in $N(v)$, $s(w)$ is not changed in this case. As the same as in (1) above, $s(x)$ is not changed for the other vertices x in V. Thus what we have to do is set $s(u) = s(v)$ in $O(1)$ time.

(2') Splitting $v \in V$ into weak twins u and v with $u \notin V$: In this case, we can observe that $s(u)$ is equal to $s(v)$ as the same as (3). On the other hand, each vertex w in $N(v)$ cannot be simplicial because u and v are not adjacent. For the other vertices x in V, $s(x)$ is not changed. Thus we have to set $s(u) = s(v)$ and $s(w) = false$ for each $w \in N(v)$ in $O(\deg_G(v))$ time.

Therefore, each update of s can be done in $O(n)$ time. □

Using the function s, the step 3.2' can be done in $O(n^2)$ time, which completes the proof of Theorem 6.

3.3 Implementation and Experimental Results

We first implemented the enumeration algorithm for distance-hereditary graphs and modified it to enumerate Ptolemaic graphs. We ran them on a supercomputer (Cray XC-40) in our university. The program is implemented to run in parallel on the supercomputer, and the running time is around 90 min to enumerate distance-hereditary graphs up to $n = 15$ and Ptolemaic graphs up to $n = 15$. Due to disk space, we publish the distance-hereditary graphs up to $n = 14$ and Ptolemaic graphs up to $n = 15$ on the web [16].

We here note that the counting problem is different from the enumeration algorithm in general. To count the number of objects, sometimes it is not necessary to enumerate all objects (see, e.g., [5]). In fact, the numbers of the graphs mentioned above have been investigated and can be found on a famous website *On-Line Encyclopedia of Integer Sequences* (oeis.org). Using these sequences, we can double-check the correctness of the outputs. (See the webpage [16] for further details.)

4   Concluding Remarks

In this paper, we show that we can enumerate non-isomorphic distance-hereditary graphs with at most n vertices in $O(n^3)$ time delay and non-isomorphic Ptolemaic graphs with at most n vertices in $O(n^3)$ time delay. The first one is slower than the previously known theoretical bound, while the second one is a new result. As described in Introduction, both algorithms can be implemented on a real computer. In fact, we have already implemented the algorithms and published their catalogs on the web [16]. When we enumerate some graph classes, it is the issue that if a theoretical algorithm can be implemented in a realistic sense. Moreover, we have to take a balance between (1) running time of the algorithm, and (2) the file size required to store the output. The running time for solving the graph isomorphism problem for the class can be a bottleneck of our framework. Future work can be efficient enumeration (in a practical sense) of some graph classes such that the graph isomorphism is GI-complete for them.

Acknowledgement. This work is partially supported by JSPS KAKENHI Grant Numbers 17H06287 and 18H04091.

References

1. Avis, D., Fukuda, K.: Reverse search for enumeration. Discret. Appl. Math. **65**, 21–46 (1996)
2. Bandelt, H.J., Mulder, H.M.: Distance-hereditary graphs. J. Comb. Theory, Ser. B **41**(2), 182–208 (1986)
3. Brandstädt, A., Le, V.B., Spinrad, J.P.: Graph Classes: A Survey. SIAM, University City (1999)
4. Golumbic, M.C.: Algorithmic graph theory and perfect graphs. In: Annals of Discrete Mathematics, 2nd ed, vol. 57. Elsevier (2004)
5. Hanlon, P.: Counting interval graphs. Trans. Am. Math. Soc. **272**(2), 383–426 (1982)
6. Harasawa, S., Uehara, R.: Efficient enumeration of connected proper interval graphs (in Japanese). IEICE Technical Report COMP2018-44, IEICE, March 2019, pp. 9–16 (2019)
7. Howorka, E.: A characterization of ptolemaic graphs. J. Graph Theory **5**, 323–331 (1981)
8. Ikeda, S.i., Uehara, R.: Implementation of enumeration algorithm for connected bipartite permutation graphs (in Japanese). IEICE Technical Report COMP2018-45, IEICE, March 2019, pp. 17–23 (2019)
9. Nakano, S.I., Uehara, R., Uno, T.: A new approach to graph recognition and applications to distance-hereditary graphs. J. Comput. Sci. Technol. **24**(3), 517–533 (2009). https://doi.org/10.1007/s11390-009-9242-3
10. Nakano, S., Uno, T.: Constant time generation of trees with specified diameter. In: Hromkovič, J., Nagl, M., Westfechtel, B. (eds.) WG 2004. LNCS, vol. 3353, pp. 33–45. Springer, Heidelberg (2004). https://doi.org/10.1007/978-3-540-30559-0_3
11. Mikos, P.: Efficient enumeration of non-isomorphic interval graphs. arXiv:1906.04094 (2019)
12. Saitoh, T., Otachi, Y., Yamanaka, K., Uehara, R.: Random generation and enumeration of bipartite permutation graphs. J. Discret. Algorithms **10**, 84–97 (2012). https://doi.org/10.1016/j.jda.2011.11.001
13. Saitoh, T., Yamanaka, K., Kiyomi, M., Uehara, R.: Random generation and enumeration of proper interval graphs. IEICE Trans. Inf. Syst. **E93–D**(7), 1816–1823 (2010)

14. Tran, D.H., Uehara, R.: Efficient enumeration of non-isomorphic ptolemaic graphs. In: Rahman, M.S., Sadakane, K., Sung, W.-K. (eds.) WALCOM 2020. LNCS, vol. 12049, pp. 296–307. Springer, Cham (2020). https://doi.org/10.1007/978-3-030-39881-1_25

15. Uehara, R., Toda, S., Nagoya, T.: Graph isomorphism completeness for chordal bipartite graphs and strongly chordal graphs. Discret. Appl. Math. **145**(3), 479–482 (2004)

16. Uehara, R.: Graph catalogs (2020). http://www.jaist.ac.jp/~uehara/graphs

17. Uehara, R., Uno, Y.: Laminar structure of ptolemaic graphs with applications. Discret. Appl. Math. **157**(7), 1533–1543 (2009)

18. Yamazaki, K., Saitoh, T., Kiyomi, M., Uehara, R.: Enumeration of nonisomorphic interval graphs and nonisomorphic permutation graphs. Theoret. Comput. Sci. **806**, 323–331 (2020). https://doi.org/10.1016/j.tcs.2019.04.017

Physical Zero-Knowledge Proof
for Ripple Effect

Suthee Ruangwises[✉][iD] and Toshiya Itoh[iD]

Department of Mathematical and Computing Science,
Tokyo Institute of Technology, Tokyo, Japan
ruangwises@gmail.com, titoh@c.titech.ac.jp

Abstract. Ripple Effect is a logic puzzle with an objective to fill numbers into a rectangular grid divided into rooms. Each room must contain consecutive integers starting from 1 to its size. Also, if two cells in the same row or column have the same number x, the space separating the two cells must be at least x cells. In this paper, we propose a physical protocol of zero-knowledge proof for Ripple Effect puzzle using a deck of cards, which allows a prover to physically show that he/she knows a solution without revealing it. In particular, we develop a physical protocol that, given a secret number x and a list of numbers, verifies that x does not appear among the first x numbers in the list without revealing x or any number in the list.

Keywords: Zero-knowledge proof · Card-based cryptography · Ripple effect · Puzzle

1 Introduction

Ripple Effect is a logic puzzle introduced by Nikoli, a Japanese company that developed many famous logic puzzles such as Sudoku, Numberlink, and Kakuro. A Ripple Effect puzzle consists of a rectangular grid of size $m \times n$ divided into polyominoes called *rooms*, with some cells already containing a number (we call these cells *fixed cells* and the other cells *empty cells*). The objective of this puzzle is to fill a number into each empty cell according to the following rules [13].

1. *Room condition*: Each room must contain consecutive integers starting from 1 to its *size* (the number of cells in the room).
2. *Distance condition*: If two cells in the same row or column have the same number x, the space separating the two cells must be at least x cells. See Fig. 1.

Suppose that Patricia, a Ripple Effect expert, created a difficult Ripple Effect puzzle and challenged her friend Victor to solve it. After a while, Victor could not solve her puzzle and began to doubt that the puzzle may have no solution. Patricia needs to convince him that her puzzle actually has a solution without showing it (which would make the challenge pointless). In this situation, Patricia needs a protocol of *zero-knowledge proof*.

© Springer Nature Switzerland AG 2021
R. Uehara et al. (Eds.): WALCOM 2021, LNCS 12635, pp. 296–307, 2021.
https://doi.org/10.1007/978-3-030-68211-8_24

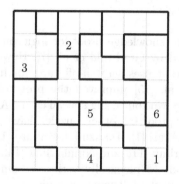 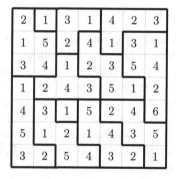

Fig. 1. An example of a Ripple Effect puzzle (left) and its solution (right)

1.1 Zero-Knowledge Proof

A zero-knowledge proof is a protocol of interactive proof between a prover P and a verifier V. Both P and V are given a computational problem x, but only P knows a solution w of x. A protocol of zero-knowledge proof enables P to convince V that he/she knows w without revealing any information of it. The protocol must satisfy the following properties.

1. **Completeness:** If P knows w, then P can convince V with high probability. (In this paper, we consider only the *perfect completeness* property where the probability to convince V is one.)
2. **Soundness:** If P does not know w, then P cannot convince V, except with a small probability called *soundness error*. (In this paper, we consider only the *perfect soundness* property where the soundness error is zero.)
3. **Zero-knowledge:** V learns nothing about w, i.e. there exists a probabilistic polynomial time algorithm S (called a *simulator*), not knowing w but having a black-box access to V, such that the outputs of S and the outputs of the real protocol follow the same probability distribution.

The concept of a zero-knowledge proof was introduced by Goldwasser et al. [6] in 1989. Goldreich et al. [5] later showed that a zero-knowledge proof exists for every NP problem. As Ripple Effect has been proved to be NP-complete [17], one can construct a computational zero-knowledge proof for it. However, such construction is not intuitive or practical as it requires cryptographic primitives.

Instead, we aim to develop a physical protocol of zero-knowledge proof with a deck of playing cards. Card-based protocols have benefit that they do not require computers and use only a small, portable deck of cards that can be found in everyday life. These protocols are also suitable for teaching purpose since they are easy to understand and verify the correctness and security, even for non-experts in cryptography.

1.2 Related Work

Development of card-based protocols of zero-knowledge proof for logic puzzles began with a protocol for Sudoku developed by Gradwohl et al. [7]. However, each of several variants of this protocol either uses scratch-off cards or has a nonzero soundness error. Later, Sasaki et al. [15] improved the protocol for Sudoku to achieve perfect soundness property without using special cards. Apart from Sudoku, protocols of zero-knowledge proof for other puzzles have been developed as well, including Nonogram [3], Akari [1], Takuzu [1], Kakuro [1,12], KenKen [1], Makaro [2], Norinori [4], Slitherlink [11], and Numberlink [14].

Many of these protocols employ methods to physically verify or compute specific number-related functions, as shown in the following examples.

- A subprotocol in [7] verifies that a list is a permutation of all given numbers in some order without revealing their order.
- A subprotocol in [2] verifies that two given numbers are different without revealing their values.
- Another subprotocol in [2] verifies that a number in a list is the largest one in that list without revealing any value in the list.
- A subprotocol in [14] counts the number of elements in a list that are equal to a given secret value without revealing that value, the positions of elements in the list that are equal to it, or the value of any other element in the list.

Observe that these four functions do not use the mathematical meaning of numbers. In these functions, numbers are treated only as symbols distinguished from one another in the sense that we can replace every number x with $f(x)$ for any function $f : \mathbb{Z}^+ \to \mathbb{Z}^+$ and the output will remain the same (for the third example, f has to be an increasing function).[1]

1.3 Our Contribution

In this paper, we propose a physical protocol of zero-knowledge proof with perfect completeness and perfect soundness properties for Ripple Effect puzzle using a deck of cards. More importantly, we extend the set of functions that are known to be physically verifiable. In particular, we develop a protocol that, given a secret number x and a list of numbers, verifies that x does not appear among the first x numbers in the list without revealing x or any number in the list.

Unlike the functions verified by many protocols in previous work, the function our protocol has to verify uses the mathematical meaning of the numbers in the sense of cardinality; it uses the value of x to determine how many elements in the list the condition is imposed on. Therefore, this function is significantly harder to verify *without revealing x*, and thus we consider this result to be an important step in developing protocols of zero-knowledge proof.

---

[1] Actually, some protocols from previous work can verify a function that uses the mathematical meaning of numbers, but still not in the sense of cardinality. For example, a subprotocol in [12] verifies that the sum of all numbers in a list is equal to a given number; for this function, we can still replace every number x with $f(x)$ for any linear function $f : \mathbb{Z}^+ \to \mathbb{Z}^+$.

2   Preliminaries

2.1   Cards

Every card used in our protocol has either ♣ or ♡ on the front side, and has an identical back side.

For $1 \leq x \leq y$, define $E_y(x)$ to be a sequence of consecutive y cards arranged horizontally, with all of them being ♣ except the x-th card from the left being ♡, e.g. $E_3(3)$ is ♣♣♡ and $E_4(2)$ is ♣♡♣♣. We use $E_y(x)$ to encode a number x in a situation where the maximum possible number is at most y. This encoding rule was first considered by Shinagawa et al. [16] in the context of using a regular y-gon card to encode each integer in $\mathbb{Z}/y\mathbb{Z}$. Additionally, we encode 0 by $E_y(0)$, defined to be a sequence of consecutive y cards, all of them being ♣.

Normally, the cards in $E_y(x)$ are arranged horizontally as defined above unless stated otherwise. However, in some situations we may arrange the cards vertically, where the leftmost card will become the topmost card and the rightmost card will become the bottommost card (hence the only ♡ will become the x-th topmost card for $x \geq 1$).

2.2   Matrix

We construct an $a \times b$ *matrix* of face-down cards. Let Row i denote the i-th topmost row, and Column j denote the j-th leftmost column. Let $M(i, j)$ denote the card at Row i and Column j of a matrix M. See Fig. 2.

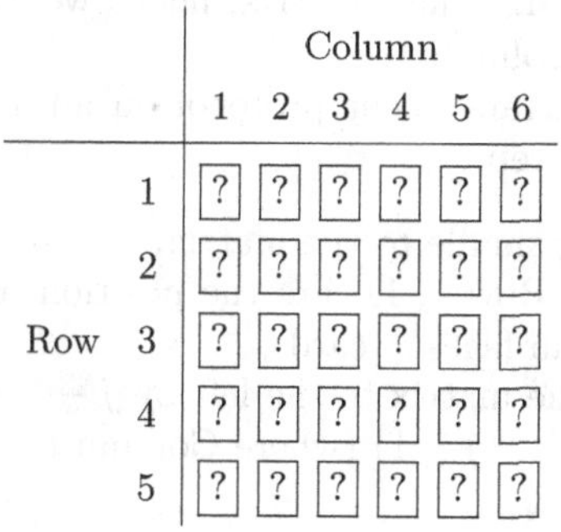

Fig. 2. An example of a 5×6 matrix

2.3   Pile-Shifting Shuffle

A *pile-shifting shuffle* was introduced by Shinagawa et al. [16]. In the pile-shifting shuffle, we rearrange the columns of the matrix by a random cyclic permutation,

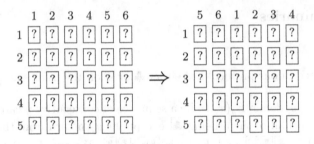

Fig. 3. An example of a pile-shifting shuffle on a 5×6 matrix with $r = 2$

i.e. move every Column ℓ to Column $\ell + r$ for a uniformly random $r \in \{0, 1, ..., b - 1\}$ (where Column ℓ' means Column $\ell' - b$ for $\ell' > b$). See Fig. 3.

One can perform the pile-shifting shuffle in real world by putting the cards in each column into an envelope and then applying a *Hindu cut*, which is a basic shuffling operation commonly used in card games [18], to the sequence of envelopes.

2.4   Rearrangement Protocol

The sole purpose of a *rearrangement protocol* is to revert columns of a matrix (after we perform pile-shifting shuffles) back to their original positions so that we can reuse all cards in the matrix without revealing them. Slightly different variants of this protocol were used in some previous work on card-based protocols [2,8,9,14,15]. Note that throughout our main protocol, we always put $E_b(1)$ in Row 1 when constructing a new matrix, hence we want to ensure that a ♡ in Row 1 moves back to Column 1.

We can apply the rearrangement protocol on an $a \times b$ matrix by publicly performing the following steps.

1. Apply the pile-shifting shuffle to the matrix.
2. Turn over all cards in Row 1. Locate the position of a ♡. Suppose it is at Column j. Turn over all face-up cards.
3. Shift the columns of the matrix to the left by $j - 1$ columns, i.e. move every Column ℓ to Column $\ell - (j - 1)$ (where Column ℓ' means Column $\ell' + b$ for $\ell' < 1$).

2.5   Uniqueness Verification Protocol

Suppose we have sequences $S_0, S_1, ..., S_a$, each consisting of b cards. S_0 encodes a positive number, while $S_1, S_2, ..., S_a$ encode nonnegative numbers. Our objective is to verify that none of the sequences $S_1, S_2, ..., S_a$ encodes the same number as S_0 without revealing any number. This protocol is a special case of the protocol developed by Ruangwises and Itoh [14] to count the number of indices i such that S_i encodes the same number as S_0. We can do so by publicly performing the following steps.

1. Construct an $(a + 2) \times b$ matrix with Row 1 consisting of a sequence $E_b(1)$ and each Row $i + 2$ $(i = 0, 1, ..., a)$ consisting of the sequence S_i.
2. Apply the pile-shifting shuffle to the matrix.
3. Turn over all cards in Row 2. Locate the position of a ♡. Suppose it is at Column j.
4. Turn over all cards in Column j from Row 3 to Row $a + 2$. If there is no ♡ among them, then the protocol continues. Otherwise, V rejects and the protocol terminates.
5. Turn over all face-up cards.

2.6   Pile-Scramble Shuffle

A *pile-scramble shuffle* was introduced by Ishikawa et al. [10]. In the pile-scramble shuffle, we rearrange the columns of the matrix by a random permutation, i.e. move every Column j to Column p_j for a uniformly random permutation $p = (p_1, p_2, ..., p_b)$ of $(1, 2, ..., b)$. See Fig. 4.

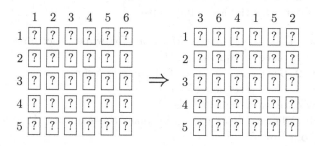

Fig. 4. An example of a pile-scramble shuffle on a 5×6 matrix

One can perform the pile-scramble shuffle in real world by putting the cards in each column into an envelope and then scrambling the envelopes together completely randomly.

3   Main Protocol

Let k be the size of the biggest room in the Ripple Effect grid. For each fixed cell with a number x, the prover P publicly puts a sequence of face-down cards $E_k(x)$ on it. Then, for each empty cell with a number x in P's solution, P secretly puts a sequence of face-down cards $E_k(x)$ on it. P will first verify the distance condition, and then the room condition.

3.1 Verification Phase for Distance Condition

The most challenging part of our protocol is to verify the distance condition, which is equivalent to the following statement: for each cell c with a number x, the first x cells to the right and to the bottom of c cannot have a number x.

First, we will show a protocol to verify that there is no number x among the first x cells to the right of c. Then, we apply the same protocol analogously in the direction to the bottom of c as well.

Suppose that cell c has a number x. Let A_0 be the sequence of cards on c. For each $i = 1, 2, ..., k$, let A_i be the sequence of cards on the i-th cell to the right of c, i.e. A_1 is on a cell right next to the right of c, A_2 is on a second cell to the right of c, and so on (if there are only $\ell < k$ cells to the right of c, we publicly put $E_k(0)$ in place of A_i for every $i > \ell$).

The intuition of this protocol is that, we will create sequences $B_1, B_2, ..., B_{k-1}$, all being $E_k(0)$, and insert them between A_x and A_{x+1} (without revealing x). Then, we will pick k sequences $A_1, A_2, ..., A_x, B_1, B_2, ..., B_{k-x}$ and use the uniqueness verification protocol introduced in Sect. 2.5 to verify that none of them encodes the same number as A_0.

P publicly performs the following steps.

1. Construct a $(k + 4) \times k$ matrix M by the following procedures. See Fig. 5.
   - In Row 1, place a sequence $E_k(1)$.
   - In Row 2, place the sequence A_0 (which is $E_k(x)$).
   - In Row 3, place a sequence $E_k(1)$.
   - In Row 4, place a sequence $E_k(0)$.
   - In each Column j ($j = 1, 2, ..., k$), place the sequence A_j arranged vertically from Row 5 to Row $k + 4$.
2. Apply the pile-shifting shuffle to M.
3. Turn over all cards in Row 2 of M. Locate the position of a $\boxed{\heartsuit}$. Suppose it is at Column j_1. Turn over all face-up cards.
4. Shift the columns of M to the right by $k - j_1$ columns, i.e. move every Column ℓ to Column $\ell + k - j_1$ (where Column ℓ' means Column $\ell' - k$ for $\ell' > k$). Observe that after this step, A_x will locate at the rightmost column.
5. Divide M into a $2 \times k$ matrix M_1 and a $(k + 2) \times k$ matrix M_2. M_1 consists of the topmost two rows of M, while M_2 consists of everything below M_1. Each cell $M(i + 2, j)$ ($i, j \geq 1$) of M will become a cell $M_2(i, j)$ of a new matrix M_2.
6. Apply the rearrangement protocol to M_1. Observe that we now have $E_k(1)$ in Row 1 and A_0 in Row 2 of M_1. From now on, we will perform operations only on M_2 while M_1 will be left unchanged.
7. Append $k - 1$ columns to the right of the matrix M_2 by the following procedures, making M_2 become a $(k + 2) \times (2k - 1)$ matrix.
   - In Row 1, place a sequence $E_{k-1}(0)$ from Column $k + 1$ to Column $2k - 1$.
   - In Row 2, place a sequence $E_{k-1}(1)$ from Column $k + 1$ to Column $2k - 1$.
   - In each Column $k + j$ ($j = 1, 2, ..., k - 1$), place a sequence $E_k(0)$ arranged vertically from Row 3 to Row $k + 2$. We call this sequence B_j.

8. Apply the pile-shifting shuffle to M_2.
9. Turn over all cards in Row 1 of M_2. Locate the position of a ♡. Suppose it is at Column j_2. Turn over all face-up cards.
10. For each $i = 1, 2, ..., k$, let S_i denote a sequence of card arranged vertically at Column $j_2 + i - 1$ (where Column ℓ' means Column $\ell' - (2k - 1)$ for $\ell' > 2k - 1$) from Row 3 to Row $k + 2$ of M_2. Observe that $(S_1, S_2, ..., S_k) = (A_1, A_2, ..., A_x, B_1, B_2, ..., B_{k-x})$. Then, construct a $(k + 2) \times k$ matrix N with Row 1 consisting of a sequence $E_k(1)$ taken from Row 1 of M_1, Row 2 consisting of the sequence A_0 taken from Row 2 of M_1, and each Row $i + 2$ ($i = 1, 2, ..., k$) consisting of the sequence S_i taken from M_2.
11. Apply the uniqueness verification protocol on N. The intuition of this step is to verify that none of the sequences $A_1, A_2, ..., A_x$ encodes the same number as A_0 (while $B_1, B_2, ..., B_{k-x}$ are all $E_k(0)$).
12. Apply the rearrangement protocol on N, put A_0 back onto c, and put $S_1, S_2, ..., S_k$ back to their corresponding columns in M_2.
13. Apply the pile-shifting shuffle to M_2.
14. Turn over all cards in Row 2 of M_2. Locate the position of a ♡. Suppose it is at Column j_3. Turn over all face-up cards.
15. Shift the columns of M_2 to the right by $k + 1 - j_3$ columns, i.e. move every Column ℓ to Column $\ell + k + 1 - j_3$ (where Column ℓ' means Column $\ell' - (2k - 1)$ for $\ell' > 2k - 1$). Then, remove Columns $k+1, k+2, ..., 2k-1$ from M_2, making M_2 become a $(k + 2) \times k$ matrix again. Observe that the columns we just removed are exactly the same $k - 1$ columns we previously appended to M_2.
16. Apply the rearrangement protocol on M_2 and put the sequences $A_1, A_2, ..., A_k$ back onto their corresponding cells on the Ripple Effect grid.

P performs these steps analogously in the direction to the right and bottom of every cell in the grid. If every cell passes the verification, P continues to the verification phase for room condition.

3.2   Verification Phase for Room Condition

The room condition of Ripple Effect is exactly the same as that of Makaro, and hence can be verified by a subprotocol in [2]. Since this is the final step of our protocol, after we finish verifying each room, we do not have to rearrange cards back to their original positions or put them back onto their cells.

P will verify each room separately. For a room R with size s, let $A_1, A_2, ..., A_s$ be the sequences of cards on the cells in R in any order. To verify room R, P publicly performs the following steps.

1. Construct a $k \times s$ matrix M by the following procedures: in each Column j ($j = 1, 2, ..., s$), place the sequence A_j arranged vertically from Row 1 to Row k.
2. Apply the pile-scramble shuffle to M.
3. Turn over all cards in M. If all columns of M are a permutation of $E_k(1)$, $E_k(2), ..., E_k(s)$ arranged vertically, then the protocol continues. Otherwise, V rejects and the protocol terminates.

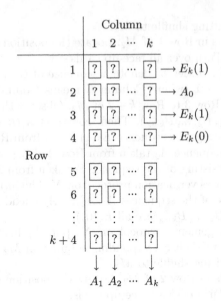

Fig. 5. A $(k+4) \times k$ matrix M constructed in Step 1

P performs these steps for every room. If every room passes the verification, then V accepts.

In total, our protocol uses $kmn + 2k^2 + 4k - 2 = \Theta(kmn)$ cards.

4   Proof of Security

We will prove the perfect completeness, perfect soundness, and zero-knowledge properties of our protocol. We omit the proofs of the verification phase for room condition as they have been shown in [2].

Lemma 1 (Perfect Completeness). *If P knows a solution of the Ripple Effect puzzle, then V always accepts.*

Proof. First, we will show that the uniqueness verification protocol will pass if none of $S_1, S_2, ..., S_a$ encodes the same number as S_0. Suppose that S_0 encodes a number $z > 0$. A ♡ in Row 2 will locate at Column z. Since none of $S_1, S_2, ..., S_a$ encodes the number z, all cards below Row 2 in the same column as that ♡ will be all ♣s. This remains true after we rearrange the columns in Step 2. Therefore, the verification in Step 4 will pass.

Now consider the main protocol. Suppose that P knows a solution of the puzzle. The verification phase for room condition will pass [2].

Consider the verification phase for distance condition. In Step 1, a ♡ in Row 2 is at the same column as A_x, and will always be. Therefore, in Step 4, the column containing A_x will move to the rightmost column.

In Step 7, the order of the sequences (which are arranged vertically from Row 3 to Row $k + 2$) from the leftmost column to the rightmost column is $A_{x+1}, A_{x+2}, ..., A_k, A_1, A_2, ..., A_x, B_1, B_2, ..., B_{k-1}$. Also, a 🂠 in Row 1 is at the same column as A_1, and a 🂠 in Row 3 is at the same column as B_1, and they will always be.

In Step 10, since A_1 locates at Column j, the sequences $S_1, S_2, ..., S_k$ will be exactly $A_1, A_2, ..., A_x, B_1, B_2, ..., B_{k-x}$ in this order. Therefore, in Step 11, the uniqueness verification protocol will pass because none of the sequences $A_1, A_2, ..., A_x$ encodes the number x, and $B_1, B_2, ..., B_{k-x}$ all encode 0.

Since this is true for every cell and every direction (to the right, left, top, and bottom), the verification phase for distance condition will pass, hence V will always accept.   □

Lemma 2 (Perfect Soundness). *If P does not know a solution of the Ripple Effect puzzle, then V always rejects.*

Proof. First, we will show that the uniqueness verification protocol will fail if at least one of $S_1, S_2, ..., S_a$ encodes the same number as S_0. Suppose that S_0 and S_d $(d > 0)$ both encode a number $z > 0$. A 🂠 in Row 2 will locate at Column z. Since S_d also encodes the number z, a card in Row d in the same column as that 🂠 will be a 🂠. This remains true after we rearrange the columns in Step 2. Therefore, the verification in Step 4 will fail.

Now consider the main protocol. Suppose that P does not know a solution of the puzzle. The numbers P puts into the grid must violate either the room condition or the distance condition. If they violate the room condition, the verification phase for room condition will fail [2].

Suppose that the numbers in the grid violate the distance condition. There must be two cells c and c' in the same row or column having the same number x, where c' locates on the right or the bottom of c with $\ell < x$ cells of space between them.

Consider when P performs the verification phase for distance condition for c in the direction towards c'. The sequence on the cell c' will be $A_{\ell+1}$

By the same reason as in the proof of Lemma 1, in Step 10, the sequences $S_1, S_2, ..., S_k$ will be exactly $A_1, A_2, ..., A_x, B_1, B_2, ..., B_{k-x}$ in this order and thus include $A_{\ell+1}$. Therefore, in Step 11, the uniqueness verification protocol will fail because A_ℓ encodes the number x, hence V will always reject.   □

Lemma 3 (Zero-Knowledge). *During the verification phase, V learns nothing about P's solution of the Ripple Effect puzzle.*

Proof. To prove the zero-knowledge property, it is sufficient to prove that all distributions of the values that appear when P turns over cards can be simulated by a simulator S without knowing P's solution.

- In the rearrangement protocol:
    • Consider Step 2 where we turn over all cards in Row 1. This occurs right after we applied the pile-shifting shuffle to the matrix. Therefore,

a $\heartsuit$ has an equal probability to appear at each of the b columns, hence this step can be simulated by S without knowing P's solution.

- In the uniqueness verification protocol:
  - Consider Step 3 where we turn over all cards in Row 2. This occurs right after we applied the pile-shifting shuffle to the matrix. Therefore, a $\heartsuit$ has an equal probability to appear at each of the b columns, hence this step can be simulated by S without knowing P's solution.
  - Consider Step 4 where we turn over all cards in Column j from Row 3 to Row $a+2$. If the verification passes, the cards we turn over must be all $\clubsuit$s, hence this step can be simulated by S without knowing P's solution.
- In the verification phase for room condition:
  - Consider Step 3 where we turn over all cards in Row 2 of M. This occurs right after we applied the pile-shifting shuffle to M. Therefore, a $\heartsuit$ has an equal probability to appear at each of the k columns, hence this step can be simulated by S without knowing P's solution.
  - Consider Step 9 where we turn over all cards in Row 1 of M_2. This occurs right after we applied the pile-shifting shuffle to M_2. Therefore, a $\heartsuit$ has an equal probability to appear at each of the $2k-1$ columns, hence this step can be simulated by S without knowing P's solution.
  - Consider Step 14 where we turn over all cards in Row 2 of M_2. This occurs right after we applied the pile-shifting shuffle to M_2. Therefore, a $\heartsuit$ has an equal probability to appear at each of the $2k-1$ columns, hence this step can be simulated by S without knowing P's solution.

Therefore, we can conclude that V learns nothing about P's solution. $\square$

5   Future Work

We developed a physical protocol of zero-knowledge proof for Ripple Effect puzzle using $\Theta(kmn)$ cards. A possible future work is to develop a protocol for this puzzle that requires asymptotically fewer number of cards, or for other popular logic puzzles.

Another challenging future work is to explore methods to physically verify other types of more complicated number-related functions.

References

1. Bultel, X., Dreier, J., Dumas, J.-G., Lafourcade, P.: Physical zero-knowledge proofs for Akari, Takuzu, Kakuro and KenKen. In: Proceedings of the 8th International Conference on Fun with Algorithms (FUN), pp. 8:1–8:20 (2016)
2. Bultel, X., et al.: Physical zero-knowledge proof for Makaro. In: Proceedings of the 20th International Symposium on Stabilization, Safety, and Security of Distributed Systems (SSS), pp. 111–125 (2018)
3. Chien, Y.-F., Hon, W.-K.: Cryptographic and physical zero-knowledge proof: from Sudoku to Nonogram. In: Proceedings of the 5th International Conference on Fun with Algorithms (FUN), pp. 102–112 (2010)

4. Dumas, J.-G., Lafourcade, P., Miyahara, D., Mizuki, T., Sasaki, T., Sone, H.: Interactive physical zero-knowledge proof for Norinori. In: Proceedings of the 25th International Computing and Combinatorics Conference (COCOON), pp. 166–177 (2019)

5. Goldreich, O., Micali, S., Wigderson, A.: Proofs that yield nothing but their validity and a methodology of cryptographic protocol design. J. ACM **38**(3), 691–729 (1991)

6. Goldwasser, S., Micali, S., Rackoff, C.: The knowledge complexity of interactive proof systems. SIAM J. Comput. **18**(1), 186–208 (1989)

7. Gradwohl, R., Naor, M., Pinkas, B., Rothblum, G.N.: Cryptographic and physical zero-knowledge proof systems for solutions of Sudoku puzzles. In: Proceedings of the 4th International Conference on Fun with Algorithms (FUN), pp. 166–182 (2007)

8. Hashimoto, Y., Shinagawa, K., Nuida, K., Inamura, M., Hanaoka, G.: Secure grouping protocol using a deck of cards. In: Proceedings of the 10th International Conference on Information Theoretic Security (ICITS), pp. 135–152 (2017)

9. Ibaraki, T., Manabe, Y.: A more efficient card-based protocol for generating a random permutation without fixed points. In: Proceedings of the 3rd International Conference on Mathematics and Computers in Sciences and Industry (MCSI), pp. 252–257 (2016)

10. Ishikawa, R., Chida, E., Mizuki, T.: Efficient card-based protocols for generating a hidden random permutation without fixed points. In: Proceedings of the 14th International Conference on Unconventional Computation and Natural Computation (UCNC), pp. 215–226 (2015)

11. Lafourcade, P., Miyahara, D., Mizuki, T., Sasaki, T., Sone, H.: A physical ZKP for Slitherlink: how to perform physical topology-preserving computation. In: Proceedings of the 15th International Conference on Information Security Practice and Experience (ISPEC), pp. 135–151 (2019)

12. Miyahara, D., Sasaki, T., Mizuki, T., Sone, H.: Card-based physical zero-knowledge proof for Kakuro. IEICE Trans. Fundamentals Electron. Commun. Comput. Sci. **E102.A**(9), 1072–1078 (2019)

13. Nikoli: Ripple Effect. https://www.nikoli.co.jp/en/puzzles/ripple_effect.html

14. Ruangwises, S., Itoh, T.: Physical zero-knowledge proof for numberlink. In: Proceedings of the 10th International Conference on Fun with Algorithms (FUN), pp. 22:1–22:11 (2020)

15. Sasaki, T., Mizuki, T., Sone, H.: Card-based zero-knowledge proof for Sudoku. In: Proceedings of the 9th International Conference on Fun with Algorithms (FUN), pp. 29:1–29:10 (2018)

16. Shinagawa, K., et al.: Multi-party computation with small shuffle complexity using regular polygon cards. In: Proceedings of the 9th International Conference on Provable Security (ProvSec), pp. 127–146 (2015)

17. Takenaga, Y., Aoyagi, S., Iwata, S., Kasai, T.: Shikaku and ripple effect are NP-complete. Congressus Numerantium **216**, 119–127 (2013)

18. Ueda, I., Nishimura, A., Hayashi, Y., Mizuki, T., Sone, H.: How to implement a random bisection cut. In: Proceedings of the 5th International Conference on the Theory and Practice of Natural Computing (TPNC), pp. 58–69 (2016)

Cyclic Shift Problems on Graphs

Kwon Kham Sai, Ryuhei Uehara, and Giovanni Viglietta[✉]

School of Information Science, Japan Advanced Institute of Science and Technology
(JAIST), Nomi, Japan
{saikwonkham,uehara,johnny}@jaist.ac.jp

Abstract. We study a new reconfiguration problem inspired by classic
mechanical puzzles: a colored token is placed on each vertex of a given
graph; we are also given a set of distinguished cycles on the graph. We
are tasked with rearranging the tokens from a given initial configuration
to a final one by using cyclic shift operations along the distinguished
cycles. We first investigate a large class of graphs, which generalizes
several classic puzzles, and we give a characterization of which final configurations can be reached from a given initial configuration. Our proofs
are constructive, and yield efficient methods for shifting tokens to reach
the desired configurations. On the other hand, when the goal is to find
a shortest sequence of shifting operations, we show that the problem is
NP-hard, even for puzzles with tokens of only two different colors.

Keywords: Cyclic shift puzzle · Permutation group · NP-hard
problem

1 Introduction

Recently, variations of reconfiguration problems have been attracting much interest, and several of them are being studied as important fundamental problems
in theoretical computer science [8]. Also, many real puzzles which can be modeled as reconfiguration problems have been invented and proposed by the puzzle
community, such as the 15-puzzle and Rubik's cube. Among these, we focus on
a popular type of puzzle based on cyclic shift operations: see Fig. 1. In these
puzzles, we can shift some elements along predefined cycles as a basic operation,
and the goal is to rearrange the pieces into a desired pattern.

In terms of reconfiguration problems, this puzzle can be modeled as fol-
lows. The input of the problem is a graph $G = (V, E)$, a set of colors $\text{COL} = \{1, 2, \ldots, c\}$, and one colored token on each vertex in V. We are also given a set
$\mathcal{C}$ of cycles of G. The basic operation on G is called "shift" along a cycle C in $\mathcal{C}$,
and it moves each token located on a vertex in C into the next vertex along C.
This operation generalizes the token swapping problem, which was introduced
by Yamanaka et al. [11], and has been well investigated recently. Indeed, when

This work is partially supported by KAKENHI grant numbers 17H06287 and
18H04091.

R. Uehara et al. (Eds.): WALCOM 2021, LNCS 12635, pp. 308–320, 2021.
https://doi.org/10.1007/978-3-030-68211-8_25

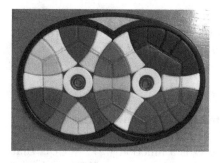

Fig. 1. Commercial cyclic shift puzzles: Turnstile (left) and Rubik's Shells (right)

we restrict each cycle in $\mathcal{C}$ to have length two (each cycle would correspond to an edge in E), the cyclic shift problem is equivalent to the token swapping problem.

In the mathematical literature, the study of permutation groups and their generators has a long history. An important theorem by Babai [1] states that the probability that two random permutations of n objects generate either the symmetric group S_n (i.e., the group of all permutations) or the alternating group A_n (i.e., the group of all even permutations) is $1 - 1/n + O(n^2)$. However, the theorem says nothing about the special case where the generators are cycles.

In [4], Heath et al. give a characterization of the permutations that, together with a cycle of length n, generate either A_n or S_n, as opposed to a smaller permutation group. On the other hand, in [7], Jones shows that A_n and S_n are the only finite primitive permutation groups containing a cycle of length $n - 3$ or less. However, his proof is non-constructive, as it heavily relies on the classification of finite simple groups (and, as the author remarks, a self-contained proof is unlikely to exist). In particular, no non-trivial upper bound is known on the distance of two given permutations in terms of a set of generators.

The computational complexity of related problems has been studied, too. It is well known that, given a set of generators, the size of the permutation group they generate is computable in polynomial time. Also, the inclusion of a given permutation π in the group is decidable in polynomial time, and an expression for π in terms of the generators is also computable in polynomial time [2].

In contrast, Jerrum showed that computing the distance between two given permutations in terms of two generators is PSPACE-complete [6]. However, the generators used for the reduction are far from being cycles.

In this paper, after giving some definitions (Sect. 2), we study the configuration space of a large class of cyclic shift problems which generalize the puzzles in Fig. 1 (Sect. 3). We show that, except for one special case, the permutation group generated by a given set of cycles is S_n if at least one of the cycles has even length, and it is A_n otherwise. This result is in agreement with Babai's theorem [1], and shows a similarity with the configuration space of the (generalized) 15-puzzle [10]. Moreover, our proofs in Sect. 3 are constructive, and yield polynomial upper bounds on the number of shift operations required to reach a given configuration. This is contrasted with Sect. 4, where we show that finding a shortest sequence of shift operations to obtain a desired configuration is NP-hard, even for puzzles with tokens of only two different colors.

2   Preliminaries

Let $G = (V, E)$ be a finite, simple, undirected graph, where V is the vertex set, with $n = |V|$, and E is the edge set. Let $\text{COL} = \{1, 2, \ldots, c\}$ be a set of colors, where c is a constant. A *token placement* for G is a function $f\colon V \to \text{COL}$: that is, $f(v)$ represents the color of the token placed on the vertex v. Without loss of generality, we assume f to be surjective.

Let us fix a set $\mathcal{C}$ of cycles in G (note that $\mathcal{C}$ does not necessarily contain all cycles of G). Two distinct token placements f and f' of G are *adjacent* with respect to $\mathcal{C}$ if the following two conditions hold: (1) there exists a cycle $C = (v_1, v_2, \ldots, v_j)$ in $\mathcal{C}$ such that $f'(v_{i+1}) = f(v_i)$ and $f'(v_1) = f(v_j)$ or $f'(v_i) = f(v_{i+1})$ and $f'(v_j) = f(v_1)$ for $1 \leq i \leq j$, and (2) $f'(w) = f(w)$ for all vertices $w \in V \setminus \{v_1, \ldots, v_i\}$. In this case, we say that f' is obtained from f by *shifting* the tokens along the cycle C. If an edge $e \in E$ is not spanned by any cycle in $\mathcal{C}$, e plays no role in shifting tokens. Therefore, without loss of generality, we assume that every edge is spanned by at least one cycle in $\mathcal{C}$.

We say that two token placements f_1 and f_2 are *compatible* if, for each color $c' \in \text{COL}$, we have $\left|f_1^{-1}(c')\right| = \left|f_2^{-1}(c')\right|$. Obviously, compatibility is an equivalence relation on token placements, and its equivalence classes are called *compatibility classes* for G and COL. For a compatibility class P and a cycle set $\mathcal{C}$, we define the *token-shifting graph of P and $\mathcal{C}$* as the undirected graph with vertex set P, where there is an edge between two token placements if and only if they are adjacent with respect to $\mathcal{C}$. A walk in a token-shifting graph starting from f and ending in f' is called a *shifting sequence between f and f'*, and the distance between f and f', i.e., the length of a shortest walk between them, is denoted as $\text{dist}(f, f')$ (if there is no walk between f and f', their distance is defined to be ∞). If $\text{dist}(f, f') < \infty$, we write $f \simeq f'$.

For a given number of colors c, we define the *c-Colored Token Shift* problem as follows. The input is a graph $G = (V, E)$, a cycle set $\mathcal{C}$ for G, two compatible token placements f_0 and f_t (with colors drawn from the set $\text{COL} = \{1, 2, \ldots, c\}$), and a non-negative integer ℓ. The goal is to determine whether $\text{dist}(f_0, f_t) \leq \ell$ holds. In the case that ℓ is not given, we consider the *c-Colored Token Shift* problem as an optimization problem that aims at computing $\text{dist}(f_0, f_t)$.

3   Algebraic Analysis of the Puzzles

For the purpose of this section, the vertex set of the graph $G = (V, E)$ will be $V = \{1, 2, \ldots, n\}$, and the number of colors will be $c = n$, so that $\text{COL} = V$, and a token placement on G can be interpreted as a permutation of V. To denote a permutation π of V, we can either use the one-line notation $\pi = [\pi(1) \ \pi(2) \ \ldots \ \pi(n)]$, or we can write down its cycle decomposition: for instance, the permutation [3 6 4 1 7 2 5] can be expressed as the product of disjoint cycles (1 3 4)(2 6)(5 7).

Note that, given a cycle set $\mathcal{C}$, shifting tokens along a cycle $(v_1, v_2, \ldots, v_j) \in \mathcal{C}$ corresponds to applying the permutation $(v_1 \ v_2 \ \ldots \ v_j)$ or its inverse $(v_j \ v_{j-1} \ \ldots \ v_1)$ to V. The set of token placements generated by shifting sequences starting from the "identity token placement" $f_0 = [1 \ 2 \ \ldots \ n]$ is therefore a permutation group with the composition operator, which we denote by $H_{\mathcal{C}}$, and we call it *configuration group generated by* $\mathcal{C}$. Since we visualize permutations as functions mapping vertices of G to colors (and not the other way around), it makes sense to compose chains of permutations from right to left, contrary to the common convention in the permutation group literature. So, for example, if we start from the identity token placement for $n = 5$ and we shift tokens along the cycles (1 2 3) and (3 4 5) in this order, we obtain the token placement

$$(1\ 2\ 3)(3\ 4\ 5) = [2\ 3\ 1\ 4\ 5]\,[1\ 2\ 4\ 5\ 3] = [2\ 3\ 4\ 5\ 1] = (1\ 2\ 3\ 4\ 5).$$

(Had we composed permutations from left to right, we would have obtained the token placement [2 4 1 5 3] = (1 2 4 5 3) as a result.)

One of our goals in this section is to determine the configuration groups $H_{\mathcal{C}}$ generated by some classes of cycle sets $\mathcal{C}$. Our choice of $\mathcal{C}$ will be inspired by the puzzles in Fig. 1, and will consist of arrangements of cycles that share either one or two adjacent vertices. As we will see, except in one special case, the configuration groups that we obtain are either the symmetric group S_n (i.e., the group of all permutations) or the alternating group A_n (i.e., the group of all even permutations), depending on whether the cycle set $\mathcal{C}$ contains at least one even-length cycle or not: indeed, observe that a cycle of length j corresponds to an even permutation if and only if j is odd.

Note that the set of permutations in the configuration group $H_{\mathcal{C}}$ coincides with the connected component of the token-shifting graph (as defined in the previous section) that contains f_0. The other connected components are simply given by the cosets of $H_{\mathcal{C}}$ in S_n (thus, they all have the same size), while the number of connected components of the token-shifting graph is equal to the index of $H_{\mathcal{C}}$ in S_n, i.e., $n!/|H_{\mathcal{C}}|$.

The other goal of this section is to estimate the diameter of the token-shifting graph, i.e., the maximum distance between any two token placements f_0 and f_t such that $f_0 \simeq f_t$. To this end, we state some basic preliminary facts, which are folklore, and can be proved by mimicking the "bubble sort" algorithm.

Proposition 1.

   1. *The n-cycle $(1\ 2\ \ldots\ n)$ and the transposition $(1\ 2)$ can generate any permutation of $\{1, 2, \ldots, n\}$ in $O(n^2)$ shifts.*
   2. *The n-cycle $(1\ 2\ \ldots\ n)$ and the 3-cycle $(1\ 2\ 3)$ can generate any even permutation of $\{1, 2, \ldots, n\}$ in $O(n^2)$ shifts.*[1]

---

[1] Of course, the two cycles generate strictly more than A_n (hence S_n) if and only if n is even; however, we will only apply Proposition 1.2 to generate even permutations.

3. *The 3-cycles* $(1\ 2\ 3)$, $(2\ 3\ 4)$, ..., $(n-2\ n-1\ n)$ *can generate any even permutation of* $\{1, 2, \ldots, n\}$ *in* $O(n^2)$ *shifts.*   □

All upper bounds given in Proposition 1 are worst-case asymptotically optimal (refer to [6] for some proofs).

3.1 Puzzles with Two Cycles

We first investigate the case where the cycle set $\mathcal{C}$ contains exactly two cycles α and β, either of the form $\alpha = (1\ 2\ \ldots\ a)$ and $\beta = (a\ a+1\ \ldots\ n)$ with $1 < a < n$, or of the form $\alpha = (1\ 2\ \ldots\ a)$ and $\beta = (a-1\ a\ a+1\ \ldots\ n)$, with $1 < a \leq n$. The first puzzle is called *1-connected* (a, b)-*puzzle*, where $n = a + b - 1$, and the second one is called *2-connected* (a, b)-*puzzle*, where $n = a + b - 2$ (so, in both cases $a > 1$ and $b > 1$ are the lengths of the two cycles α and β, respectively). See Fig. 2 for some examples. Note that the Turnstile puzzle in Fig. 1 (left) can be regarded as a 2-connected $(6, 6)$-puzzle.

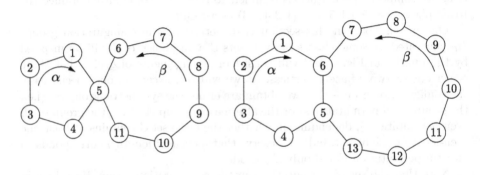

Fig. 2. A 1-connected $(5, 7)$-puzzle (left) and a 2-connected $(6, 9)$-puzzle (right)

Theorem 1. *The configuration group of a 1-connected* (a, b)-*puzzle is* A_n *if both* a *and* b *are odd, and it is* S_n *otherwise. Any permutation in the configuration group can be generated in* $O(n^2)$ *shifts.*

Proof. Observe that the commutator of α and β^{-1} is the 3-cycle $\alpha^{-1}\beta\alpha\beta^{-1} = (a-1\ a\ a+1)$. So, we can apply Proposition 1.2 to the n-cycle $\alpha\beta = (1\ 2\ \ldots\ n)$ and the 3-cycle $(a-1\ a\ a+1)$ to generate any even permutation in $O(n^2)$ shifts. If a and b are odd, then α and β are even permutations, and therefore cannot generate any odd permutation.

On the other hand, if a is even (the case where b is even is symmetric), then the a-cycle α is an odd permutation. So, to generate any odd permutation $\pi \in S_n$, we first generate the even permutation $\pi\alpha$ in $O(n^2)$ shifts, and then we do one extra shift along the cycle α^{-1}.   □

Our first observation about 2-connected (a, b)-puzzles is that the composition of α^{-1} and β is the $(n-1)$-cycle $\alpha^{-1}\beta = (a-2\ a-3\ \ldots\ 1\ a\ a+1\ \ldots\ n)$, which excludes only the element $a-1$. Similarly, $\alpha\beta^{-1} = (1\ 2\ \ldots\ a-1\ n\ n-1\ \ldots\ a+1)$, which excludes only the element a. We will write γ_1 and γ_2 as shorthand for $\alpha^{-1}\beta$ and $\alpha\beta^{-1}$ respectively, and we will use the permutations γ_1 and γ_2 to conjugate α and β, thus obtaining different a-cycles and b-cycles.[2]

Lemma 1. *In a 2-connected $(3, b)$-puzzle, any even permutation can be generated in $O(n^2)$ shifts.*

Proof. If we conjugate the 3-cycle α^{-1} by the inverse of γ_1, we obtain the 3-cycle $\gamma_1\alpha^{-1}\gamma_1^{-1} = (2\ 3\ 4)$. By applying Proposition 1.2 to the $(n-1)$-cycle β and the 3-cycle $(2\ 3\ 4)$, we can generate any even permutation of $V \setminus \{1\}$ in $O(n^2)$ shifts.

Let $\pi \in A_n$ be an even permutation of V. In order to generate π, we first move the correct token $\pi(1)$ to position 1 in $O(n)$ shifts, possibly scrambling the rest of the tokens: let σ be the resulting permutation. If σ is even, then $\sigma^{-1}\pi$ is an even permutation of $V \setminus \{1\}$, and we can generate it in $O(n^2)$ shifts as shown before, obtaining π.

On the other hand, if σ is odd, then one of the generators α and β must be odd, too. Since α is a 3-cycle, it follows that β is odd. In this case, after placing the correct token in position 1 via σ, we shift the rest of the tokens along β, and then we follow up with $\beta^{-1}\sigma^{-1}\pi$, which is an even permutation of $V \setminus \{1\}$, and can be generated it in $O(n^2)$ shifts. Again, the result is $\sigma\beta\beta^{-1}\sigma^{-1}\pi = \pi$. □

Lemma 2. *In a 2-connected (a, b)-puzzle with $a \geq 4$ and $b \geq 5$, any even permutation can be generated in $O(n^2)$ shifts.*

Proof. As shown in Fig. 3, the conjugate of β by γ_1 is the b-cycle

$$\delta_1 = \gamma_1^{-1}\beta\gamma_1 = (a\ a+1\ \ldots\ n-1\ a-1\ 1),$$

and the conjugate of β^{-1} by γ_2 is the b-cycle

$$\delta_2 = \gamma_2^{-1}\beta^{-1}\gamma_2 = (n\ n-1\ \ldots\ a+2\ a\ a-2\ a-1).$$

Their composition is $\delta_1\delta_2 = (1\ a\ a-2)(a-1\ n)(a+1\ a+2)$, and therefore $(\delta_1\delta_2)^2$ is the 3-cycle $(1\ a-2\ a)$. Conjugating this 3-cycle by α^{-1}, we finally obtain the 3-cycle $\tau = \alpha(\delta_1\delta_2)^2\alpha^{-1} = (1\ 2\ a-1)$; note that τ has been generated in a number of shifts independent of n. Now, since the 3-cycle τ and the $(n-1)$-cycle γ_2 induce a 2-connected $(3, n-1)$-puzzle on V, we can apply Lemma 1 to generate any even permutation of V in $O(n^2)$ shifts. □

Theorem 2. *The configuration group of a 2-connected (a, b)-puzzle is:*

   1. *Isomorphic to $S_{n-1} = S_5$ if $a = b = 4$.*

---

[2] If g and h are two elements of a group, the *conjugate* of g by h is defined as $h^{-1}gh$. In the context of permutation groups, conjugation by any h is an automorphism that preserves the cycle structure of permutations [9, Theorem 3.5].

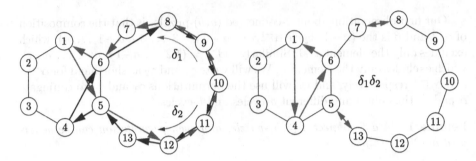

Fig. 3. Some permutations constructed in the proof of Lemma 2

2. A_n *if both a and b are odd.*
3. S_n *otherwise.*

Any permutation in the configuration group can be generated in $O(n^2)$ shifts.

Proof. By the symmetry of the puzzle, we may assume $a \leq b$. The case with $a = 2$ is equivalent to Proposition 1.1, so let $a \geq 3$. If $a \neq 4$ or $b \neq 4$, then Lemmas 1 and 2 apply, hence we can generate any even permutation in $O(n^2)$ shifts: the configuration group is therefore at least A_n. Now we reason as in Theorem 1: if a and b are odd, then α and β are even permutations, and cannot generate any odd one. If a is even (the case where b is even is symmetric), then α is an odd permutation. In this case, to generate any odd permutation $\pi \in S_n$, we first generate the even permutation $\pi\alpha$ in $O(n^2)$ shifts, and then we do one more shift along the cycle α^{-1} to obtain π.

The only case left is $a = b = 4$. To analyze the 2-connected $(4,4)$-puzzle, consider the outer automorphism $\psi \colon S_6 \to S_6$ defined on a generating set of S_6 as follows (cf. [9, Corollary 7.13]):

$$\psi((1\ 2)) = (1\ 5)(2\ 3)(4\ 6), \qquad \psi((1\ 3)) = (1\ 4)(2\ 6)(3\ 5),$$
$$\psi((1\ 4)) = (1\ 3)(2\ 4)(5\ 6), \qquad \psi((1\ 5)) = (1\ 2)(3\ 6)(4\ 5),$$
$$\psi((1\ 6)) = (1\ 6)(2\ 5)(3\ 4).$$

Because ψ is an automorphism, the subgroup of S_6 generated by α and β is isomorphic to the subgroup generated by the permutations $\psi(\alpha)$ and $\psi(\beta)$. Since $\alpha = (1\ 2\ 3\ 4) = (1\ 2)(1\ 3)(1\ 4)$ and $\beta = (3\ 4\ 5\ 6) = (1\ 3)(1\ 4)(1\ 5)(1\ 6)(1\ 3)$, and recalling that $\psi(\pi_1\pi_2) = \psi(\pi_1)\psi(\pi_2)$ for all $\pi_1, \pi_2 \in S_6$, we have:

$$\psi(\alpha) = \psi((1\ 2))\psi((1\ 3))\psi((1\ 4)) = [1\ 5\ 6\ 4\ 3\ 2] = (2\ 5\ 3\ 6) \text{ and}$$
$$\psi(\beta) = \psi((1\ 3))\psi((1\ 4))\psi((1\ 5))\psi((1\ 6))\psi((1\ 3)) = [3\ 1\ 5\ 4\ 2\ 6] = (1\ 3\ 5\ 2).$$

Note that the new generators $\psi(\alpha)$ and $\psi(\beta)$ both leave the token 4 in place, and so they cannot generate a subgroup larger than S_5 (up to isomorphism). On the other hand, we have $\psi(\alpha)\psi(\beta) = (1\ 6\ 2)$. This 3-cycle, together with the 4-cycle $\psi(\alpha)$, induces a 2-connected $(3,4)$-puzzle on $\{1,2,3,5,6\}$: as shown before, the configuration group of this puzzle is (isomorphic to) S_5. We conclude that the configuration group of the 2-connected $(4,4)$-puzzle is isomorphic to S_5, as well. A given permutation $\pi \in S_6$ is in the configuration group if and only if $\psi(\pi)$ leaves the token 4 in place.   □

3.2   Puzzles with Any Number of Cycles

Let us generalize the (a,b)-puzzle to larger numbers of cycles. (As far as the authors know, there are commercial products that have 2, 3, 4, and 6 cycles.) We say that two cycles are *properly interconnected* if they share exactly one vertex, of if they share exactly two vertices which are consecutive in both cycles. Note that all 1-connected and 2-connected (a,b)-puzzles consist of two properly interconnected cycles. Given a set of cycles $\mathcal{C}$ in a graph $G = (V,E)$, let us define the *interconnection graph* $\hat{G} = (\mathcal{C}, \hat{E})$, where there is an (undirected) edge between two cycles of $\mathcal{C}$ if and only if they are properly interconnected.

Let us assume $|V| > 6$ (to avoid special configurations of small size, which can be analyzed by hand), and let $\mathcal{C}$ consist of k cycles of lengths n_1, n_2, ..., n_k, respectively. We say that $\mathcal{C}$ induces a *generalized* $(n_1, n_2, \ldots, n_k)$-*puzzle* on V if there is a subset $\mathcal{C}' \subseteq \mathcal{C}$ such that:

(1) $\mathcal{C}'$ contains at least two cycles;
(2) the induced subgraph $\hat{G}[\mathcal{C}']$ is connected;
(3) each vertex of G is contained in at least one cycle in $\mathcal{C}'$.

When we fix such a subset $\mathcal{C}'$, the cycles in $\mathcal{C}'$ are called *relevant cycles*, and the vertices of G that are shared by two properly interconnected relevant cycles are called *relevant vertices* for those cycles. See Fig. 4 for an example of a generalized puzzle.

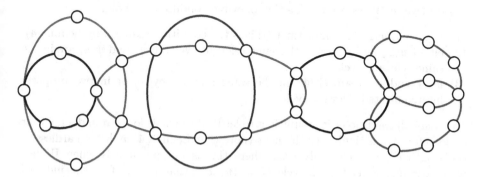

Fig. 4. A generalized puzzle where any permutation can be generated in $O(n^5)$ shifts, due to Theorem 3. Note that the blue cycle is the only cycle of even length, and is not properly interconnected with any other cycle. Also, the two red cycles and the two green cycles intersect each other but are not properly interconnected. (Color figure online)

The next two lemmas are technical; their proof is found in the full version of the paper: https://arxiv.org/abs/2009.10981.

Lemma 3. *In a generalized puzzle with three relevant cycles, $C' = \{C_1, C_2, C_3\}$, such that C_1 and C_2 induce a 2-connected $(4,4)$-puzzle, any permutation involving only vertices in C_1 and C_2 can be generated in $O(n^2)$ shifts.*   □

Lemma 4. *Let $V = \{1, \ldots, n\}$, and let $W = (w_1, \ldots, w_m) \in V^m$ be a sequence such that each element of V appears in W at least once, and any three consecutive elements of W are distinct. Then, the set of 3-cycles $C = \{(w_{i-1}\ w_i\ w_{i+1}) \mid 1 < i < m\}$ can generate any even permutation of V in $O(n^3)$ shifts.*   □

Theorem 3. *The configuration group of a generalized $(n_1, n_2, \ldots, n_k)$-puzzle is A_n if $n_1, n_2, \ldots, n_k$ are all odd, and it is S_n otherwise. Any permutation in the configuration group can be generated in $O(n^5)$ shifts.*

Proof. Observe that it suffices to prove that the given cycles can generate any even permutation in $O(n^5)$ shifts. Indeed, if all cycles have odd length, they cannot generate any odd permutation. On the other hand, if there is a cycle of even length and we want to generate an odd permutation π, we can shift tokens along that cycle, obtaining an odd permutation σ, and then we can generate the even permutation $\sigma^{-1}\pi$ in $O(n^5)$ shifts, obtaining π.

Let us fix a set of $k' \geq 3$ relevant cycles $C' \subseteq C$: we will show how to generate any even permutation by shifting tokens only along relevant cycles. By properties (2) and (3) of generalized puzzles, there exists a walk W on G that visits all vertices (possibly more than once), traverses only edges of relevant cycles, and transitions from one relevant cycle to another only if they are properly interconnected, and only through a relevant vertex shared by them. We will now slightly modify W so that it satisfies the hypotheses of Lemma 4, as well as some other conditions. Namely, if w_{i-1}, w_i, w_{i+1} are any three vertices that are consecutive in W, we would like the following conditions to hold:

(1) w_{i-1}, w_i, w_{i+1} are all distinct (this is the condition required by Lemma 4);
(2) either w_{i-1} and w_i are in the same relevant cycle, or w_i and w_{i+1} are in the same relevant cycle;
(3) w_{i-1} and w_{i+1} are either in the same relevant cycle, or in two properly interconnected relevant cycles.

To satisfy all conditions, it is sufficient to let W do a whole loop around a relevant cycle before transitioning to the next (note that Lemma 4 applies regardless of the length of W). The only case where this is not possible is when W has to go through a relevant 2-cycle $C = (u_1\ u_2)$ that is a leaf in the induced subgraph $\hat{G}[C']$, such that C shares exactly one relevant vertex, say u_1, with another relevant cycle $C' = (v_0\ u_1\ v_1\ v_2\ \ldots)$. To let W cover C in a way that satisfies the above conditions, we set either $W = (\ldots, v_0, u_1, u_2, v_1, \ldots)$ or $W = (\ldots, v_1, u_1, u_2, v_0, \ldots)$: that is, we skip u_1 after visiting u_2. After this modification, W is no longer a walk on G, but it satisfies the hypotheses of Lemma 4, as well as the three conditions above.

We will now show that the 3-cycle $(w_{i-1}\ w_i\ w_{i+1})$ can be generated in $O(n^2)$ shifts, for all $1 < i < |W|$. By Lemma 4, we will therefore conclude that any even permutation of V can be generated in $O(n^2) \cdot O(n^3) = O(n^5)$ shifts. Due to conditions (2) and (3), we can assume without loss of generality that w_{i-1} and w_i are both in the same relevant cycle C_1, and that w_{i+1} is either in C_1 or in a different relevant cycle C_2 which is properly interconnected with C_1. In the first case, by property (1) of generalized puzzles, there exists another relevant cycle C_2 properly interconnected with C_1. So, in all cases, C_1 and C_2 induce a 1-connected or a 2-connected $(|C_1|, |C_2|)$-puzzle.

That the 3-cycle $(w_{i-1}\ w_i\ w_{i+1})$ can be generated in $O(n^2)$ shifts now follows directly from Theorems 1 and 2, except if $|C_1| = |C_2| = 4$ and C_1 and C_2 share exactly two vertices: indeed, the 2-connected $(4, 4)$-puzzle is the only case where we cannot generate any 3-cycle. However, since we are assuming that $V > 6$, there must be a third relevant cycle C_3, which is properly interconnected with C_1 or C_2. Our claim now follows from Lemma 3.   □

4   NP-Hardness for Puzzles with Two Colors

In this section, we show that the 2-Colored Token Shift problem is NP-hard. That is, for a graph $G = (V, E)$, cycle set $\mathcal{C}$, two token placements f_0 and f_t for G, and a non-negative integer ℓ, it is NP-hard to determine if $\text{dist}(f_0, f_t) \le \ell$.

Theorem 4. *The 2-Colored Token Shift problem is NP-hard.*

Proof. We will give a polynomial-time reduction from the NP-complete problem 3-Dimensional Matching, or 3DM [3]: given three disjoint sets X, Y, Z, each of size m, and a set of triplets $T \subseteq X \times Y \times Z$, does T contain a matching, i.e., a subset $M \subseteq T$ of size exactly m such that all elements of X, Y, Z appear in M?

Given an instance of 3DM (X, Y, Z, T), with $n = |T|$, we construct the instance of the 2-Colored Token Shift problem illustrated in Fig. 5.

The vertex set of $G = (V, E)$ includes the sets X, Y, Z (shown with a green background in the figure: these will be called *green vertices*), as well as the vertex u. Also, for each triplet $\hat{t}_i = (x, y, z) \in T$, with $1 \le i \le n$, the vertex set contains three vertices $t_{i,1}, t_{i,2}, t_{i,3}$ (shown with a yellow background in the figure: these will be called *yellow vertices*), and the cycle set $\mathcal{C}$ has the three cycles $(u, t_{i,1}, t_{i,2}, t_{i,3}, x)$, $(u, t_{i,1}, t_{i,2}, t_{i,3}, y)$, and $(u, t_{i,1}, t_{i,2}, t_{i,3}, z)$ (drawn in blue in the figure). Finally, we have the vertex w, and the vertices $v_1, v_2, \ldots, v_{3n-3m}$; for each $i \in \{1, 2, \ldots, n\}$, the cycle set $\mathcal{C}$ contains the cycle $(t_{i,3}, t_{i,2}, t_{i,1}, v_1, v_2, \ldots, v_{3n-3m}, w)$ (drawn in red in the figure). In the initial token placement f_0, there are black tokens on the $3n$ vertices of the form $t_{i,j}$, and white tokens on all other vertices. In the final token placement f_t, there is a total of $3m$ black tokens on all the vertices in X, Y, Z, plus $3n - 3m$ black tokens on $v_1, v_2, \ldots, v_{3n-3m}$; all other vertices have white tokens. With this setup, we let $\ell = 3n$.

It is easy to see that, if the 3DM instance has a matching $M = \{\hat{t}_{i_1}, \hat{t}_{i_2}, \ldots, \hat{t}_{i_m}\}$, then $\text{dist}(f_0, f_t) \le \ell$. Indeed, for each $\hat{t}_{i_j} = (x_j, y_j, z_j)$, with

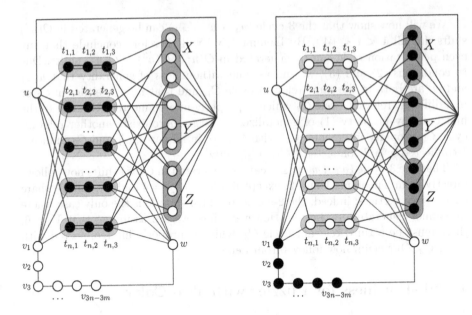

Fig. 5. The initial token placement f_0 (left) and the final token placement f_t (right) (Color figure online)

$1 \leq j \leq m$, we can shift tokens along the three blue cycles containing the yellow vertices $t_{i_j,1}$, $t_{i_j,2}$, $t_{i_j,3}$, thus moving their three black tokens into the green vertices x_j, y_j, and z_j. Since M is a matching, these $3m$ shifts eventually result in X, Y, and Z being covered by black tokens. Finally, we can shift the $3n - 3m$ black tokens corresponding to triplets in $T \setminus M$ along red cycles, moving them into the vertices $v_1, v_2, \ldots, v_{3n-3m}$. Clearly, this is a shifting sequence of length $3n = \ell$ from f_0 to f_t.

We will now prove that, assuming that $\mathrm{dist}(f_0, f_t) \leq \ell$, the 3DM instance has a matching. Note that each shift, no matter along which cycle, can move at most one black token from a yellow vertex to a non-yellow vertex. Since in f_0 there are $\ell = 3n$ black tokens on yellow vertices, and in f_t no token is on a yellow vertex, it follows that each shift must cause exactly one black token to move from a yellow vertex to a non-yellow vertex, and no black token to move back into a yellow vertex.

This implies that no black token should ever reach vertex u: if it did, it would eventually have to be moved to some other location, because u does not hold a black token in f_t. However, the black token in u cannot be shifted back into a yellow vertex, and therefore it will be shifted into a green vertex along a blue cycle. Since every shift must cause a black token to leave the set of yellow vertices, such a token will move into u: we conclude that u will always contain a black token, which is a contradiction. Similarly, we can argue that the vertex w should never hold a black token.

Let us now focus on a single triplet of yellow vertices $t_{i,1}$, $t_{i,2}$, $t_{i,3}$. Exactly three shifts must involve these vertices, and they must result in the three black tokens leaving such vertices. Clearly, this is only possible if the three black tokens are shifted in the same direction. If they are shifted in the direction of $t_{i,3}$ (i.e., rightward in Fig. 5), they must move into green vertices (because they cannot go into w); if they are shifted in the direction of $t_{i,1}$ (i.e., leftward in Fig. 5), they must move into v_1 (because they cannot go into u).

Note that, if a black token ever reaches a green vertex, it can no longer be moved: any shift involving such a token would move it back into a yellow vertex or into u. It follows that the only way of filling all the green vertices with black token is to select a subset of exactly m triplets of yellow vertices and shift each of their black tokens into a different green vertex. These m triplets of yellow vertices correspond to a matching for the 3DM instance.   □

In the above reduction, we can easily observe that the final token placement f_t can always be reached from the initial token placement f_0 in a polynomial number of shifts. Therefore, for this particular set of instances, the 2-Colored Token Shift problem is in NP. The same is also true of the puzzles introduced in Sect. 3, due to the polynomal upper bound given by Theorem 3. However, we do not know whether this is true for the c-Colored Token Shift problem in general, even assuming $c = 2$. A theorem of Helfgott and Seress [5] implies that, if $f_0 \simeq f_t$, the distance between f_0 and f_t has a quasi-polynomial upper bound; this, however, is insufficient to conclude that the problem is in NP. On the other hand, it is not difficult to see that the c-Colored Token Shift problem is in PSPACE; characterizing its computational complexity is left as an open problem. It would also be interesting to establish if the problem remains NP-hard when restricted to planar graphs or to graphs of constant maximum degree.

References

1. Babai, L.: The probability of generating the symmetric group. J. Comb. Theory Ser. A **52**, 148–153 (1989)
2. Furst, M., Hopcroft, J., Luks, E.: Polynomial-time algorithms for permutation groups. In: Proceedings of the 21st Annual Symposium on Foundations of Computer Science, pp. 36–41 (1980)
3. Garey, M.R., Johnson, D.S.: Computers and Intractability: A Guide to the Theory of NP-Completeness. W. H. Freeman and Company, San Francisco (1979)
4. Heath, D., Isaacs, I.M., Kiltinen, J., Sklar, J.: Symmetric and alternating groups generated by a full cycle and another element. A. Math. Mon. **116**(5), 447–451 (2009)
5. Helfgott, H.A., Seress, Á.: On the diameter of permutation groups. Anna. Math. **179**(2), 611–658 (2014)
6. Jerrum, M.R.: The complexity of finding minimum-length generator sequences. Theor. Comput. Sci. **36**, 265–289 (1985)
7. Jones, G.A.: Primitive permutation groups containing a cycle. Bull. Aust. Math. Soc. **89**(1), 159–165 (2014)
8. Nishimura, N.: Introduction to rconfiguration. Algorithms **11**(4), 1–25 (2018)

9. Rotman, J.J.: An Introduction to the Theory of Groups, 4th edn. Springer, New York (1995). https://doi.org/10.1007/978-1-4612-4176-8
10. Wilson, R.M.: Graph puzzles, homotopy, and the alternating group. J. Comb. Theory Ser. B **16**, 86–96 (1974)
11. Yamanaka, K., et al.: Swapping colored tokens on graphs. Theor. Comput. Sci. **729**, 1–10 (2018)

Mathematical Characterizations and Computational Complexity of Anti-slide Puzzles

Ko Minamisawa[1(✉)], Ryuhei Uehara[1], and Masao Hara[2]

[1] School of Information Science, Japan Advanced Institute of Science and Technology (JAIST), Ishikawa, Japan
{minamisawa,uehara}@jaist.ac.jp
[2] Department of Mathematical Science, Tokai University, Kanagawa, Japan
masao@tokai-u.jp

Abstract. For a given set of pieces, an anti-slide puzzle asks us to arrange the pieces so that none of the pieces can slide. In this paper, we investigate the anti-slide puzzle in 2D. We first give mathematical characterizations of anti-slide puzzles and show the relationship between the previous work. Using a mathematical characterization, we give a polynomial time algorithm for determining if a given arrangement of polyominoes is anti-slide or not in a model. Next, we prove that the decision problem whether a given set of polyominoes can be arranged to be anti-slide or not is strongly NP-complete even if every piece is x-monotone. On the other hand, we show that a set of pieces cannot be arranged to be anti-slide if all pieces are convex polygons.

Keywords: Anti-slide puzzle · Interlock puzzle · Strongly NP-completeness · x-monotone · Convexity

1 Introduction

A *silhouette puzzle* is a puzzle where, given a set of polygons, one must place them on the plane in such a way that their union is a target polygon. One of the most famous ones is the *Tangram*, which is the set of seven polygons (Fig. 1). Of anonymous origin, their first known reference in literature is from 1813 in China [11]. While we are given a silhouette as the goal in a silhouette puzzle, many puzzles are invented such that the goal is not explicitly given recently. For example, puzzles that ask to make a "symmetric shape" are quite popular in puzzle society, and its computational complexity has been investigated [7].

In this paper, we investigate *anti-slide* puzzles, which is one of the puzzles that have a characteristic difficulty. Typically, we are given a set $\mathcal{P}$ of pieces and a frame F. The goal of this puzzle is finding an arrangement of all pieces of $\mathcal{P}$ in F so that no piece slides even if we tilt or shake the frame in any direction. (In this paper, we also investigate similar puzzles known as *interlock* puzzles; in

© Springer Nature Switzerland AG 2021
R. Uehara et al. (Eds.): WALCOM 2021, LNCS 12635, pp. 321–332, 2021.
https://doi.org/10.1007/978-3-030-68211-8_26

Fig. 1. Silhouette puzzle "Tangram"   **Fig. 2.** Commercial product of the
first anti-slide puzzle [12]

an interlock puzzle, only a set of pieces are given without frame and the goal is interlocking all pieces.)

In *packing* puzzles, we are also given a set $\mathcal{P}$ of pieces and a frame F. The goal of this puzzle is packing all pieces in $\mathcal{P}$ into the frame F. In packing puzzles, the difficulty comes from the fact that we cannot pack them into the frame without a neat way, e.g., since the total area of the pieces is equal to the vacant area of F. On the other hand, it is easy to pack the pieces into the frame in an anti-slide puzzle, while it is difficult to support with each other. That is, the difficulty of anti-slide puzzles comes from the other aspect from packing puzzles.

Originally, the first anti-slide puzzle was invented by William Strijbos, who is a famous puzzle designer, in 1994 [12]. It was the second place winner at the 1994 Hikimi Puzzle Competition in Japan, and it was sold as a commercial product by Hanayama in 2007 (Fig. 2). This anti-slide puzzle is a 3D version; it consists of 15 "caramels" of size $1 \times 2 \times 2$ and a case of size $4 \times 4 \times 4$. The puzzle asks us to put 15, 14, 12, or 13 pieces into the case so that no caramel can slide in any direction in the case. Since then, especially in 2D, several dozen of anti-slide puzzles have been invented in the puzzle society. It is too many to follow all these puzzles, however, we introduce an impressive puzzle *Lock Device* invented by Hiroshi Yamamoto, who is also a famous puzzle designer, in 2012 (Fig. 3). This is an interlock puzzle that consists of seven distinct L-shape pieces: We have no frame and we can arrange them so that no piece can slide from the other pieces (it is a quite counterintuitive difficult puzzle).

While anti-slide puzzles and interlock puzzles are common in puzzle society, little research has been done on them from the viewpoint of computer science. Amano et al. gave an IP formulation of the problem in 2015 [1]. They investigated a generalization of the puzzle proposed by Strijbos; that is, they investigated a set of caramels of size $1 \times 2 \times 2$ into a case of size $n \times m \times \ell$. On the other hand, in 2019, Takenaga et al. investigated anti-slide puzzles in 2D [13]; they formalized the case that the pieces are *polyominoes* (see [10]) and solved them by using OBDD (Ordered Binary Decision Diagram). In this previous work, they investigated quite restricted cases, and general properties of anti-slide puzzles are not yet formalized.

In this paper, we first give some mathematical models of the notion of anti-slide. The formulations are given to some different models, which correspond to previous research and natural anti-slide puzzles. As an application of the models, we consider anti-slide puzzles of the set of polyominoes. We can give a characterization of anti-slide (and interlock) polyominoes by using two directed graphs. In this characterization, strong connectivity of these graphs coincides with the property of anti-slide. Using the result, we can determine whether a given arrangement of polyominoes is anti-slide or not in polynomial time.

Next, for a given set $\mathcal{P}$ of pieces and frame F, we consider the decision problem that asks if $\mathcal{P}$ can be anti-slide in F or not. In a sense, the Rectangle Packing Puzzle problem in [7], a packing problem of rectangle pieces into a rectangular frame, gives a proof of NP-completeness. However, this is a difficulty of a packing problem, which does not give us any difficulty of anti-slide puzzles. Therefore, we add one more condition that we consider the set $\mathcal{P}$ of pieces such that a packing of $\mathcal{P}$ into the frame F is trivial (always we can and it is easy to find). Under this condition, we show that the decision problem of an anti-slide puzzle is strongly NP-complete. Moreover, even if we remove F from the puzzle, it is still strongly NP-complete for the decision problem of an interlock puzzle. Especially, these NP-completeness hold even if all pieces in $\mathcal{P}$ are x-monotone.

Lastly, we consider the interlock puzzle for the convex pieces. Although the decision problem of an interlock puzzle is strongly NP-complete even if all pieces are x-monotone, any set of convex polygons cannot be interlocked. That is, when all pieces are convex, we prove that we cannot interlock them.

2   Preliminaries

We follow the basic notations and notions in [5]: We assume that each polygon P is given by a sequence of vertices in counterclockwise. A polygon P is *convex* if the line segment $\overline{pq}$ between any pair of points p and q in P is included in P. A polygon is a *simple polygon* when it is connected, no self intersections of its edges, and no holes. For a line l, a polygon P is said to be *l-monotone* if the intersection of P and l' is connected (i.e., a line segment, a point, or empty) for any l' perpendicular to l. If the line l can be the x-axis, we call it *x-monotone* (Fig.). ∂P denote the boundary of a polygon P. Two polygons P and Q *touch* if $\partial P \cap \partial Q \neq \emptyset$. A *feasible arrangement* of a set $\mathcal{P}$ of polygons is an arrangement of $\mathcal{P}$ such that for any pair of $P, Q \in \mathcal{P}$, they do not share any points except on ∂P and ∂Q. Intuitively, no pair of polygons overlaps in a feasible arrangement. Hereafter, we only consider feasible arrangements of polygons, therefore, we say just arrangement.

When we consider slide of polygons, we have to be conscious of the direction of gravity. When a direction is given as a vector $\boldsymbol{d}$, we consider the vector indicates "down" and fix an xy-coordinate following $\boldsymbol{d}$. Then, we denote the coordinate of a point p by $(x_d(p), y_d(p))$ with respect to $\boldsymbol{d}$. A polygon is said to be *orthogonal* when every edge of P is parallel to the x-axis or y-axis.

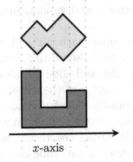

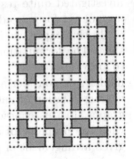

Fig. 3. 2D interlock puzzle "Lock Device" [14]

Fig. 4. Two x-monotone polyominoes

Fig. 5. Complete list of twelve pentominoes

The input of an anti-slide puzzle consists of a set $\mathcal{P}$ of polygons and a frame F. Each element in $\mathcal{P}$ is called *piece* and F is a polygon that has holes inside it. Without loss of generality, the total area of the holes in F is equal to or greater than the total area of the pieces in $\mathcal{P}$. Each polygon and frame are given as a sequence of the points and the total number of the vertices is n.

A *polyomino* is a polygon obtained by aligning squares of equal size arranged with coincident sides [10]. Depending on its area, it is called monomino, domino, and so on. A complete list of pentominoes (of size 5) is shown in Fig. 5.

A *directed graph* $D = (V, A)$ is defined by a finite set V and a subset A of $V \times V$. Each element in $V = V(D)$ is a *vertex* and each element in $A = A(D) \subseteq V(D) \times V(D) = \{(u, v) : u, v \in V(D)\}$ is an *arc*. On a directed graph D, a vertex v is *reachable* from u if we have a *directed path* $(u, \ldots, v)$ starting from u to v in D. A directed graph D is *strongly connected* if v is reachable from u and u is reachable from v for any pair of vertices u and v in $V(D)$ (see [2] for further details).

3   Mathematical Characterizations of Anti-slide Puzzles

In this section, we give some models of anti-slide puzzles for polygons in 2D and show some basic properties. When two polygons P and Q touch, a point q in ∂Q *supports* a point p in ∂P *with respect to a direction* $\boldsymbol{d}$ when the following conditions hold (Fig. 6):

1. both points p and q are in $\partial P \cap \partial Q$ and
2. we have the following for any $\epsilon > 0$:
   (a) let $C_\epsilon(q)$ be a circle centered at q with radius ϵ and $C_\epsilon(p)$ a circle centered at p with radius ϵ,
   (b) let $(q_{\epsilon-}, q_{\epsilon+})$ be two intersection points of $C_\epsilon(q)$ and ∂Q with $x_d(q_{\epsilon-}) < x_d(q_{\epsilon+})$ and $(p_{\epsilon-}, p_{\epsilon+})$ be two intersection points of $C_\epsilon(p)$ and ∂P with $x_d(p_{\epsilon-}) < x_d(p_{\epsilon+})$, and

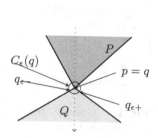

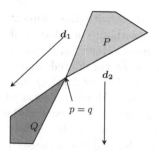

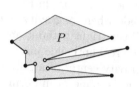

Fig. 6. Q supports P from below

Fig. 7. Q supports P with respect to a direction d_1, however, Q does not support P with respect to d_2

Fig. 8. Lower envelope of a polygon P; black endpoints are included in the envelope, while white endpoints are not

(c) we have $x_d(q_{\epsilon-}) < x_d(p) < x_d(q_{\epsilon+})$ or $x_d(p_{\epsilon-}) < x_d(q) < x_d(p_{\epsilon+})$.

Intuitively, the neighbor of the points p or q has a positive width for supporting the other point. We note that it depends on the direction d (Fig. 7).

For a direction d, the *lower envelope* of a polygon P is the set of points p in ∂P such that there is no point p' inside of P with $dist(p,p') < \epsilon$, $x_d(p) = x_d(p')$, and $y_d(p) > y_d(p')$ for any $\epsilon > 0$, where $dist(p,p')$ is the distance between p and p' (Fig. 8). We define the *upper envelope* in the same manner.

For a given arrangement of a set of polygons, a polygon P is *weakly anti-slide with respect to d* when there is a polygon Q ($P \neq Q$) such that a point on the lower envelope of P is supported by a point on the upper envelope of Q. In short, we say Q *supports* P in this case (Fig. 9).

	(a) 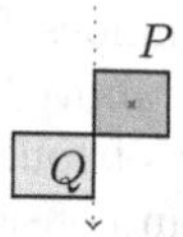	(b) 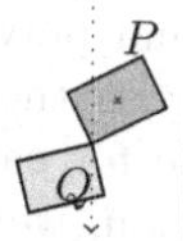	(c) 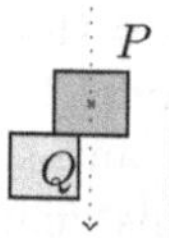	(d) 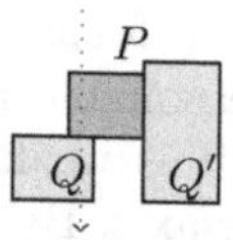	(e) 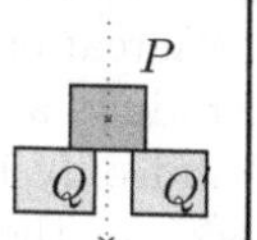
Weak anti-slide	N	Y	Y	Y	Y
Strong anti-slide	N	N	N	N	Y
Anti-slide for polyominoes	N	undefined	N	Y	Y

Fig. 9. Yes/No cases for each anti-slide model when d indicates below (each red $\times$ indicates the gravity center of P)

In [13], Takenaga et al. dealt with the problems in the weak anti-slide model (without considering the case in Fig. 9(b)) in this context. However, in puzzle society, the case in Fig. 9(c) is not considered as anti-slide when its gravity center of P is not "on" Q since P tilts right without any support. When we consider the real anti-slide puzzles, we have to be aware of the gravity center of a polygon. We suppose that a polygon is made of uniform material and has a constant thickness. We introduce its *gravity center* in the standard way of physics. We note that some polygons have their gravity centers outside of them. Therefore we introduce a notion of balance point (Fig. 10): For a direction d, the *balance point* of a polygon P is the minimum point (taking of $y_d(p)$) of the intersection points of P and the line ℓ that is parallel to d and passes through the gravity center of P. We have the following observation for the balance point:

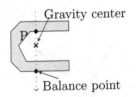

Fig. 10. Gravity center and balance point

Observation 1. *For any polygon P and a direction d, P has a unique balance point with respect to d. Let g and b be the gravity center and the balance point of P, respectively. Then we have $x_d(g) = x_d(b)$.*

Now we are ready to introduce another anti-slide model. For a given arrangement of a set of polygons, a polygon P is *strongly anti-slide with respect to d* when there are two points p_r and p_l in the lower envelope of P such that;

1. for the balance point p of P, we have $x_d(p_l) \leq x_d(p)$ and $x_d(p) \leq x_d(p_r)$ and
2. there are two polygons Q and Q' such that p_r (and p_l) is supported by a point q (and q') in the upper envelope of Q (and Q', respectively).

We here note that we may have $p_r = p_l$ or $Q = Q'$. Intuitively, the gravity center of P is supported from below by Q and Q'. Some representative examples are shown in Fig. 9. Especially, we note that the case 9(d) is not the case of strongly anti-slide. In real puzzles, if P has some thickness, it is considered as an anti-slide case since P is supported by Q' from the right.

Hereafter, we focus on orthogonal polyominoes for fixed xy-axes. We now introduce a reasonable model between them above. A polyomino is *anti-slide* if and only if it is weakly anti-slide for each of four directions parallel to the x-axis or the y-axis. By this definition, we can consider the case in Fig. 9(d) is anti-slide. This model fits to the anti-slide puzzles in 3D version by Amano et al. [1], and it also fits to the real anti-slide puzzles for the set of polyominoes, which are the most popular ones in the puzzle society.

4   Decision Problem for Interlocking of Polyominoes

In this section, we show a polynomial time algorithm for the decision problem of whether a given arrangement of orthogonal polyominoes is anti-slide or not.

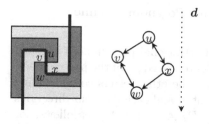

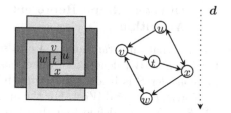

Fig. 11. Each of four pieces is anti-slide, however, we can slide by combining two of them along the blue thick lines in figure.

Fig. 12. Five pieces are interlocking; by the central square, we cannot slide any set of pieces in 2D

To simplify, we consider the decision problem for the interlock problem without frame for the set of polyominoes. (Its extension to the anti-slide puzzle with frame is easy, and omitted here.)

First, we remark that the arrangement may not be interlocked even if every polyomino is anti-slide. Each polyomino in the arrangement of four polyominoes in Fig. 11(left) is anti-slide, however, if we consider two pieces as a "block", we can slide the block and separate into two blocks of two pieces. On the other hand, a similar arrangement of five polyominoes in Fig. 12(left) is interlocking and we cannot slide any block of polyominoes from the other. (The puzzle "Lock Device" designed by Hiroshi Yamamoto consists of five L-shape polyominoes similar to ones in Fig. 11 [14]. That is, without square piece as in Fig. 12, we can construct the interlock structure by using just five distinct L-shape polyominoes. In this paper, we do not give the sizes in honor to the designer.)

In order to deal with such cases, we introduce some notions. When two polyominoes share at least one edge in an arrangement, a *merge* of two polyominoes is a replacement of these two polyominoes by one polyomino which is obtained by gluing these two polyominoes. Let $\mathcal{A}$ be an arrangement of a set $\mathcal{P}$ of polyominoes. Let $\mathcal{A}'$ be a partial arrangement induced by any subset $\mathcal{P}'$ of $\mathcal{P}$. We repeat merging of two polyominoes in $\mathcal{A}'$ as many times as we can. We call this operation a *merge* of the subset $\mathcal{P}'$. We suppose that we obtain a single polyomino P' after merging all elements in the subset $\mathcal{P}'$. Then $\mathcal{A}'$ is said to be *anti-slide* when the corresponding polyomino P' is anti-slide after replacing the partial arrangement $\mathcal{A}'$ by P' in $\mathcal{A}$. We define that the arrangement $\mathcal{A}$ is *interlocking* if and only if $\mathcal{P}'$ is anti-slide for any subset $\mathcal{P}'$ of $\mathcal{P}$.

By the definition, we can decide that the arrangement in Fig. 11 is not anti-slide, while the other one in Fig. 12 is interlocking. However, it seems that we have to check exponentially many subsets $\mathcal{A}'$ of $\mathcal{A}$ whether the corresponding $\mathcal{P}'$ is anti-slide or not. It is not the case; we show a polynomial-time algorithm for solving this problem.

4.1 Directed Graph Representation and Polynomial-Time Algorithm

Assume that an arrangement $\mathcal{A}$ of a set $\mathcal{P}$ of orthogonal polyominoes is given. Let $\{N, E, S, W\}$ denote four directions parallel to the edges of the polyominoes. We consider S is below (South in a map), and the other letters corresponding to North, East, and West. For the arrangement $\mathcal{A}$ and each d of four directions $\{N, E, S, W\}$, we define four directed graphs $D_d(\mathcal{A}) = (\mathcal{P}, A_d)$ as follows; for $P, Q \in \mathcal{P}$, we have an arc $(P, Q) \in A_d$ if and only if Q supports P in direction d. Then we have the following lemma.

Lemma 1. *Let $\mathcal{P}$ and $\mathcal{A}$ be a set of polyominoes and its arrangement. Then $\mathcal{A}$ interlocks if and only if the four graphs $D_N(\mathcal{A}), D_E(\mathcal{A}), D_S(\mathcal{A})$, and $D_W(\mathcal{A})$ are strongly connected.*

Proof. (Outline) We here consider the directed graph $D_S(\mathcal{A})$ for $d = S$. Without loss of generality, we can assume that the underlying graph of $D_S(\mathcal{A})$ is connected.

We first show that when $\mathcal{A}$ is interlocking, $D_S(\mathcal{A})$ is strongly connected. To derive contradictions, we suppose $D_S(\mathcal{A})$ is not strongly connected. Since $D_S(\mathcal{A})$ is not strongly connected, there is a partition $\mathcal{P}_1$ and $\mathcal{P}_2$ of $\mathcal{P}$ such that, even while P_1 is reachable to P_2 in $D_S(\mathcal{A})$, any $P_2 \in \mathcal{P}_2$ is not reachable to any $P_1 \in \mathcal{P}_1$. For the set $\mathcal{P}_1$, we merge the elements in $\mathcal{P}_1$ as many times as we can. Then the obtained polyomino (if there are two or more, we can pick up any of them) can be slid to the direction N from $\mathcal{P}_2$. This contradicts the definition of interlock. Thus $D_S(\mathcal{A})$ is strongly connected.

We next show that $\mathcal{A}$ is interlocking when $D_S(\mathcal{A})$ is strongly connected. The transitivity of supporting, when there is a directed path from P_1 to P_2, P_1 supports P_2 from below through some polyominoes between P_1 and P_2. On the other hand, if there is a directed path from P_2 to P_1, P_2 supports P_1 from below in the same manner. Therefore, if we have two directed paths between them, they are supporting with each other, thus we cannot slide P_1 from P_2 to any direction of S and N. Therefore, when $D_S(\mathcal{A})$ is strongly connected, any pair of polyominoes are in the same relationship, or $\mathcal{A}$ is interlocking with respect to this direction.

By the symmetric arguments, an arrangement $\mathcal{A}$ is interlocking if and only if the four graphs $D_N(\mathcal{A}), D_E(\mathcal{A}), D_S(\mathcal{A}), D_W(\mathcal{A})$ are strongly connected. □

We note that the definitions are symmetric with respect to the pair of directions (S, N) and (E, W). Therefore, when we reverse all edges in $D_N(\mathcal{A})$, we can obtain $D_S(\mathcal{A})$, and so $D_E(\mathcal{A})$ and $D_W(\mathcal{A})$ are. This implies the following corollary:

Corollary 1. *Let $\mathcal{P}$ and $\mathcal{A}$ be a set of polyominoes and its arrangement. Then $\mathcal{A}$ interlocks if and only if the two graphs $D_N(\mathcal{A})$ and $D_E(\mathcal{A})$ are strongly connected.*

Theorem 1. *For a given arrangement $\mathcal{A}$ of a set $\mathcal{P}$ of polyominoes, we can determine whether $\mathcal{A}$ is interlocking or not in $O(n^2)$ time, where n is the total number of the vertices of the polyominoes.*

Proof. By Corollary 1, it is sufficient to construct two graphs $D_N(\mathcal{A})$ and $D_E(\mathcal{A})$. Since the total number of vertices in $\mathcal{P}$ is n, we have $|\mathcal{P}| = O(n)$. The decision problem for strong connectivity can be done in $O(|\mathcal{P}|^2)$ time (see, e.g., [4]), and hence we have the theorem.   □

In Fig. 11 and Fig. 12, the corresponding $D_S(\mathcal{A})$s are shown in the right side. In Fig. 11, the graph indicates that contracting u and x, we obtain a merged vertex on v, w. On the other hand, in Fig. 12, we have no such a vertex set.

We note that since the underlying graph of $D_d(\mathcal{A})$ is a planar graph, the running time in Theorem 1 can be improved to linear time if the graph is explicitly given as an input.

5   Decision Problem for Anti-slide Puzzle on Polygons

In this section, for a given instance of the anti-slide puzzle, we consider the decision problem that asks whether we can arrange the given pieces anti-slide or not. We first prove that this problem is NP-complete even if the pieces are x-monotone polyominoes. Next, we show that any set of convex polygons is never anti-slide.

5.1   NP-completeness of the Anti-slide Problem on x-monotone Polyominoes

In this section, we deal with polyominoes. We reduce the following *3-Partition* problem to our problem.

Input: A set $\hat{A} = \{a_1, a_2, a_3, \ldots, a_{3m}\}$ and a positive integer B, where each a_i is a positive integer with $\frac{1}{4}B < a_i < \frac{1}{2}B$.

Output: Determine whether we can partition $\hat{A}$ into m 3-tuples $A_1, A_2, \ldots, A_m$ so that $a_i + a_j + a_k = B$ for each $A_l = \{a_i, a_j, a_k\}$.

It is well known that the 3-Partition problem is strongly NP-complete [9]. Without loss of generality, we also assume that each a_i is an odd number (by replacing each number a_i by $2a_i + 1$ and B by $2B + 3$), and $\sum_i a_i = mB$.

Theorem 2. *Let $\mathcal{P}$ and F be an instance of the anti-slide puzzle. It is strongly NP-complete to find an anti-slide arrangement for $\mathcal{P}$ and F.*

We note that the instance of the anti-slide puzzle should be easy for packing.

Proof. We first remark that each piece in $\mathcal{P}$ is a polyomino and so F is. Then, by Theorem 1, we can determine whether an arrangement is anti-slide or not in polynomial time. Therefore, the anti-slide puzzle is in NP. Next, we show NP-hardness. We reduce the 3-Partition problem to our problem.

For a given instance $\hat{A} = \{a_1, \ldots, a_{3m}\}$, we construct a piece P_i for each a_i as shown in Fig. 13. The frame F is a rectangular polyomino with m rectangular holes of size $B \times 2B$. The reduction can be done in polynomial time.

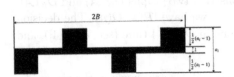

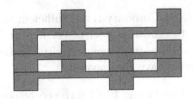

Fig. 13. A piece P_i for an odd integer a_i

Fig. 14. Packing arrangement of any three pieces in $\mathcal{P}$

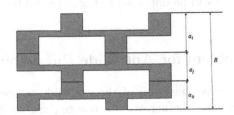

Fig. 15. Anti-slide arrangement of $\{P_i, P_j, P_k\}$

We first confirm that the packing problem for the $\mathcal{P}$ and F is easy to solve. We divide $\mathcal{P}$ into m 3-tuples in any way and pack each 3 pieces of a 3-tuple into a hole of size $B \times 2B$ of F as shown in Fig. 14. Since we have $\frac{1}{4}B < a_i < \frac{1}{2}B$, for any three pieces, by putting the shortest one among three pieces at the center, the total height of them is less than $B/2 + B/2 + 1$, or at most B. Therefore, the packing problem is trivial.

Thus we show that the instance $\hat{A}$ and B of the 3-Partition problem has a solution if and only if we can make an anti-slide arrangement of $\mathcal{P}$ into F.

First we assume that the instance $\hat{A}$ and B has a solution of the 3-Partition. Then we have m 3-tuples $A_l = \{a_i, a_j, a_k\}$ such that $a_i + a_j + a_k = B$ for each l. Using these 3-tuples, we can fill each hole of size $B \times 2B$ in F by corresponding three pieces P_i, P_j, P_k as shown in Fig. 15. Clearly, they are anti-slide. Next we assume that $\mathcal{P}$ and F are anti-slide. Then, by the condition $\frac{1}{4}B < a_i < \frac{1}{2}B$, each hole has exactly three pieces, and it is easy to see that the directions of three pieces should be the same as shown in Fig. 15 (or their up-side-down direction).

Therefore, the instance $\hat{A}$ and B of the 3-Partition has a solution if and only if the instance $\mathcal{P}$ and F of the anti-slide puzzle has an anti-slide arrangement. Thus the generalized anti-slide puzzle is strongly NP-complete. □

We note that each piece in $\mathcal{P}$ in the proof of Theorem 2 is x-monotone. We can give a stronger result as follows:

Theorem 3. *Let $\mathcal{P}$ be an instance of the interlock puzzle. It is strongly NP-complete to find an interlocking arrangement of $\mathcal{P}$ even if all pieces in $\mathcal{P}$ are x-monotone.*

Proof. We divide the frame F in the proof of Theorem 2 into five x-monotone polyominoes as shown in Fig. 16. Making the size of unit square small enough compared to the other elements in $\mathcal{P}$, the way of interlocking of F is uniquely determined. Therefore, the proof of Theorem 2 also proves this theorem.   □

Fig. 16. Interlocking frame F

5.2   Convex Polygons Cannot Interlock

The pieces of "Lock Device" and ones in the proof of Theorem 3 are all x-monotone. That is, the set of x-monotone polyominoes can interlock. In this section, we consider general polygons, and we show that the set of convex polygons never interlock.

Theorem 4. *Let* $\mathcal{P} = \{P_1, \ldots, P_k\}$ *be a set of convex polygons for some positive integer k. Then $\mathcal{P}$ does not interlock for any arrangement.*

Proof. (Outline) It is easy to see that the claim holds for $k = 1, 2$. We assume that k is the minimum integer such that $\mathcal{P}$ interlocks for some arrangement $\mathcal{A}$. Then we apply similar techniques in [8]: Roughly, we "grow" convex polygons in $\mathcal{P}$ to inscribe a triangle T without changing their geometric properties, especially, convexity. After growing and fitting to T, the polygon at the corner of T can be slid farther from the other polygons. This implies that the corresponding polygon in $\mathcal{P}$ in $\mathcal{A}$ can also be slid farther, which contradicts the minimality of k.   □

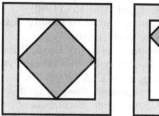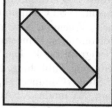

Fig. 17. Are they anti-slide?

6 Concluding Remarks

In this paper, we proposed some 2D models for anti-slide puzzles and interlock puzzles. Extensions to 3D are future work. In puzzle society, there are some puzzles called *coordinate-motion* in which all pieces have to be moved smoothly and simultaneously to assemble or disassemble [3]. These puzzles are decided to be anti-slide or interlocking in our framework, however, they actually can be slid and disassembled in neat ways. In order to deal with such movements, we may have to consider *infinitesimal rigidity* in terms of architecture (see, e.g., [6, Sec. 4.4]). However, it is not so simple as the first impression. For example, in Fig. 17, we feel that the left tilted square in the frame is *not* anti-slide, while the right tilted rectangle in the frame *is* anti-slide because the square can be rotated, but the rectangle cannot be. Therefore, even for polyominoes, it is not straightforward to extend it to general (non-orthogonal) case.

Acknowledgements. This work is partially supported by JSPS KAKENHI Grant Numbers 17H06287 and 18H04091. The authors thank Prof. Yoshio Okamoto, who mentioned that the techniques in [8] can be used to prove Theorem 4.

References

1. Amano, K., Nakano, S., Yamazaki, K.: Anti Slide. J. Inf. Process. **23**(3), 252–257 (2015)
2. Chartrand, G., Lesniak, L., Zhang, P.: Graphs and Digraphs, 6th edn. CRC Press, Boca Raton (2015)
3. Coffin, S.T.: The Puzzling World of Polyhedral Dissections: Chapter 12 - Coordinate-Motion Puzzles (1990–2012). https://puzzleworld.org/PuzzlingWorld/chap12.htm. Accessed September 2020
4. Cormen, T.H., Leiserson, C.E., Rivest, R.L., Stein, C.: Introduction to Algorithms, 3rd edn. MIT Press, Cambridge (2009)
5. de Berg, M., Cheong, O., van Kreveld, M., Overmars, M.: Computational Geometry: Algorithms and Applications, 3rd edn. Springer-Verlag, Heidelberg (2008). https://doi.org/10.1007/978-3-540-77974-2
6. Demaine, E.D., O'Rourke, J.: Geometric Folding Algorithms: Linkages, Origami, Polyhedra. Cambridge University Press, Cambridge (2007)
7. Demaine, E.D., et al.: Symmetric assembly puzzles are hard, beyond a few pieces. Comput. Geom. Theory Appl. **90**(101648), 1–11 (2020)
8. Edelsbrunner, H., Robison, A.D., Shen, X.-J.: Covering convex sets with non-overlapping polygons. Discrete Math. **81**, 153–164 (1990)
9. Garey, M.R., Johnson, D.S.: Computers and Intractability–A Guide to the Theory of NP-Completeness. Freeman, New York (1979)
10. Golomb, S.W.: Polyominoes. Princeton University Press, Princeton (1994)
11. Slocum, J.: The Tangram Book: The Story of the Chinese Puzzle with Over 2000 Puzzle to Solve. Sterling Publishing, New York (2004)
12. Strijbos, W.: Anti-Slide. (Commertial product was sold by Hanayama, 2007) (1994)
13. Takenaga, Y., Yang, X., Inada, A.: Anti-slide placements of pentominoes. In: The 22nd Japan Conference on Discrete and Computational Geometry, Graphs, and Games (JCDCGGG), pp. 121–122 (2019)
14. Yamamoto, H.: Lock Device (2012). https://puzzleworld.org/DesignCompetition/2012/. Accessed September 2020

Author Index

Printed in the United States
By Bookmasters